D0929213

A Practical Guide to BOX-JENKINS Forecasting

A Practical Guide to BOX-JENKINS Forecasting

JOHN C. HOFF

LIFETIME LEARNING PUBLICATIONS
Belmont, California
A division of Wadsworth, Inc.
London, Singapore, Sydney, Toronto, Mexico City

Jacket and Text Designer: R. Kharibian

Content Editor: Carol Beal

Illustrator: John Foster

Compositor: Graphic Typesetting Service

Printed in the United States of America

1 2 3 4 5 6 7 8 9 10—87 86 85 84 83

Library of Congress Cataloging in Publication Data

Hoff, John C.
A practical guide to Box-Jenkins forecasting.

Bibliography: p.
Includes index.
1. Economic forecasting—Mathematical models.
2. Time-series analysis. I. Title. II. Title: Box-Jenkins forecasting.
HB3730.H64 1983 338.5'442'0724 83-16210
ISBN 0-534-02719-9

Contents

Preface

In recent years the Box-Jenkins method of forecasting has enjoyed a large degree of popularity in the business and economic forecasting community—at least in name, if not in actual application. Since Box and Jenkins first introduced the method in 1970 in their book *Time Series Analysis, Forecasting and Control,* it has been included on the roster of time series forecasting methods in nearly every practical and theoretical book on the subject of time series forecasting. Moreover, a large percentage of forecasting analysts—preparers of forecasts—in the business and economic environment have either read about it, inquired about it, or attempted to use it. However, most forecasting practitioners never move beyond these initial steps. This is unfortunate because the Box-Jenkins method is one of the few time series forecasting methods that is both *theoretically sound* and *practically applicable.*

The problem is really threefold. First, most descriptions of the method are written in an academic style replete with an imposing body of statistical jargon and theory, and complicated mathematical formulas and derivations, which tend to overwhelm even those who have some statistical background. Second, the practical aspects—the problems, pitfalls, and rules of thumb—in applying the method in real-life situations are usually only lightly treated, if at all. Third, the seemingly complex nature of the method raises gigantic barriers in communicating to the person who needs the forecast the methods by which it was obtained. In point of fact, it is not really necessary to have a complete understanding of the underlying statistical theory in order to successfully apply the Box-Jenkins method.

Purpose of the Book

The purpose of this book is to provide an easy-to-understand *explanation* of the important concepts behind the Box-Jenkins method, and to provide practical *guidance* in applying the method in real-life forecasting situations. The explanation provided will minimize or strip away the jargon and mystique associated with the statistical and mathematical nature of the method, and the guidance provided will show you how to deal with the practical aspects of applying the method, and precisely where and when that method becomes applicable.

Audience for and Organization of the Book

This book is written with both the forecasting practitioner (analyst, preparer of forecasts) and the forecast user (manager, planner, etc.) in mind. In particular, the material in Part 1 provides the forecasting practitioner and the forecast user with common ground on which to meet. This material should help create an atmosphere of mutual understanding that is so often lacking between these parties. Part 1 introduces the basic concepts of time series and time series forecasting and it indicates where the Box-Jenkins method becomes applicable (Chapters 1 and 2). The requirements for the successful application of the method are also outlined (Chapter 3), and some observations are made on the overall problem of forecasting and the context in which the Box-Jenkins method, or any method, should be used (Chapter 4).

Parts 2 and 3 are directed primarily to the forecasting practitioner. The forecast user, however, can gain some insight into what is involved in the Box-Jenkins method and what Box-Jenkins models actually look like by reading Chapters 5 and 6 in Part 2 and Chapters 12 and 14 in Part 3. Part 2 itself describes the basic concepts and tools that form the foundation of the Box-Jenkins method. The basic Box-Jenkins models are described and the tools for identifying and constructing these models are introduced. Part 3 builds on the foundation of Part 2 to develop the complete repertoire of Box-Jenkins models. Many examples for identifying and constructing Box-Jenkins models are given in both Parts 2 and 3.

Part 4 is directed to both the forecasting practitioner and the forecast user. This part describes the use of Box-Jenkins models to generate forecasts, and discusses the problems of assessing forecast accuracy and updating forecasts and forecasting models.

For the analyst who desires more statistical and mathematical explanations, appendices A, B, and D are provided to help satisfy that need.

Acknowledgments

I would like to thank the TimeWare* Corporation for its support in writing this book, George Ledin for his insistence that it should be written, Alex Kugushev and the editorial staff at Lifetime Learning Publications for their help and encouragement, Sarah Fisher for typing the manuscript—and, of course, my wife, Linda, and Nicholas, Andrew, and Christopher for enduring many fatherless weekends without complaint.

John C. Hoff

*TimeWare is the Software developer of the forecasting and graphics computer systems, TIMEPACK and PICTURE-PAC, which were used to produce many of the examples and illustrations in this book. (TIMEPACK is a registered trademark of TimeWare Corporation, Palo Alto, California. PICTURE-PAC is a registered trademark of Control Data Corporation, Minneapolis, Minnesota.)

A Practical Guide to BOX-JENKINS Forecasting

Part 1

Introduction

The purpose of Part 1 is to provide you with the necessary background and information in order to understand and successfully apply the Box-Jenkins forecasting method. The material in Part 1 will introduce you to some of the basic concepts and terminology used in forecasting and in time series forecasting, in particular.

In Part 1 you will learn:

- What a time series and time series forecasting are.
- What self-projecting time series forecasting models are.
- What type of time series forecasting method the Box-Jenkins method is.
- When the Box-Jenkins forecasting method can be used.
- What you should do before applying the Box-Jenkins method.
- How to use the Box-Jenkins method within the framework of an overall forecasting process.

If you don't know what time series forecasting is or have only a passing acquaintance with the subject, Chapter 1 will help you understand the nature of the problem, the basic approaches and methodologies used in time series forecasting, and when it is appropriate to use the Box-Jenkins method.

If you already know what time series forecasting is, but are a little rusty on some of the details and concepts, Chapters 2 and 3 will refresh your memory. Even if you are quite familiar with time series forecasting, you may still find it helpful to review Chapters 2 and 3 since the basic notation and definitions used throughout the book are introduced here.

Specifically, Chapter 1 introduces the concept of time series and time series forecasting and describes briefly where the Box-Jenkins method is applicable. The two basic approaches to forecasting time series are described.

Chapter 2 deals with the concept of time series forecasting models and describes the process of building such models. Examples of traditional models and their disadvantages are discussed and compared with the advantages of Box-Jenkins models.

Chapter 3 discusses what you should be aware of, or what you should do, in order to be successful in using the Box-Jenkins method.

Chapter 4 provides an overview of the forecasting process in general, and indicates where time series forecasting and Box-Jenkins fit into the overall scheme of forecasting.

CHAPTER 1

Box-Jenkins and Time Series Forecasting

The Box-Jenkins forecasting method is a method for forecasting time series. This statement, of course, will have little meaning for you unless you know what a time series is and in what forecasting situations time series arise. The purposes of this chapter are to introduce the basic concepts of time series and time series forecasting and to indicate when the Box-Jenkins method can be used.

In this chapter you will learn:

- What a time series is.
- What time series forecasting is.
- When the Box-Jenkins method can be used.
- What the basic approaches to forecasting time series are.
- What type of time series forecasting method the Box-Jenkins method is.

1.1 WHEN CAN THE BOX-JENKINS FORECASTING METHOD BE USED?

Forecasting is a ubiquitous activity in the operation of any enterprise or institution in business and government. Anyone in a position of planning, controlling, and managing projects, people, finances, and operations has a constant requirement for making estimates of the future state of the environment. On the basis of these estimates, plans are drawn up and acted upon, and management decisions are made.

For example, product-line managers need to estimate return on investment when proposing the introduction of a new product. Marketing managers need to estimate future consumer buying trends in order to plan their marketing strategies. Sales managers need to estimate future sales when setting sales quotas. Production managers need to know future demand in order to efficiently plan and control the production process. Financial planners need to project cash flows in order to manage and plan for future cash requirements. Corporate executives need to project overall economic conditions and trends in order to develop long-term corporate strategies.

Obviously, the areas and situations in which forecasting is required are widely diverse. In most of these situations, however, there is almost always a need to forecast certain processes or activities that have *quantified track records*. That is, the numerical level of the activity—whether we're talking about unit sales, company sales revenue, personnel requirements, cash flows, expenditures, product demand, shipments, inventory, economic indicators, etc.—has been recorded at various time intervals in the past. For example, product sales may be known in terms of the number of units sold in each of the past 60 months. Data in this form is generally called a *time series,* i.e., a series of data values associated with a particular set of time intervals, such as months or years.

It turns out that time series data can contain a lot of useful information that is helpful in developing estimates of the future behavior of the activity or process it represents. The process of obtaining this information and developing forecasts from it is called *time series forecasting*. It is within this context that the Box-Jenkins method of forecasting is applicable.

We will, of course, be more precise about the nature of time series data, time series forecasting, and the Box-Jenkins method later in this chapter. In broadest terms, however, you can use the Box-Jenkins forecasting method if you have the following:

- Data representing the historical behavior of what you want to forecast (a time series).
- Enough data (an established track record).
- Short- to medium-term forecasting requirements (not-too-distant forecasts are required).

What is meant by "enough data" and "short- to medium-term forecasts" is addressed in later chapters.

1.2 WHAT IS A TIME SERIES?

A *time series* is a set of numbers that measures the status of some ongoing process or activity over time. The measurements are assumed to be taken at equally spaced time intervals or periods.

For example, a monthly time series for unit sales represents the number of units sold for each month in the past, as shown in Table 1.1. In this time series the *process* or activity is product sales, the *measurement* is in number of units sold, and the *time periods* are months.

Table 1.1 A Monthly Time Series

Time Period	Unit Sales
Jan 78	13,727
Feb 78	16,109
Mar 78	15,589
Apr 78	14,741
May 78	17,205
Jun 78	16,373
Jul 78	15,286
Aug 78	17,364
Sep 78	16,376
Oct 78	16,225
Nov 78	18,600
Dec 78	17,570

Most time series forecasting methods assume that a time series has the following characteristics:

1. The time periods at which each measurement, or data value, is recorded are of equal length. The most common time periods encountered are years, months, weeks, and days (months are a special case since they are not quite equal in length; see Chapter 3, Section 3.3.2, for more details on this topic).
2. The data values are always arranged in order from the earliest time period to the latest. For self-projecting methods it is assumed there is a data value for each time period; i.e., there are no missing values.
3. The process or activity and the method of measuring the status of the activity remain consistent over time (see Chapter 3, Section 3.2, for more details on this topic).

The business and economic environment is an especially fertile source of time series. Indeed, many of the processes or activities that businesspeople are interested in forecasting have been tracked historically and recorded as time series. The accompanying table lists some examples of these types of activities.

Marketing	Production	Financial Planning	Corporate Planning
Product sales	Product demand	Sales revenue	Company sales
Sales revenue	Shipments	Cash flows	Earnings
Prices	Material costs and requirements	Expenses	General economic indicators
Market share	Labor costs and requirements	Budgets	Capital expenditures
Consumer trends	Inventory levels	Inventory levels	
	Plant utilization	Prices	
		Interest rates	

The most convenient way to understand the behavior of a time series is to display it, not in tabular form, as shown previously, but in a graph. Graphs of time series provide visual insight into how the process or activity it represents has behaved historically—whether there is a consistent upward or downward trend in the level of the activity or whether certain patterns repeat themselves, and so on.

Figure 1.1 shows a graph of the unit sales time series data from Table 1.1 plotted against time. In this graph time is marked off on the horizontal axis and the unit sales level on the vertical axis. The time series values are then represented by points on the graph. The points are connected to visually reinforce the change in level of the time series from one time period to the next. A casual inspection of the graph, for example, shows a simple recurring pattern every three months with a slight overall upward movement.

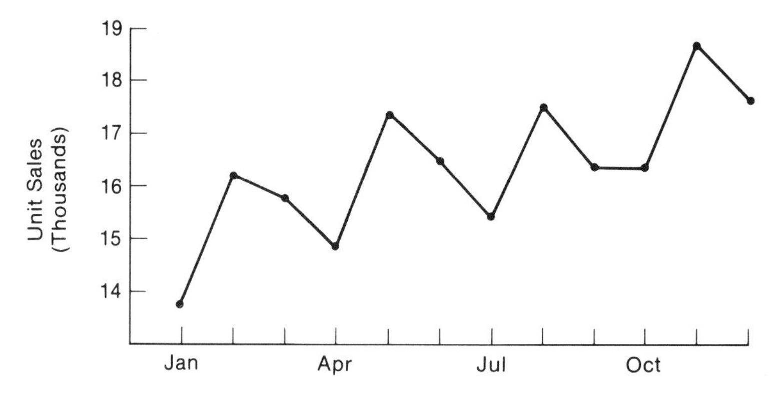

Figure 1.1 A Time Series Graph

Graphs of time series are commonplace. Newspapers and magazines covering business and economic news use this format to demonstrate the historical behavior of some business or economic activity. The front page of every edition of the *Wall Street Journal,* for example, shows a time series graph of some business or economic activity.

Figures 1.2 through 1.6 illustrate some additional examples of time series displayed in graphical form.

1.3 WHAT ARE THE BASIC APPROACHES TO FORECASTING TIME SERIES?

The Box-Jenkins method is one of many time series forecasting methods that have been developed over the years. These methods are analytical in nature; i.e., they use a variety of mathematical and statistical concepts and techniques to extract

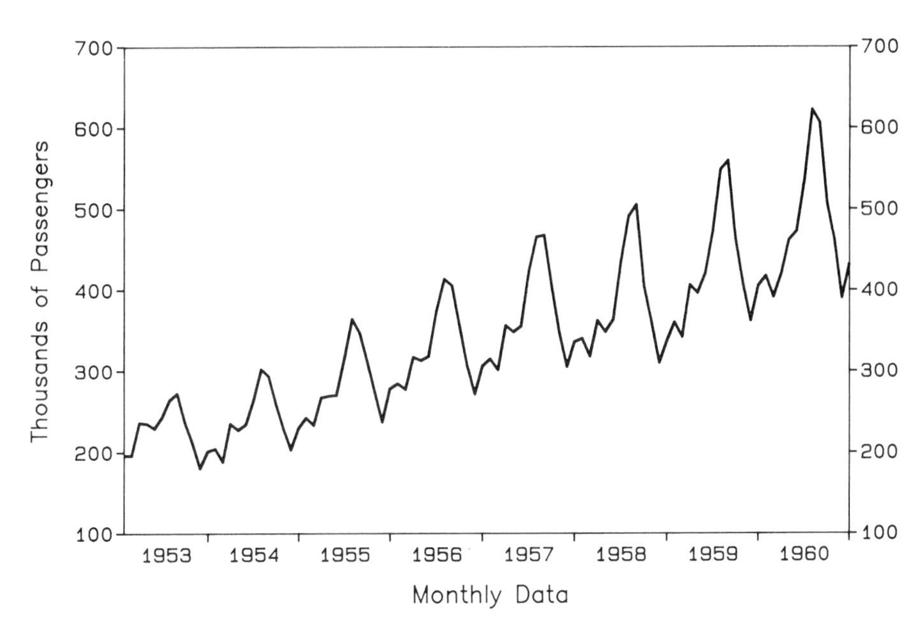

Figure 1.2 Monthly International Airline Passengers

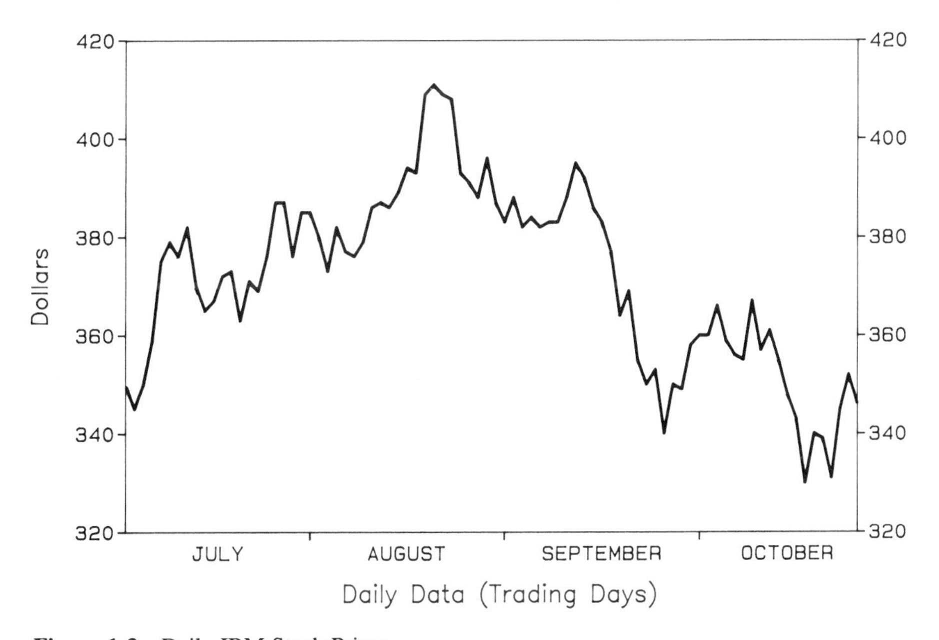

Figure 1.3 Daily IBM Stock Prices

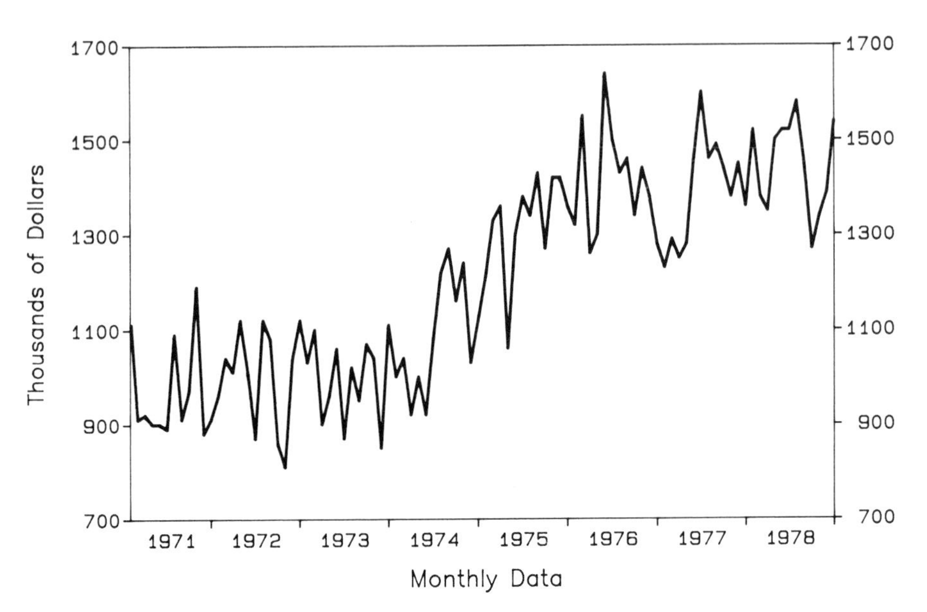

Figure 1.4 Monthly Company Sales Revenue

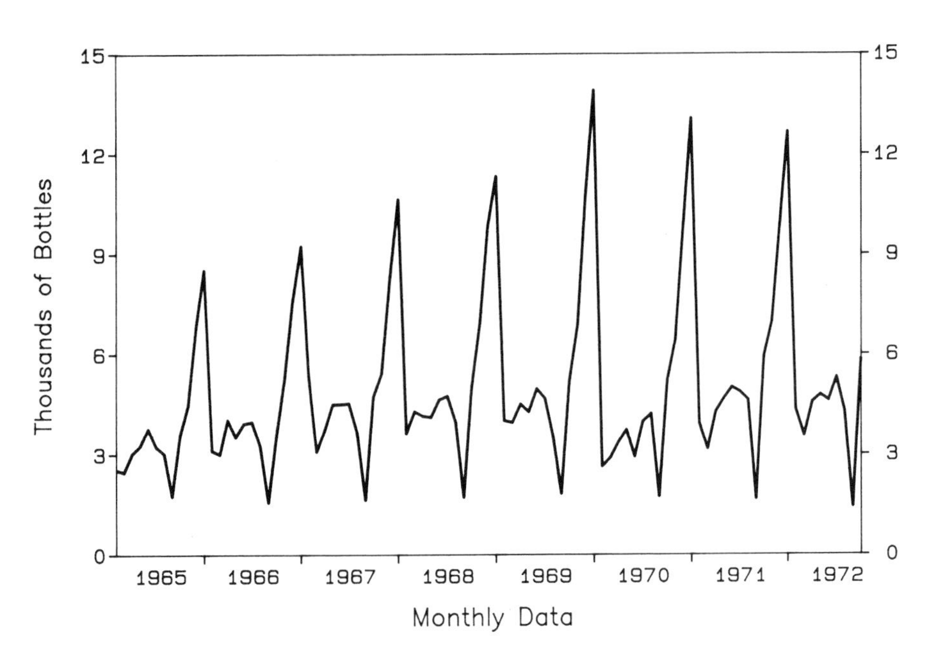

Figure 1.5 Monthly Champagne Sales

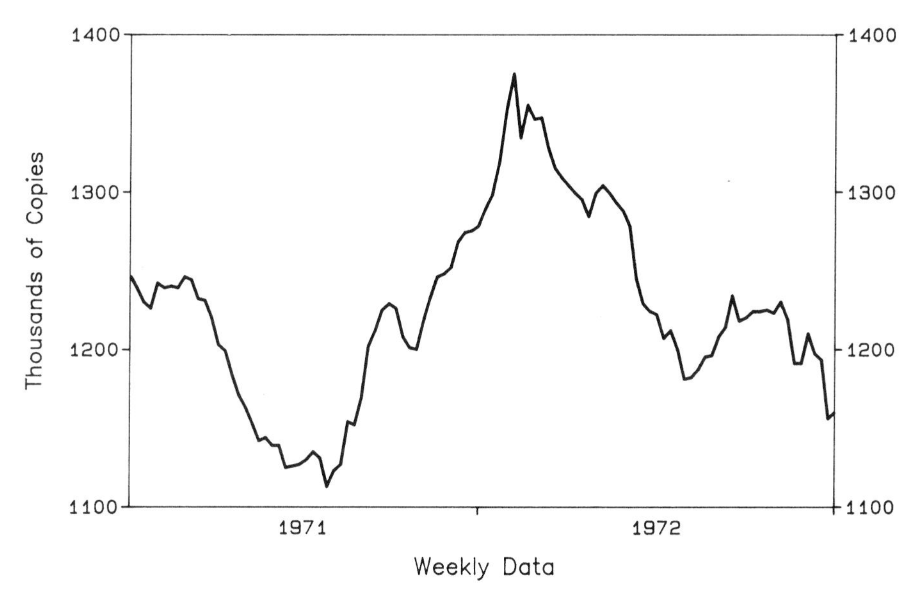

Figure 1.6 Weekly Magazine Distribution

pertinent information from time series data, establish relationships among relevant factors, and extrapolate past behavior into the future.

Time series forecasting methods vary widely in degree of sophistication and application. Some are based on simple or intuitive ideas, while others are based on more complex statistical concepts. All of them, however, follow one of two basic approaches to forecasting time series: the *self-projecting approach* or the *cause-and-effect approach*.

1.3.1 The Self-projecting Approach

Self-projecting methods derive forecasts of a time series solely on the basis of the historical behavior of the series itself. This approach is illustrated in Figure 1.7. Self-projecting methods are often referred to as *univariate methods* (i.e., only "one variable" is forecasted).

Self-projecting methods can range from simple naive methods, such as trend projections and moving averages, to sophisticated methods, such as Box-Jenkins. We will look at some of these self-projecting methods in more detail in Chapter 2.

Self-projecting methods are useful for a number of reasons:

- They can be applied quickly and easily. In situations where hundreds of time series need to be forecasted (e.g., forecasting unit sales for hundreds of product lines), this approach may be the only reasonable one.

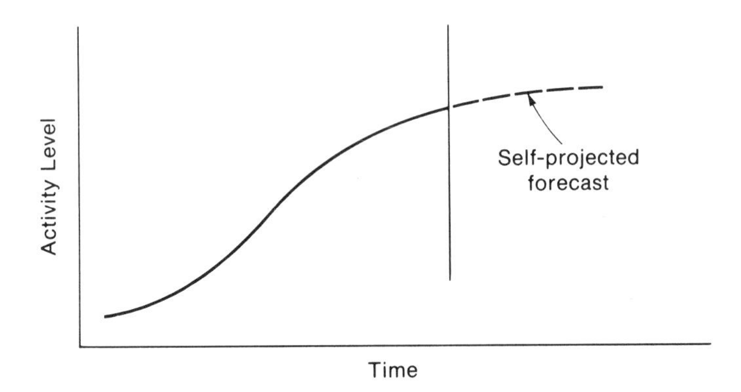

Figure 1.7 The Self-projecting Approach

- They provide a basis by which forecasts developed through other methods can be measured against.
- They require a minimum of data, and they are usually inexpensive to apply.
- They can provide reasonably accurate short- to medium-term forecasts for reasonably well behaved time series.

Self-projecting methods, however, are generally not useful for forecasting into the far distant future. Moreover, by their very nature they do not take into account any external factors that may have influence on the activity or process to be forecasted.

1.3.2 The Cause-and-Effect Approach

In many forecasting situations the activity or process to be forecasted is affected by outside influences. For example, product sales may be highly influenced by the level of advertising expenditures and personal disposable income. Cause-and-effect methods of time series forecasting try to take these influences into account by establishing a mathematical relationship between the series to be forecasted and one or more other time series representing the influencing factors. This approach is illustrated in Figure 1.8.

Cause-and-effect methods are often called *causal* or *explanatory methods* since the related series can be thought of as "causing" or "explaining" the behavior of the series to be forecasted.

Cause-and-effect methods can range from simple regression techniques to highly complex methods for solving large econometric models.

Cause-and-effect methods are useful for the following reasons:

- They bring more information to bear on the forecasting problem.
- They take into account interrelationships among various factors in the business environment and determine the extent of those relationships.

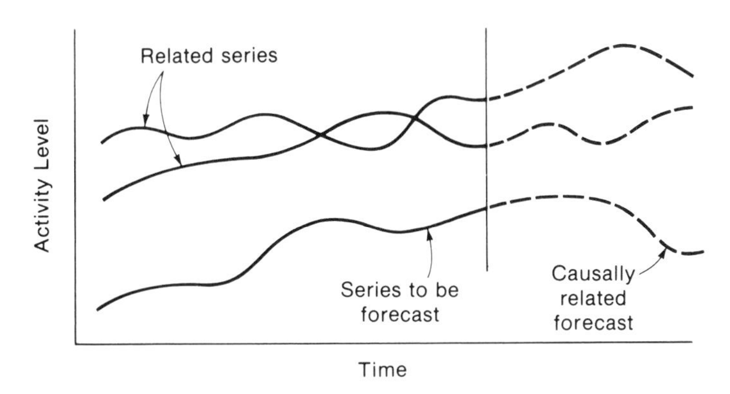

Figure 1.8 The Cause-and-Effect Approach

- They can exploit the forecasting power of leading indicator relationships.
- They can provide more accurate medium- to long-range forecasts.

Cause-and-effect methods, however, require larger amounts of data and are generally more time-consuming and expensive to apply than other methods. In addition, cause-and-effect methods require forecasts of the explanatory time series themselves for computing forecasts of the time series to be forecasted. For some methods, such as regression, these forecasts must be obtained by some other means (e.g., forecasts of personal disposable income might be obtained from government sources, while "forecasts" of advertising expenditures are actually predetermined, or controllable). This requirement, of course, increases the scope of the forecasting problem and introduces another degree of uncertainty into the forecast. Other, more sophisticated cause-and-effect methods simultaneously generate forecasts of the explanatory time series and the series to be forecasted (much like a combination of a self-projecting and a regression-type cause-and-effect method). Such methods are often referred to as *multivariate methods* (i.e., "multiple variables" are forecasted simultaneously).

1.4 WHAT IS THE BOX-JENKINS FORECASTING METHOD?

The Box-Jenkins method discussed in this book is a self-projecting (univariate) time series forecasting method. It is often referred to as univariate Box-Jenkins. The methodology and concepts employed in univariate Box-Jenkins, however, have also been extended to the cause-and-effect approach. The resulting method is known as multivariate Box-Jenkins.* Although this book only deals with the univariate Box-

*The multivariate Box-Jenkins method is also known as the Box-Jenkins transfer function method.

Jenkins method, it also provides the starting point for working with the multivariate Box-Jenkins method, since multivariate Box-Jenkins requires, among other things, repeated applications of univariate Box-Jenkins.

The Box-Jenkins method is founded on statistical concepts and principles, which is the basis of its power and wide applicability. It is also the source, unfortunately, of the difficulty encountered by many would-be users of the method. The purpose of this book is to eliminate that difficulty. In Chapter 2 and Part 2 you will learn why the Box-Jenkins method is so powerful and so unique among time series forecasting methods.

1.5 REVIEW OF KEY CONCEPTS

To understand and successfully apply the Box-Jenkins forecasting method, you should know what a time series is, what time series forecasting is, and what the basic approaches to time series forecasting are. With this information you will then know when you can use the Box-Jenkins forecasting method.

A *time series* is a set of numbers that measures the status of some activity (e.g., unit sales) over time. In other words, the historical behavior of the activity has been numerically recorded at various time intervals in the past. In this book we assume the following:

- Measurements are taken at equally spaced intervals (months are an exception to this rule).
- There are no missing measurements between the first and last time periods.
- The method of measurement and the activity being measured remain consistent over time.

The most common time periods in the business environment are years, months, quarters, and weeks. Time series are commonplace in the business and economic environment. Activities such as product sales, sales revenue, shipments, orders placed, inventory levels, and cash flow are routinely tracked and recorded as time series.

A *time series graph* is one of the best ways to get a feeling for the historical behavior of a given activity represented by the series. Such a graph is simply a plot of the series values against time. Time series graphs show the period-to-period changes in the level of an activity as well as its overall movement.

Time series forecasting is the process of obtaining information from time series data for forecasting purposes. There are two basic approaches to forecasting time series: the self-projecting approach and the cause-and-effect approach.

The *self-projecting time series forecasting approach* uses only the time series data of the activity to be forecasted to generate forecasts. Self-projecting methods have the following characteristics:

- They can be applied quickly and easily.
- They require little data to work with.
- They are usually inexpensive to apply.
- They are good for short- to medium-term forecasting.

The *cause-and-effect time series forecasting approach* derives forecasts on the basis of establishing relationships between the time series to be forecasted and one or more other series that are assumed to influence, or cause, the behavior of the first time series. Cause-and-effect methods have the following characteristics:

- They bring more information to bear on the forecasting problem.
- They exploit leading indicator relationships.
- They are usually more time-consuming and expensive to apply.
- They are useful for longer-term forecasting.
- They require forecasts of the explanatory time series.

The Box-Jenkins forecasting method described in this book is a self-projecting time series forecasting method. Thus you can use the Box-Jenkins forecasting method described here if you have the following:

- A time series representing the activity you want to forecast.
- Short- to medium-term forecasting requirements.
- An established track record, i.e., enough data.

The self-projecting Box-Jenkins forecasting method is often referred to as the univariate Box-Jenkins method. The Box-Jenkins methodology has also been used to develop a cause-and-effect method called multivariate Box-Jenkins. The multivariate method is beyond the scope of this book.

CHAPTER 2

Self-projecting Time Series Forecasting Models

Since the Box-Jenkins method described in this book is a self-projecting forecasting method, a basic understanding of how self-projecting methods are derived is vital for understanding how the Box-Jenkins method works. In general, all self-projecting methods are based on the concept of a mathematical model. The main purpose of this chapter, therefore, is to describe what self-projecting models are and how they are constructed.

In this chapter you will learn:

- What a self-projecting forecasting model is.
- How a self-projecting model is represented mathematically.
- What the model-building process is.
- What some of the traditional self-projecting models and their disadvantages are.
- What the advantages of Box-Jenkins models are.

Since self-projecting models depend entirely on the time series data to be forecasted, the historical behavior of the time series plays a central role in the construction of self-projecting forecasting models. In particular, the identification of consistent patterns of behavior is of prime importance. We therefore begin our discussion of self-projecting time series forecasting models with a description of some special time series patterns.

2.1 TIME SERIES PATTERNS

In the self-projecting forecasting approach the only thing we have to work with is the past behavior of the time series itself. Fortunately, most time series usually exhibit some consistent patterns of behavior. To forecast a time series, then, we must be able to identify what these patterns are and, by some means, extend them into the future. This is the basic philosophy behind all self-projecting forecasting methods.

Some of the more common patterns encountered in business and economic time series are described below:

1. The *overall trend pattern* is the overall movement or direction in the time series from beginning to end. While the trend often will be simply rising, falling, or horizontal, it is possible to consider more complex overall trends. A typical overall trend pattern is shown in Figure 2.1.
2. The *seasonal pattern* is a repetition of the same basic pattern at regular intervals. The time interval is called the season (e.g., a year), and the time periods within this interval are called the periods per season (e.g., 12 months or 4 quarters). Figure 2.2 shows a typical seasonal pattern.
3. A *cyclic pattern* is a more or less consistently rising and falling pattern over an extended period of time. The distances between peaks and valleys are generally not equal. Often the overall trend and cyclic behaviors of a series are considered as one pattern, called the *trend cycle*. A typical cyclic pattern is illustrated in Figure 2.3.

The above patterns (trend, seasonal, and cyclic) are visual patterns; i.e., they can generally be recognized when a time series is displayed in graphical form. Some time series may contain only one of these patterns, while others may contain all of them combined. Such combinations can occur in a wide variety of ways, leading to more and more complex patterns. Most traditional self-projecting forecasting methods are based on the attempt to generate these types of patterns or combinations of patterns.

It is possible, however, that patterns that are not visually perceptible might exist in time series data. In this case, without the proper tools it would be very difficult to identify such patterns, let alone generate them. Such patterns we will call *statistical patterns,* since it is with statistical tools that they can be identified:

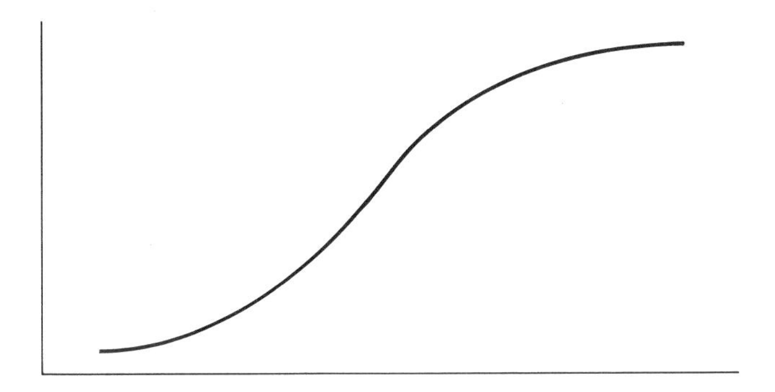

Figure 2.1 Overall Trend Pattern

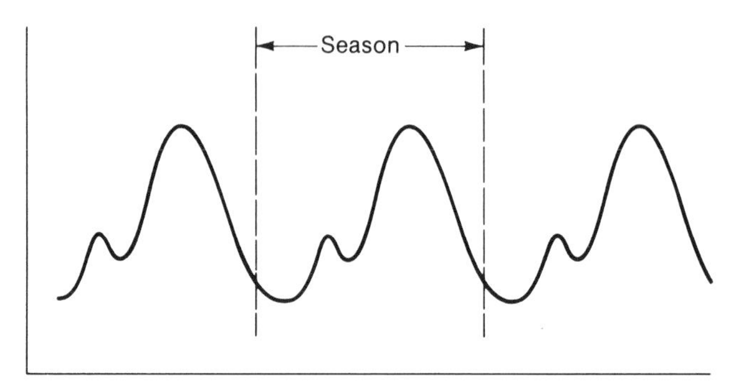

Figure 2.2 Seasonal Pattern

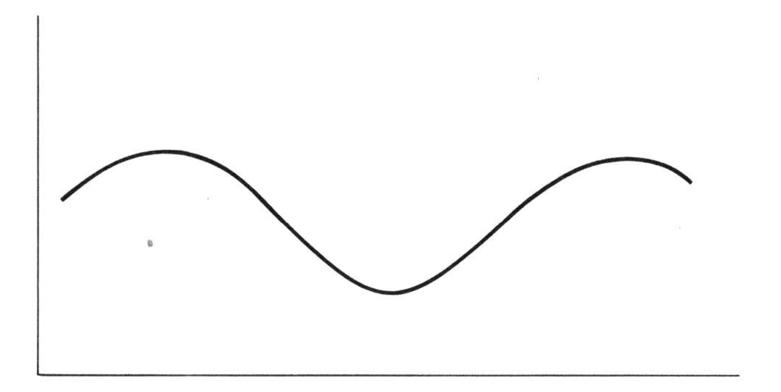

Figure 2.3 Cyclic Pattern

4. A *statistical pattern* is a consistent relationship among time series values (e.g., an increase in value from one period to the next may, more often than not, be followed by a decrease in value in the next period). Recognition of statistical patterns forms the basis of the Box-Jenkins forecasting method. Note that the concept of statistical patterns includes visual patterns as well.

2.2 SELF-PROJECTING FORECASTING MODELS

Once we have identified the patterns in a time series, how do we extend these patterns into the future? We do so by developing a *mathematical model*. The main part of a model, in simplest terms, is nothing more than a formula. When certain values are substituted into the formula, a series of values is generated. If the formula is a good one, the values it generates will follow the patterns in the original time series. If such a formula is found, it can then be used to generate additional values (forecasts) that extend the patterns into the future.

2.2.1 The Pattern and Error Components of a Series

Intuitively, we know that no mathematical formula will exactly reproduce each and every value in a time series. That is, there will always be a discrepancy, or error, between a formula-generated value and the actual time series value at a given period. Even if the process we are interested in forecasting could be precisely described by some mathematical formula, the time series values representing the process are only *inexact measurements* of the process. Consequently, the formula would not reproduce the time series values exactly.

For example, the speed of a falling object can be precisely represented by the formula *speed* = *distance* × *time*. Yet if we were to measure the speed of the object at a given time by using some measuring device, the recorded value would necessarily be in error because of the inexact nature of the measurement process; and if several measurements were made at the same time, different errors would result. Unless our measurement device was biased in some way, the errors would tend to be random; i.e., some measurements would be larger and some smaller than the true speed value, and the average of all the measurement errors would tend to be something close to zero.

What this discussion means is that any time series is assumed to have two major components: a *pattern component* and a nonpattern, or *random error, component;* i.e.,

Time series = Patterns + Random error

By its very nature, the random error component cannot be forecasted. The best we can do is provide some estimate of how big it might be in future time series values. Thus the greater the presence of patterns in a series, the more accurate the forecasts will be (assuming we can find a formula to reproduce the pattern component). And likewise, the larger the random errors in time series data are, the less accurate a forecast of the series is likely to be, i.e., the greater the degree of uncertainty in the forecast.

2.2.2 Model Representation

To summarize what we have said so far, suppose we have N periods of time series data and these periods are numbered 1, 2, 3, . . . , N. The time series value for the first period can then be denoted by the symbol X_1 (read "X sub 1"), the second period by X_2, and so on, down to X_N. In general, we have the series values X_t for $t = 1, 2, \ldots, N$. In addition, we will represent the pattern component in a series by F_t and the random error component by E_t. We can then write our forecasting model as follows:

$$X_t = F_t + E_t$$

This expression is the general representation of all self-projecting forecasting models.

Now suppose we find an actual formula that we think will reproduce the patterns in the series; i.e., we use the formula to generate a value for F_t at each period t. The *computed* F_t are then called *fitted values* for t less than or equal to N ($t \leq N$) and *forecast values* for t greater than N ($t > N$). The F_t are then subtracted from the X_t to obtain E_t, for $t \leq N$. The *computed* E_t are called *residuals*. If our formula is the correct one, then the residuals should behave like a random error component. Note that we don't know what E_t is for $t > N$. It turns out, though, that sometimes we can make a statement about how big the E_t are likely to be in the future.

The notation introduced above is summarized in Table 2.1.

Table 2.1 Summary of Notation

Period, t	1	2	3	...	N	$N + 1$	$N + 2$	$N + 3$ ...
Actual Series, X_t	X_1	X_2	X_3	...	X_N	?	?	?
Fitted/Forecast Values, F_t	F_1	F_2	F_3	...	F_N	F_{N+1}	F_{N+1}	F_{N+3} ...
Residuals, E_t	E_1	E_2	E_3	...	E_N	?	?	?

Forecast periods

2.3 MODEL FORMULAS

What is the nature of a formula that generates the fitted and forecast values F_t? First of all, a formula indicates how certain values are to be substituted and combined arithmetically to produce values that conform to the desired patterns. For example, the formula

$$F_t = \tfrac{2}{3}X_{t-1} + \tfrac{1}{3}X_{t-2}$$

indicates that the fitted (or forecast) value F_t, at period t, is produced by multiplying the time series values at period $t - 1$ and $t - 2$ (i.e., the two immediately preceding periods) by $\frac{2}{3}$ and $\frac{1}{3}$, respectively, and adding these results together. The two numbers $\frac{2}{3}$ and $\frac{1}{3}$ are called *parameters*. For example, if $X_{21} = 12.9$ and $X_{22} = 18.0$, then

$$F_{23} = \tfrac{2}{3}X_{22} + \tfrac{1}{3}X_{21} = \tfrac{2}{3}(18.0) + \tfrac{1}{3}(12.9) = 16.3$$

There are therefore three elements in a model formula:

1. The *data* that is substituted into the formula.
2. The *parameters*.

3. The *form* of the formula, i.e., the indication of how the data and parameters are to be combined arithmetically.

For example, the form of the above formula could be expressed more generally as

$$F_t = AX_{t-1} + BX_{t-2}$$

where A and B represent the parameters. A particular formula of the above form is then obtained by assigning specific values to the parameters A and B, as in

$$F_t = \tfrac{2}{3}X_{t-1} + \tfrac{1}{3}X_{t-2}$$

2.4 THE PROCESS OF BUILDING MODELS

Given a time serie,s how do you proceed in determining a forecasting model for your series? In general the procedure is a fourfold process:

1. *Identify the correct model form.* This step is done by identifying the past patterns and then selecting the appropriate formulas that can generate the kind of patterns identified.
2. *Determine specific values for the parameters in the model.* For example, if the model form is $F_t = AX_{t-1} + BX_{t-2}$, values for A and B are computed from the given time series data so that the generated F_t are as close as possible to the original X_t, i.e., so that the E_t are minimized in some sense (the concept of minimizing the E_t is discussed later in Chapter 10). If the model form is the correct one, the residuals E_t should also not exhibit any patterns, since they are supposed to be unrelated random errors. Depending on the model form, one of a number of known mathematical procedures can be used to find the appropriate parameter values. No matter what procedure is used, though, the process of computing the parameter values is called *estimation.* This term is used because the parameters are computed from real-life time series data, which, you'll recall, are inexact measurements of the activity to be forecasted. Thus we can only approximate, or estimate, the true parameter values.
3. *Forecast the time series* by using the specific model formula determined in steps 1 and 2. That is, generate values from the formula for periods greater than N, *and* provide a measure of the uncertainty in these forecasts.
4. *Monitor the accuracy of the forecasts* when new, actual time series values are obtained. Forecasts that are off the mark more than expected may suggest a change in the historical patterns (and therefore the model form) or a change in the model parameter values, which must therefore be reestimated. In this sense building a forecasting model is a dynamic process.

The process outlined above, however, raises many questions:

- How do you identify the patterns in the series?
- What mathematical formulas can be used to generate these patterns?
- How do you estimate the specific parameter values?
- How do you measure the uncertainty in the forecasts?

Many forecasting methods have been developed over the years to address these questions. Let's take a look at some of these traditional methods and discuss their shortcomings.

2.5 EXAMPLES OF SOME TRADITIONAL SELF-PROJECTING MODELS

2.5.1 Overall Trend Models

One of the most common historical patterns in time series data is *overall trend.* Most series covering long periods of time generally tend to exhibit a relatively simple overall growth or decline trend pattern. Figure 2.4 shows some typical overall trend patterns.

It turns out that these overall trends can be expressed mathematically in terms of the time period t. For example, the straight-line trend has the form

$$\text{Trend}_t = A + Bt$$

where A and B are parameters and t represents the values 1, 2, ... The formula-generated value at period t is, therefore, A plus B times the value t.

For example, the only pattern demonstrated in the time series plotted in Figure 2.4, is an overall, increasing straight-line trend. Since we have a formula that can generate straight lines, a possible model for this series would be

$$X_t = \overbrace{A + B_t}^{F_t} + E_t$$

Values for the parameters A and B can then be determined for a particular series by using a well-known procedure, called the least squares method, that minimizes the E_t. (Actually, it is the sum of all the squared residuals E_t^2 that is minimized—hence the name *least squares*.) The results of applying this procedure to the series in Figure 2.5 are shown in Figure 2.6 (results have been rounded to the nearest unit).

The problem with overall trend models, though, is a critical one: It is not always clear what the best trend model should be for forecasting. For example, in Figure

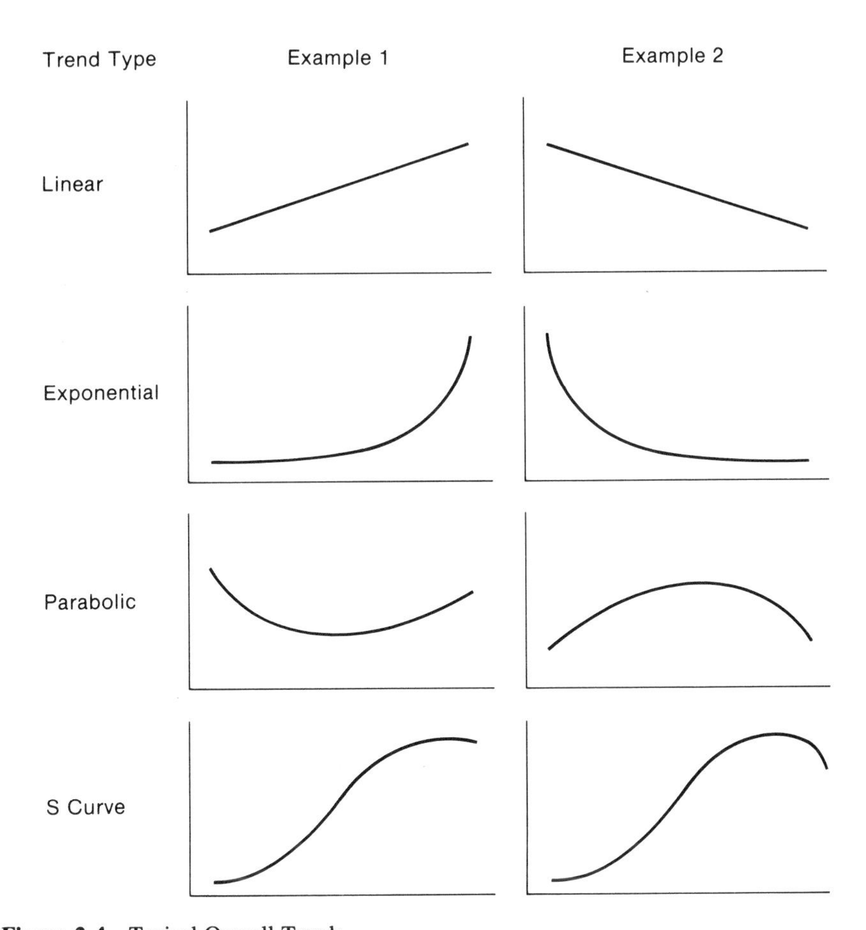

Figure 2.4 Typical Overall Trends

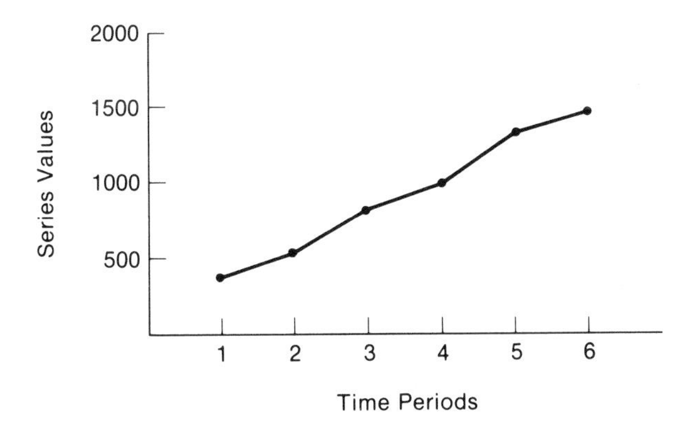

Figure 2.5 A Series with an Overall Straight-Line Trend Pattern

Model Formula: $F_t = 152 + 225t$			
t	X_t	F_t	E_t
1	395	377	18
2	575	602	−27
3	850	827	23
4	1000	1052	−52
5	1325	1277	48
6	1490	1502	−12
7	—	1727	—
8	—	1752	—
⋮	⋮	⋮	⋮

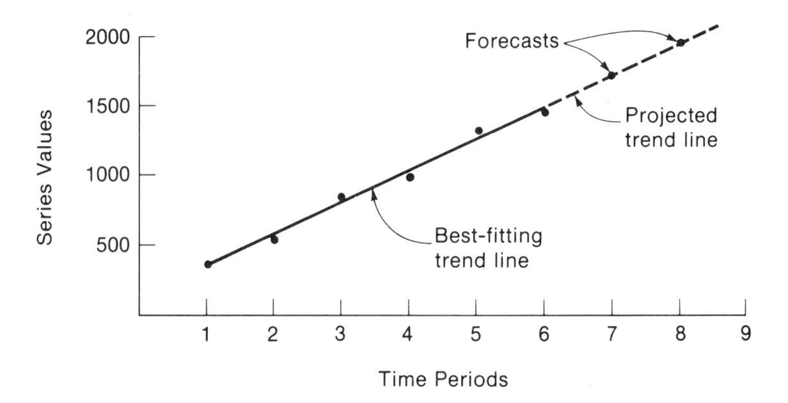

Figure 2.6 Results for the Straight-Line Trend Model

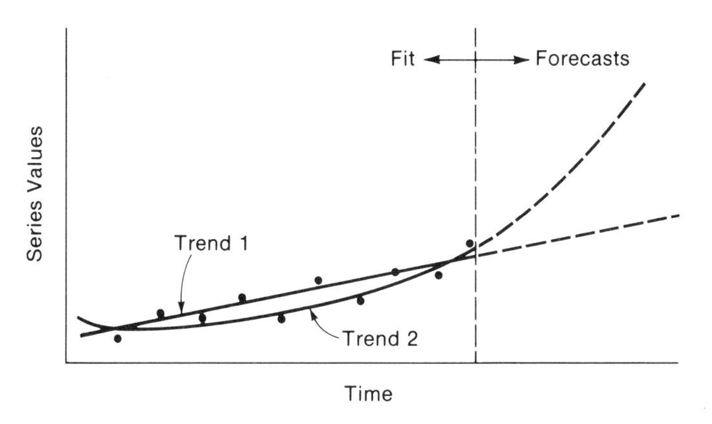

Figure 2.7 Unclear Choice of a Trend Model

2.7 two trend models were used to fit the same time series. Both models produce similar fits to the time series data, yet the forecasts generated by the two models are radically different.

Generally, there is no systematic way to go about identifying and selecting the right overall trend for forecasting. Often the final choice is arbitrary.

2.5.2 Smoothing Models

An overall trend model, by nature, has problems in tracking short-term changes in a time series. If more short-term forecasting is desired, an overall trend model by itself may not be appropriate. Consequently, a wide variety of methods, classified as *smoothing methods,* have evolved over the years that attempt to model intermediate trending and/or cyclic behavior. In other words, these models place more emphasis on responding to the most recent behavior of the series to produce near-term forecasts.

Most smoothing methods are based on intuitive and commonsense considerations. In general, they employ the idea of computing averages or weighted averages of past time series values. The main idea behind smoothing is based on two intuitive feelings held by most people: (1) that there usually is some relationship between what happened in the last few periods and what is likely to happen in the next period, and (2) that there are, in fact, unexplainable irregularities in the past data values that should be ignored, if possible. Using an average of past values as a forecast for the next period seems to be a reasonable mathematical translation of these intuitive feelings, since it does establish a relationship, and the irregularities tend to cancel each other out in the averaging process. In other words, the random errors in the data are "smoothed out," leaving only the intrinsic pattern of the series. The amount of smoothness obtained depends on how many past values are averaged and the weights (if any) that are assigned to the past values.

For example, the series in Figure 2.8 doesn't exhibit an overall trend, but there are definite short-term trend and horizontal patterns that are partially obscured by the random fluctuations in the data.

Some of the random fluctuations in this series can be smoothed out by taking a simple average of past values to produce a fitted value for the current period. One possible model (among many) that can do this is as follows:

$$
\begin{aligned}
X_t &= 1/3X_{t-1} + 1/3X_{t-2} + 1/3X_{t-3} + E_t \\
&= 1/3(X_{t-1} + X_{t-2} + X_{t-3}) + E_t
\end{aligned}
$$

The results obtained from this simple-average smoothing model are shown in the graph in Figure 2.9.

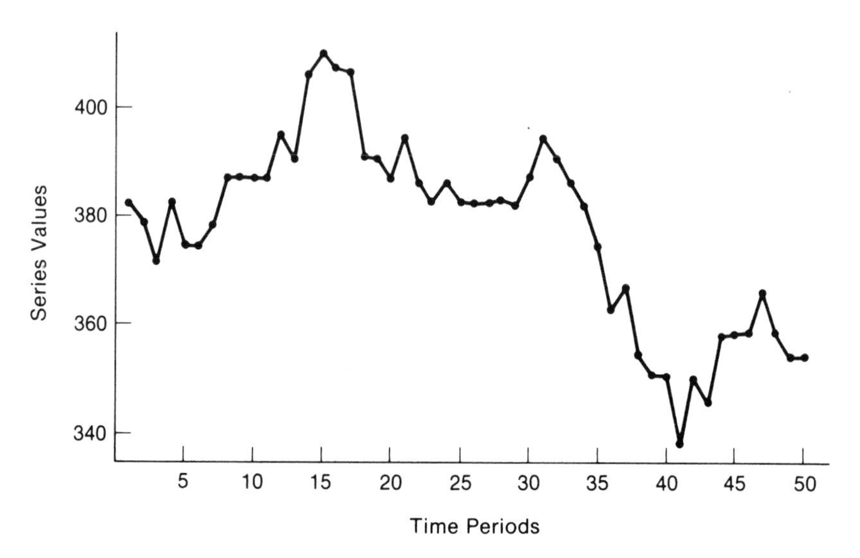

Figure 2.8 A Series with No Overall Trend Pattern

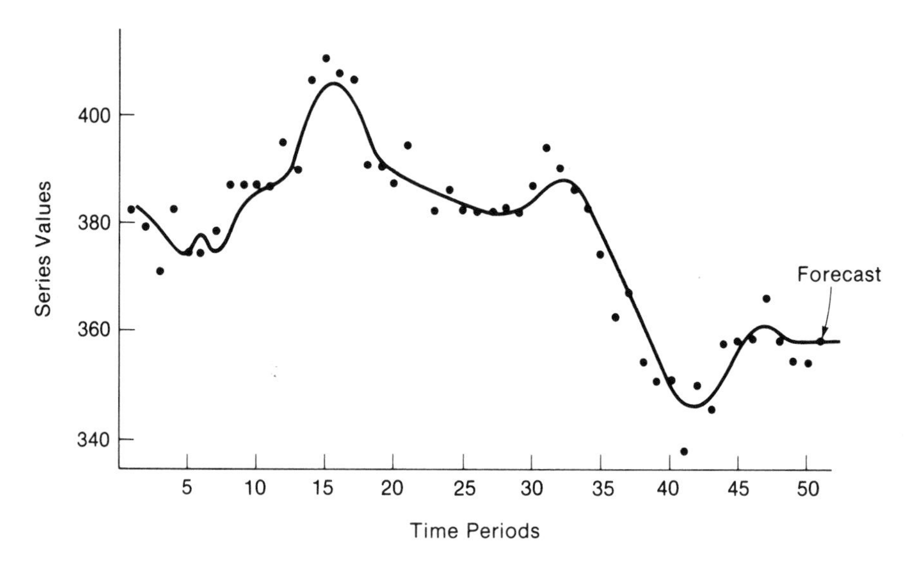

Figure 2.9 Results for the Simple-Average Smoothing Model

There is, of course, a huge repertoire of smoothing models, ranging from the simple averaging technique previously illustrated to sophisticated weighting schemes. The well-known exponential smoothing method falls into this latter category. For example, simple exponential smoothing* has the following model representation:

*It turns out that the simple exponential smoothing model is one example of a simple Box-Jenkins model.

$$X_t = \overbrace{AX_{t-1} + (1 - A)F_{t-1}}^{F_t} + E_t$$

where A is called the smoothing parameter, which can be determined by known optimization procedures (note the use of F_{t-1} in the computation of F_t).

The problem with smoothing models, as with overall trend models, is that there is no systematic means by which to select the appropriate model. Most smoothing models were developed in an *ad hoc* manner to support intuitive notions and simplicity in computation. Thus model selection is generally an arbitrary, trial-and-error process. In addition, long-term or medium-term forecasting is normally not possible with smoothing models.

2.5.3 Seasonal and Decomposition Models

Many series, of course, exhibit patterns other than long- and short-term trend patterns. Seasonal patterns are extremely prevalent in business and economic time series, and many models have been developed to fit and forecast such patterns.

In general, however, most seasonal time series also contain long- and short-term trend patterns. Consequently, *seasonal models* have been combined with trend and smoothing models to produce composite models that fit all these patterns simultaneously. One general representation of such a composite model is

$$X_t = \overbrace{(\text{Overall trend}_t) \times (\text{Seasonal}_t) \times (\text{Cycle}_t)}^{F_t} + E_t$$

Because the series is, in essence, decomposed into its separate patterns, and each pattern is modeled separately, the above model is called a **decomposition model.**

2.5.4 Disadvantages of Traditional Models

The traditional models just described can often be successful forecasting models in certain circumstances. In general, however, the use of traditional models has some serious drawbacks:

- There are a *limited number of models:* The models available for consideration often are unable to reproduce many of the patterns (both visual and nonvisual) that exist in a given time series.
- There is *difficulty in finding the right model:* There is no systematic approach for the identification and selection of an appropriate model. In practice, the model identification process is a combination of visual recognition of patterns and trial-and-error.

- There is *difficulty in verifying the validity of the model:* There are little or no statistical guidelines for judging whether or not the model obtained is a good one or for assessing the uncertainty in the forecasts produced. This problem results because most traditional methods were developed from intuitive and practical considerations rather than from a statistical foundation.

2.6 BOX-JENKINS MODELS

The disadvantages associated with traditional forecasting methods are, for the most part, not present in Box-Jenkins forecasting models. This happy circumstance is primarily due to the origin and development of the models used in the Box-Jenkins method. Actually, the types of models employed in the Box-Jenkins method have been around for many years, having been developed for purposes and disciplines other than business or economic forecasting.* It wasn't until the 1960s that Box and Jenkins recognized the importance of these models in the area of economic forecasting and developed the Box-Jenkins forecasting methodology to take advantage of them.

The important characteristic of these models is that they were developed in terms of statistical concepts under the assumption that the processes or activities being modeled are dynamic and subject to statistical fluctuations. This assumption, of course, is quite compatible with the behavior of economic and business activities represented by time series. Because of the statistical nature of these models and their wide applicability, Box-Jenkins forecasting models have the following advantages over traditional models:

- There is a *large class of models*. The generalized Box-Jenkins model form can model a wide variety of time series behavior and patterns.
- There is a *systematic approach to model identification*. The Box-Jenkins method provides statistical tools for identifying the particular model form required for a given series.
- The *validity of models can be verified*. After the specific model parameter values have been determined, the Box-Jenkins method provides many statistical tests for verifying the validity of the chosen model.
- *Forecast accuracy can be measured*. The statistical theory behind Box-Jenkins models allows statistical measurements of the uncertainty in a forecast to be made.

*These models are more generally known as ARIMA models. The origin of this acronym will be described in Part 3.

2.7 REVIEW OF KEY CONCEPTS

The basic idea behind self-projecting time series forecasting models is to find a mathematical formula that will approximately generate the historical patterns in a time series. *Time series patterns* come in a variety of forms such as overall trends, seasonal fluctuations, and cyclic movements. Most business and economic time series contain one or more of these patterns, as well as other statistical patterns not visible to the eye.

Self-projecting forecasting models can be stated in symbolic terms as

$$X_t = F_t + E_t$$

where t stands for the period numbers 1, 2, . . . N. The F_t represents the pattern component in a series and the E_t represents the random error component. When the F_t are generated by some formula, they are called *fitted values* for $t \leq N$ and *forecast values* for $t > N$. The E_t are then computed as the differences between X_t and F_t for $t \leq N$ and are called the *residuals*. The idea is to find an appropriate formula so that the residuals are as small as possible and exhibit no pattern, i.e., are not related to each other.

The *model formulas* used to generate the F_t consist of three elements: (1) the form of the formula, which indicates how certain values are to be combined arithmetically, (2) data that is dynamically substituted into the formula (i.e., the data substituted into the formula changes for each period t), and (3) parameters that remain constant for a specific formula.

Examples of model formulas are

$$F_t = A + Bt \qquad \text{(straight-line trend)}$$

$$F_t = AX_{t-1} + BX_{t-2} + CX_{t-3} \qquad \text{(weighted average)}$$

where A, B, and C in the above formulas are the parameters. The data substituted into the straight-line trend formula are the period numbers t, and the data substituted into the weighted average formula are past series values.

The *model-building process* is a four-step process. The object of this process is to end up with a specific formula for F_t that replicates the patterns in the series as closely as possible and also produces accurate forecasts. The steps in this process are as follows:

1. Identify the model form.
2. Compute appropriate values for the model parameters from the time series data.
3. Forecast the time series by using the specific formula obtained in step 2.
4. Monitor the accuracy of the forecasts in real time, and repeat steps 1 through 3 if necessary.

Traditional self-projecting forecasting models have attempted to model the common patterns that normally exist in business and economic time series, such as overall trends (trend models), short-term trends or horizontal movement (smoothing models), and seasonal fluctuations (seasonal models). Models that address combinations of seasonal and other patterns are called decomposition models.

The *problems with traditional forecasting models* are as follows:

- There are a limited number of models to choose from, relative to the complex behavior of many time series.
- There is no systematic approach for identifying the right model form.
- There are little or no guidelines for verifying that a selected model is a valid one.

Box-Jenkins forecasting models are based on statistical concepts and principles and are able to model a wide spectrum of time series behavior. The Box-Jenkins approach to model building provides the following advantages:

- There is a large class of models to choose from.
- There is a systematic approach for identifying the correct model form.
- There are statistical tests for verifying model validity.
- There are statistical measures of forecast uncertainty.

One of the virtues of many traditional methods, however, is that they are easy to understand, apply, and communicate to others, while the Box-Jenkins method, to many people, appears complicated and difficult to understand. This difficulty is due primarily to the statistical underpinnings of the method and the corresponding body of jargon normally used to describe and derive it.

As a practical matter, it is really not necessary to have a complete understanding of the underlying statistical theory or the various mathematical derivations that are used to develop the Box-Jenkins method. The basic concepts behind Box-Jenkins model building are actually easy to understand, and the process of building a Box-Jenkins forecasting model is easy to grasp and apply. Learning this process is what you will accomplish in Parts 2 and 3 of this book.

CHAPTER **3**

What You Need to Do to Successfully Use Box-Jenkins

Before you can successfully apply the Box-Jenkins forecasting method, you will need to have good data to work with, and you will need to be able to quickly and easily perform the required Box-Jenkins computations on this data.

In this chapter you will learn:

- How to obtain access to a Box-Jenkins computer program.
- How to collect and qualify your data.
- How to prepare and adjust your data, when necessary.

3.1 OBTAIN ACCESS TO A COMPUTER PROGRAM

The computations involved in the Box-Jenkins method are much too laborious and time-consuming to do by hand. It is absolutely necessary to gain access to a computer and a computer program to perform these computations for you.

Many Box-Jenkins programs may be accessed through commercial computer time-sharing services or are available for sale from computer software distributors or developers (commercial and noncommercial). The latter may be installed on your company's computer if you have an appropriate computer and computer system. Most of these programs are similar in terms of the type of output provided but may differ in how they are operated. A partial list of some of the currently available Box-Jenkins programs is given in Appendix E; some of the attributes of these programs are also discussed there.

The program and environment you should choose will depend on cost factors, the availability of computer resources at your company, and your own personal preferences. Probably the best way to start is by signing up on a time-sharing service, even if you have the necessary computer resources in-house to buy a program. When you become familiar with the Box-Jenkins modeling process and operation of the program on that system, you will then be in a better position to compare other programs and environments and decide which is best for you.

3.2 COLLECT AND QUALIFY YOUR DATA

One of the biggest reasons for failure in the application of time series forecasting methods is a misunderstanding of the nature of the data you are dealing with. Before building a Box-Jenkins model, you should therefore ask some of the following questions about the data you intend to collect or have collected:

1. Does the definition of the data have more than one interpretation? For example, is the data for product demand recorded on the basis of orders placed or of shipments made? Depending on what the forecast is to be used for, an incorrect interpretation of the data can lead to misinterpreted forecasts.
2. Does the data truly represent what you think it does? For example, if product demand is recorded as orders placed, does the data also include adjustments for orders lost?
3. Where does the data come from and how is it produced? Data is produced and recorded by people who often "filter" the data for various reasons (not necessarily devious). For example, changes in accounting practices for tax reasons can produce changes in the way financial information is recorded and, therefore, in the definition of what the data represents.
4. Is the process or activity represented by the data constant over time? For example, is product A still really product A as originally defined, or have sales of various related models or products been lumped into "sales of product A" over the years?
5. Do you have enough data? Most methods require some minimum amount of data before you can theoretically apply them. But generally, the practical minimums are much larger than the theoretical ones. In Box-Jenkins the rule of thumb is to have at least 40 to 50 periods of data, or 4 to 5 seasons of seasonal data, whichever is larger.

Asking questions like these can help you make sure you are collecting the right data and can point out what might be done to improve the quality of the data. Knowing what your data really represents and how it was recorded can eliminate common misuses of forecasts and failures in forecasting.

3.3 CLEAN AND ADJUST YOUR DATA

Almost all data needs to be altered in some way before you try to develop any forecasting model. Problems such as actual errors made in recording the data, one-time events (strikes, bad weather) that cause abnormal or extreme behavior, trading-day variations in monthly or quarterly data, or missing data interfere with or disrupt what would otherwise be normal series behavior. To the extent possible, these problems should be identified and cured.

3.3.1 Extreme Values

Extreme values in time series data are usually the result of one-time abnormal events such as strikes or bad weather or errors in data recording. There is no standard way to identify extreme values other than by noting them visually from graphs of the series or through special knowledge of what has happened in the past. Often the application of a forecasting method will uncover problem values when unusually large residuals are computed. If you have monthly or quarterly data, the X11 seasonal adjustment method [38] used by the U.S. government provides a statistical technique for identifying extreme values and replacing them. This method is also available as a computer program. In any event, extreme values should be replaced by some more representative value. This replacement may be done with the X11 technique described above (if applicable) or by some *ad hoc* method.

3.3.2 Trading-Day Variations in Monthly Data

Trading-day variations are a source of inconsistent behavior in certain business-oriented, *monthly or quarterly* time series arising from unequal-length months and the uneven distribution of days of the week in the different months. For example, a time series of retail sales revenue for a store that is only open Monday through Friday will tend to have higher sales in months that have fewer weekend days, i.e., more business or trading days. The most common method in use today for adjusting monthly time series data for trading-day variations is also found in the X11 seasonal adjustment method noted previously. Fortunately, however, most time series do not suffer from trading-day variations.

3.3.3 Missing Values

The Box-Jenkins method assumes (as do most other self-projecting methods) that all time periods have data recorded for them. Values must be supplied for periods with no recorded data, or the portion of the series prior to and including the last missing value must be thrown away. Often these values can be generated by some *ad hoc* method or simple interpolative techniques.

3.3.4 Skimpy Data

Time series data collected for a detailed activity, such as the inventory level for a single product, may at times be skimpy or incomplete because of low sales levels, for example. The historical data for this one period may not contain enough information to determine a reasonable forecasting model. But if the time series data for similar products were combined, the resultant time series—the *aggregate series*—might then exhibit more recognizable patterns than the individual series do. The aggregate series can then be adequately forecasted. Forecasts for the constituent series may not be possible, however, unless the aggregate forecast can be broken down in some way.

In a similar manner, time series data can be *compacted* across time intervals instead of across activities. For example, suppose monthly cash flow has some seasonal regularity but is inconsistent with regard to the exact months in which highs and lows occur; i.e., cash flow may always be high in the first quarter of the year, but the monthly highs may occur in either January, February, or March. This type of erratic pattern can be overcome by compacting the data into larger time intervals, such as quarters, to obtain a series with a more consistent seasonal pattern.

3.3.5 Buried Information

On the other side of the coin, it is sometimes desired to forecast an aggregate activity, such as total product sales, where the sales data for the individual products contain a lot of information. This information can often be buried in the aggregate. In many cases then, more accurate forecasts of the aggregate can be obtained if each of the individual constituent activities is forecasted and their forecasts totaled to obtain a forecast of the aggregate.

In other cases more accurate forecasts may be obtained by using smaller time intervals. For example, yearly data might be broken down into monthly data. The pattern contained in the same data broken down into months may produce better monthly forecasts, which, in turn, may produce a better composite yearly forecast by summing the forecasts of the monthly series.

3.4 BE PRACTICAL

Often a forecaster will get carried away with a technique or method and begin to rely solely on the mechanical and technical aspects of the procedure. But the Box-Jenkins method is not a black box. The application of the method, as you will see, requires a commonsense approach and has many judgmental aspects to it. Sometimes, any statistical or mathematical technique can provide information that doesn't make sense in real life (recall the trend model examples in Chapter 2). The danger with Box-Jenkins is that because of its structure and systematic approach, there is a tendency to rely completely on the numbers and statistics that are printed out and to interpret everything strictly by the numbers. Don't fall into this trap!

3.5 REVIEW OF KEY CONCEPTS

To be successful in using the Box-Jenkins forecasting method, you will need to do the following:

1. *Obtain access to a Box-Jenkins computer program.* A computer program is required since the mathematical computations that need to be performed are much too complex and laborious to be done by hand. Access to such programs can be obtained in a variety of ways, such as through commercial time-sharing services or by buying a computer program that can be installed on your organization's computer.
2. *Collect and qualify your data.* The misuse, misunderstanding, and inaccuracy of forecasts is often the result of not fully understanding the nature of the data you are working with. Asking the following questions about your data can help you make sure you are collecting the right data and that you will interpret forecasts properly.
    - Does the data really represent what you think it does?
    - Does the data have more than one interpretation?
    - How was the data recorded and who recorded it?
    - Is the data consistent across time?
    - Do you have enough data? (The rule of thumb for Box-Jenkins is to have at least 40 to 50 periods of data.)
3. *Clean and adjust your data.* One or more of the following problems often exist in data and should be corrected before applying any time series forecasting technique:
    - *Extreme values,* caused by abnormal or unexpected events. Extreme values should be replaced by more reasonable estimates.
    - *Missing values.* Missing values should be filled in with reasonable estimates, or the time periods prior to and including the last missing value should be deleted from the series.
    - *Trading-day variations* in monthly or quarterly data caused by the unequal distribution of the days of the week in different months. Trading-day var-

iations should be eliminated if possible (about the only practical way to do so is with the X11 seasonal adjustment program).

- *Skimpy data* due to low levels of activity of the item to be forecasted. Aggregation of several time series representing similar activities (e.g., total sales of all products in a product group) or compaction of the series into larger time intervals can provide some help in this case.
- *Buried information* due to unnecessary aggregation or compaction. The component series in an aggregate series, for example, can often contain a lot of useful information, so that forecasting the component series individually can yield better forecasts of the aggregate.

4. *Be practical.* Never treat any forecasting method as a black box. Always exercise judgment and common sense in the process of forecasting as well as in the interpretation of forecasting results.

CHAPTER 4

How to Increase Your Chances of Being Successful in Forecasting

There is much more to forecasting than simply applying a particular method to some data and producing forecasts. To be successful, you should do your forecasting within the context of some overall process where everyone involved has a clear understanding of what forecasting is and what the forecasting situation is. It is therefore important to understand what some of the basic concepts and issues are in forecasting.

In this chapter you will learn:

- What a major source of problems in forecasting is.
- What the makeup of a forecast is.
- What the basic ingredients in forecasting are.
- What forecasting requirements and constraints need to be determined.
- What a typical formalized forecasting process is.

4.1 A MAJOR SOURCE OF PROBLEMS IN FORECASTING

Forecasting is not an end in itself. Its purpose is to provide input to the planning and decision-making process from which the forecasting requirement arose. The tendency on the part of many preparers of forecasts, however, is to become enamored with their forecasting methods and models to the exclusion of any consideration of the problem or situation that gave rise to the forecasting requirements in the first place. On the other hand, many forecast users tend to be suspicious of any forecasting method, particularly if it is based on statistical analyses they don't understand, and they would rather rely on intuition and gut feelings. The net result is forecasts produced with little understanding of what they are intended for and forecasts used with little confidence, or inappropriately, or not at all. The best forecasting methods in the world are of little value unless those who use the forecasts and those who prepare them share the same understanding of how and why the forecasts were derived.

The truth is that good forecasting is a result of using all available information in a rational and logical manner, whether the source of information is historical data and statistical analyses or the experience and intuition of individuals. Clearly, then, good forecasting can only be accomplished if there is a mutual understanding among all involved about what forecasting is and what the forecasting situation is. In the remainder of this chapter you will learn about some of the concepts and issues that should be understood in order to accomplish this objective.

4.2 THE MAKEUP OF A FORECAST

The end product of forecasting is, of course, a forecast. Many people think of a forecast as a simple prediction of what will happen in the future. To be of value, however, a real forecast should provide the following information:

1. A prediction of what will happen in the future (usually numerical): "Unit sales are projected to be 5000 units next month."
2. An assessment, or measurement, of the uncertainty in the prediction: "There is a 20% chance that sales will be under 4500 units."
3. A description of how the prediction was derived and what assumptions were made in deriving it: "Sales will continue to grow at the same constant rate, but they will need to be adjusted for the seasonal fluctuations demonstrated in past years and a planned promotion in the middle of the month."

Unfortunately, many forecasters (preparers of forecasts) and managers (users of forecasts) often think of a forecast as simply a one-number prediction of what will happen in the future. A prediction, however, is not really a forecast without an assessment, or measurement, of the uncertainty in the prediction. This assessment allows the planner or decision maker to weigh the risks involved in making a wrong decision based on a possibly inaccurate prediction. Such an assessment of uncertainty can drastically alter the decisions and plans that would have otherwise been made without such an assessment.

Likewise, a description of the assumptions used in obtaining the prediction and a description of how the prediction was derived are necessary elements in providing a forecast. These descriptions allow the user of the forecast to understand what the forecast really means and to use the forecast information in the proper way and proper place. They are also useful in bringing the preparers and users of forecasts together in an atmosphere of mutual understanding.

4.3 THE BASIC INGREDIENTS IN FORECASTING

In any forecasting situation there are three basic ingredients that we use to produce forecasts:

- Historical data
- Experience and judgment
- Forecasting methodologies

Historical data is simply quantified information about the past, such as time series. Monthly sales revenue, personnel statistics, demographic data, weekly production statistics, and market survey data are all examples of historical data that can provide valuable information about how certain activities have behaved in the past or about the nature of the environment. The analysis and interpretation of historical data is the foundation upon which most forecasts are built.

Experience and judgment provide the human factors in forecasting. In some cases there is little or no historical data to work with, so the experience, judgment, and expertise of certain individuals are all that's available. But even when forecasts are generated from historical data by using methods such as Box-Jenkins, they must be reviewed in the light of the experience and special knowledge of those involved in the forecasting problem. For example, knowledge of impending government regulations, the understanding of the company's political environment, or a special insight into future economic conditions must be factored into any forecast. The main assumption behind forecasts developed strictly from historical data is that the behavior of the past will continue into the future. Taking these forecasts as a foundation and then factoring in the experience, judgment, and special knowledge of individuals will produce more reliable forecasts and forecasts that will be used with greater confidence.

Forecasting methodologies provide the means by which historical data and/or experience and judgment are used and combined to produce forecasts. The appropriate methodology depends on the requirements and constraints of a particular forecasting problem.

4.4 UNDERSTANDING THE FORECASTING PROBLEM

There are many different kinds of forecasting problems. Forecasting production requirements for next week, for example, is a much different problem than forecasting return on investment for a proposed new product. These two problems vary widely in terms of the forecasting requirements, the purpose of the forecast, the environmental constraints, the data availability, and so on. Such differences suggest the use of different approaches and methodologies for solving each problem. It is therefore important to thoroughly understand the unique characteristics of your forecasting problem in terms of requirements and constraints in order to make the best possible use of the three basic ingredients in forecasting. Some of the most important considerations are the following:

- *What is the purpose of the forecast?* How is the forecast to be used? What decisions are to be made? Using sales forecasts to set sales quotas and using forecasts to prepare a budget have different implications.
- *What information is required?* What factors influence the activity or event being forecasted? Which factors are under your control and which are not? Which factors can be quantified? What data is available?
- *What is the quality of the data?* What is the relevance of the available data? Where did it come from? How was it recorded? What are the possibility of errors?
- *What is the forecast horizon?* How far in the future must a forecast be provided? Usually, forecasters divide the forecast horizon into short-term (e.g., 1 to 3 months), medium-term (e.g., 3 to 24 months), and long-term (e.g., over 2 years) forecast horizons. The terms *short, medium,* and *long,* of course, are relative, since what may be considered short-term in one forecasting situation may be considered long-term in another.
- *What accuracy is required?* What is the impact of a bad forecast? If there is low sensitivity to a bad forecast, there is no sense in trying to produce the most accurate forecast possible.
- *What resources are available?* How much time, money, and expertise are available to develop the forecast? If there is very little lead time, for example, the use of a complicated or sophisticated method is probably out of the question.
- *Who will use the forecast?* Who will act upon the forecast? What are their assumptions and perceptions of the situation? If the users of a forecast don't understand how the forecast was derived, they probably won't have confidence in it, let alone use it.

Finding answers to these questions is the necessary first step in the forecasting process.

4.5 THE FORECASTING PROCESS

Most companies have a wide range of forecasting needs. Moreover, there is usually a high degree of interdependence among the forecasts required in different areas of the business. Sales forecasts, for example, are an important component of the forecasts required in most other areas.

Because of this wide range of needs and the interdependence of forecasts, forecasting cannot be effective unless it is done within the framework of some overall formalized process. The implementation of such a process is also necessary to ensure a meeting of the minds between preparers and users of forecasts, where everyone understands the assumptions, requirements, and purpose involved in the forecasting situation. Only in this way can the decision maker accept and use the forecasts provided and the analyst select the correct data and appropriate methods to produce the forecast. Only in this way, too, can forecasts be critically reviewed and the effectiveness of the forecasting process appraised and improved on a regular basis.

Figure 4.1 shows a simplified view of the main steps involved in a typical formal forecasting process. These steps are discussed below.

Step 1. Define the forecasting problem. This step is probably the most important since understanding the nature of the forecasting situation affects how all succeeding steps in the process are carried out. From this understanding, information and data requirements are determined, accuracy and forecast horizon requirements are set, and time and cost constraints are revealed.

Step 2. Collect and prepare the data. This step is often a difficult task since historical data is rarely in an immediately usable form and may be difficult to find. Also, you may not know precisely what it represents. As you learned in Chapter 3, almost all data need to be adjusted in some way in order to successfully apply most forecasting methods. The accuracy and consistency of the data must also be assessed and understood. (See Chapter 3 for more details on data collection and preparation.)

Step 3. Select and apply a forecasting method. The selection of an appropriate method depends on what was determined in steps 1 and 2. The type and quantity of data available, the accuracy and the forecast horizon requirements, the time and cost constraints, the expertise available, and management's understanding and acceptance are all factors in the choice of an appropriate method.

Step 4. Review and adjust preliminary forecasts. This step provides the means by which historical data and management experience and judgment are combined to produce a final forecast. Once a preliminary forecast has been produced by applying a particular method, the results should be carefully reviewed by both the preparer and the user of the forecast. Are the forecasts reasonable? Are they consistent with the assumptions made? The judgment of those familiar with the business environment in which the forecast is made must be brought to bear on preliminary forecasts. For example, special events such as impending strikes or government regulations cannot be predicted from the historical data, or conditions in the current business environment may not be reflected in the historical data; but such events or conditions must be factored into any forecast. In short, the results of applying forecasting methods provide a starting point to guide the decision maker in producing a final forecast.

Steps 5 and 6. Track the forecast accuracy and update the forecasts and the forecasting system. These steps provide the means by which the effectiveness of

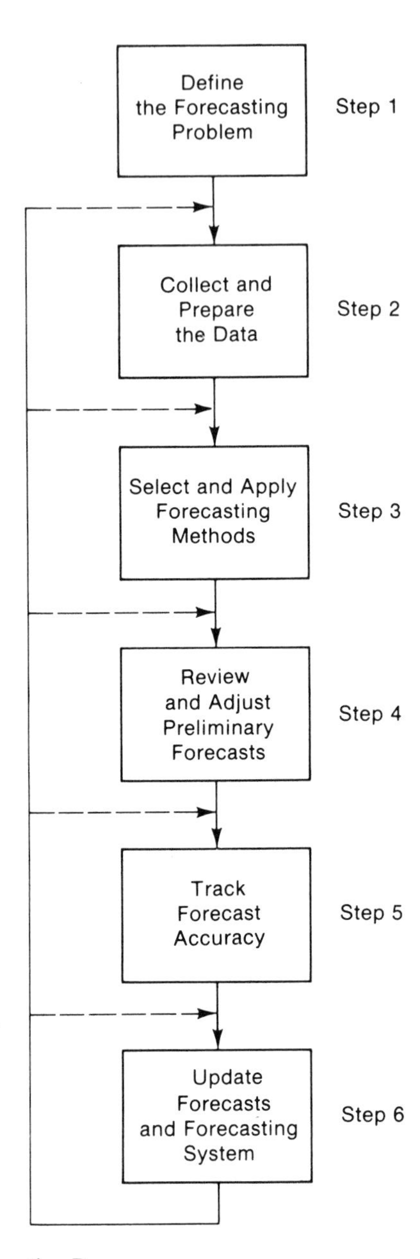

Figure 4.1 The Forecasting Process

the process is evaluated and updated. Final forecasts should constantly be compared with what actually happens in order to determine the effectiveness of the forecasting methods used. When accuracy requirements are not met, the entire process should be reviewed and updated. As additional data becomes available, new and revised forecasts should be made, and the validity of these forecasts should be checked on a regular basis. Forecasting is rarely a one-time activity. As time moves on, the assumptions, requirements, and constraints that were initially developed will probably change.

4.6 REVIEW OF KEY CONCEPTS

The application of any forecasting method cannot be done in a vacuum. For forecasting to be successful, there must be a mutual understanding between the forecasting practitioner and the forecast user about what the makeup of a forecast is, what the ingredients in forecasting are, what the requirements and constraints of the specific forecasting problem are, and what the forecasting process should be. Misunderstandings between these two people on these issues is one of the major sources of problems in forecasting. Only if a mutual understanding between the forecast preparer and user is attained will forecasts be used with confidence and more accurate forecasts made in the long run. The application of a specific forecasting method is only one part of an overall process.

A *forecast* consists of three components:

1. A prediction, which is usually numerical.
2. An assessment of the uncertainty in the prediction.
3. A description of the method and assumptions used in deriving the forecast.

The *three basic ingredients in forecasting* are as follows:

1. Historical data
2. Experience and judgment
3. Forecasting methodologies

The way in which these ingredients are combined and used depends on the nature of the forecasting problem.

The *nature of a forecasting problem* is assessed by determining and understanding the following elements:

- What the purpose of the forecast is.
- What information is required.
- What the quality of the data is.
- What the forecast horizon is.

- What accuracy is required.
- What the resource constraints are.
- Who the user of the forecast is.

Forecasting methods by themselves are of little value unless they are used within the framework of an overall *forecasting process*. A forecasting process involves several steps:

1. Define the forecasting problem (purpose, requirements, constraints, assumptions).
2. Prepare the data (collection, interpretation, adjustment).
3. Select and apply the appropriate methods.
4. Review and adjust the forecasts generated by these methods.
5. Track the accuracy of the forecasts.
6. Update the process (new data, revised forecasts, new methods, new assumptions and requirements).

Part 2

The Foundation

Before continuing with Part 2, you should already be familiar with the following concepts:

- Time series.
- The self-projecting time series forecasting approach.
- The basic structure of forecasting models: model form, parameters, fitted values, forecasts, and residuals (random error).
- The model-building process: identification of model form, determination of parameter values, forecasting with the model, and monitoring forecasts.

If you are not familiar with the above concepts or you feel you need a brief refresher, please read Chapters 1 and 2 in Part 1.

The material in Part 2 lies at the heart of the Box-Jenkins modeling process. Once you have mastered this material, you will know how to:

- Recognize the basic Box-Jenkins models.
- Identify and select the correct model for some simple series.
- Determine the specific parameters in a selected model.
- Evaluate the specific model.

Models for time series with more complex behavior are developed in Part 3. The models described in Part 2, however, provide the foundation for building all Box-Jenkins models, and the techniques and tools presented in Part 2 for the selection and evaluation of the basic models are the same as those for the more complex models.

CHAPTER **5**

The Box-Jenkins Model-Building Process

In Chapter 2 you learned about self-projecting forecasting models and the process of building such models for a particular series. Since an understanding of the Box-Jenkins forecasting method depends heavily on understanding these two concepts, it is worthwhile to review the basic ideas involved.

In this chapter, therefore, we will review:

- What the basic structure of self-projecting forecasting models is and what the objectives in building these models are.
- What the model-building process is from the point of view of the Box-Jenkins method.

5.1 A REVIEW OF SELF-PROJECTING FORECASTING MODELS

As you learned in Chapter 2, a self-projecting time series forecasting model can be stated in the general form

$$X_t = F_t + E_t, \qquad t = 1, 2, \ldots$$

where X_t represents the original time series, F_t represents the *pattern component* of the series, and E_t represents the *random error component* of the series. The objective, then, is to find some mathematical formula that can reproduce the pattern component in the series. This formula is also used to extend the pattern component into the future. The values generated by the formula are called *fitted values* for $t \leq N$ and *forecast values* for $t > N$, where N is the number of time periods. The differences between X_t and F_t are called the *residuals;* they represent the random error component.

Note that since the random error component, by definition, cannot be forecasted, a forecast consists only of an extension of the pattern component. This fact implies that a *forecast is always expected to be in error.* Thus, one question that eventually needs to be answered is, how big is the forecast error expected to be?

In summary, given a formula for reproducing the pattern component in a series, we have the following definitions:

$$X_t = \text{Time series value}, \qquad t = 1, \ldots, N$$

$$F_t = \begin{cases} \text{Fitted values}, & t \leq N \\ \text{Forecast values}, & t > N \end{cases}$$

$$E_t = X_t - F_t = \text{Residuals}, \qquad t = 1, \ldots, N$$

The random error component is assumed to have one additional property that may seem obvious but must be stressed—namely, that each individual random error *E_i is independent of, and unrelated to, any other E_j* when j is not equal to i. This property will be important to remember when we describe how models are identified (Chapters 8 and 9) and evaluated (Chapters 10 and 11).

The objective, then, is to find some formula that will generate the fitted values so that the computed residuals are as small as possible but are unrelated to each other. Precisely what is meant by "small as possible" and "unrelated to each other" will be discussed in subsequent chapters.

5.2 THE THREE-PHASE BOX-JENKINS MODEL-BUILDING PROCESS

Box-Jenkins models are put together by using up to seven types of parameters, or model-building blocks. These building blocks may be combined in several different ways to construct a rich variety of forecasting model formulas capable of accounting for a wide range of time series behavior.

Since the Box-Jenkins approach to model building is based on statistical concepts, certain statistical analyses of the historical data provide the necessary information for selecting the particular building blocks required to model a given time series. Once the building blocks have been selected, the best values for the parameters are determined by using a known procedure. Finally, after a check of the validity of the model has been performed by using certain statistical tests, the model may be used to generate forecasts of the series. Further statistics are also provided that indicate how confident you can be in these forecasts, i.e., that measure the degree of uncertainty in the forecast.

The above description suggests a three-phase process for building Box-Jenkins models that follows the general process outlined in Chapter 2:

1. *Model identification phase*. The type of Box-Jenkins model building blocks (parameters) needed to construct the model for the time series are identified by using various statistics computed from the historical data.

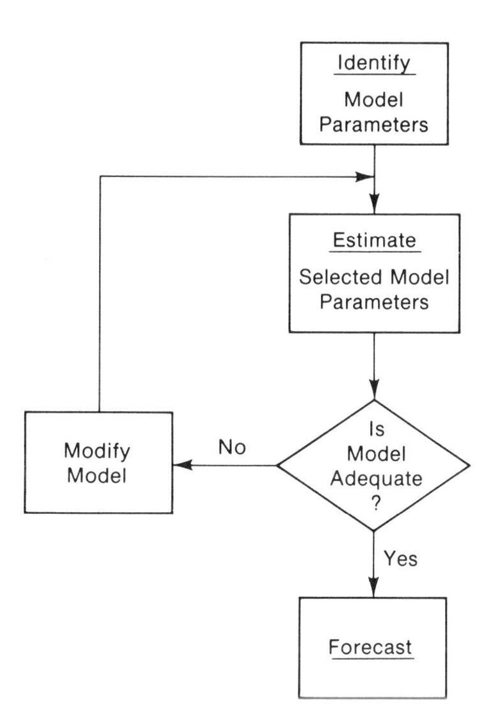

Figure 5.1 The Box-Jenkins Model-Building Process

2. *Model estimation and validation phase.* The best, or optimum, numerical values for the selected model parameters are determined (estimated) so that the model generates fitted values that are as close as possible to the original series values and so that the computed residuals are unrelated to each other. Certain diagnostics are provided to check the validity of the model and suggest alternative models.
3. *Model forecasting phase.* The selected and estimated model is used to generate forecasts of the time series. Forecast confidence limits (measures of uncertainty) are also obtained.

In practice, this process is usually iterative, as suggested by the diagram in Figure 5.1. That is, you will normally go back and forth a few times between the first two phases, each time altering the model, until you are satisfied. Once a final model has been chosen, estimated, and validated, you are ready to generate forecasts.

To be complete, any model-building process should also include a fourth phase, namely tracking and updating. In this phase the accuracy of forecasts is monitored as new data is made available. These results may then indicate that forecasts need to be updated or the model rebuilt.

CHAPTER 6

The Basic Box-Jenkins Models

As noted in the previous chapter, Box-Jenkins models can contain up to seven different building blocks or parameters. In this chapter you will learn about two of these building blocks from which the most basic Box-Jenkins models can be constructed. These basic models provide the foundation for developing the more generalized models described in Parts 3 and 4.

In particular, in this chapter you will learn:

- What series can be modeled by the basic models.
- What a basic autoregressive model is.
- What a basic moving-average model is.

6.1 SERIES THAT CAN BE MODELED WITH THE BASIC MODELS

The basic models cannot be used for the majority of business and economic time series you are likely to encounter. For example, a series that exhibits a constant growth pattern, or overall trend, or a series that moves back and forth from one established level to another cannot be modeled by the basic models.

For application of the basic models the series you are working with must have the property that it is stationary. In practical terms, a series is *stationary* if it tends to wander more or less uniformly about some fixed level. It may take short trips away from this fixed level, but it will eventually gravitate back to it. The fixed level about which it wanders is generally the mean, or average, of the series. An example of a stationary series is shown in Figure 6.1.

Limiting the application of the basic models to stationary series may seem awfully restrictive—and it is. But you need to know how to model stationary series first, before tackling series with more complex patterns. It is interesting to note, however,

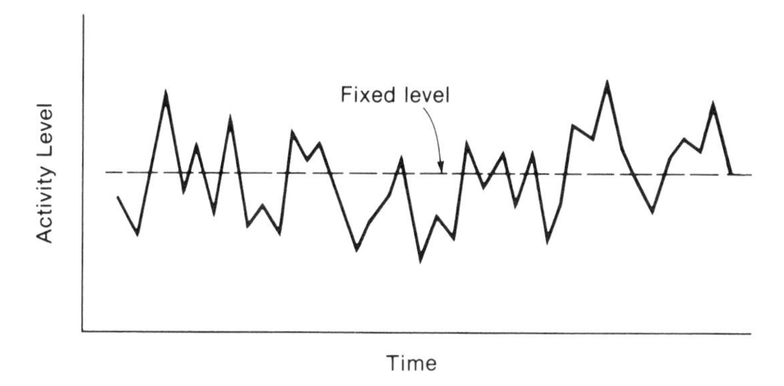

Figure 6.1 A Stationary Series

that some economic time series whose values represent period-to-period *changes* in some activity (e.g., the increase or decrease in earnings from one quarter to the next) are stationary series.

In this chapter, and in the rest of Part 2, then, we will assume that all series are stationary. It is also convenient, for the time being, to assume that the average, or mean value, of the series is zero; i.e., the fixed level, about which the series fluctuates, is zero. This assumption is made only for notational convenience. All conclusions and results presented in the following chapters also apply to stationary series with nonzero averages. In Chapter 12 you will learn how to determine whether a series is stationary and what you can do if it isn't.

6.2 AUTOREGRESSIVE MODELS (AR MODELS)

The first type of basic model is built with parameters called *autoregressive parameters* (AR parameters).

To understand how AR parameters work, consider a Box-Jenkins model that contains only one AR parameter. This model is written in the following form:*

$$X_t = \overbrace{A_1 X_{t-1}}^{F_t} + E_t$$

where X_t is the stationary series. The term A_1X_{t-1} represents the fit to the series value X_t, and A_1 is called an *AR parameter of order 1*. The term E_t, of course, represents the assumed random error in the data at period t.

*The parameter notation used in this book is not the standard notation normally used in the Box-Jenkins literature. The standard notation uses Greek letters and a more abstract symbolic representation of the models themselves. See Appendix D for a complete mathematical description of Box-Jenkins models.

The above model is simple enough, but what does it mean? It simply says that any given value X_t in the series is directly proportional to the previous value X_{t-1} plus some random error E_t. That is, what happens this period is only dependent on what happened last period, plus some current random error.

It is possible, of course, that X_t could be directly related to more than just one past value. For example, an autoregressive model with two AR parameters is written as follows:

$$X_t = \overbrace{A_1X_{t-1} + A_2X_{t-2}}^{F_t} + E_t$$

where A_1 is an AR parameter of order 1 and A_2 is an AR parameter of order 2. This model indicates that X_t is related to a combination of the two immediately preceding values, X_{t-1} and X_{t-2}, plus some current random error E_t.

Extending this idea further, we may write a general autoregressive model with p AR parameters as follows:

$$X_t = \overbrace{A_1X_{t-1} + A_2X_{t-2} + \cdots + A_pX_{t-p}}^{F_t} + E_t$$

where $A_1, A_2, \ldots, A_p$ are the AR parameters. The subscripts on the A's are called the *orders* of the AR parameters. The highest order p is referred to as the *order of the model*. In this generalized autoregressive* model any series value X_t is expressed as an arithmetic combination of p past series values plus some random error E_t.

Note that there is nothing to say we can't exclude some of the lower-order terms in the above generalized model. For example, the model $X_t = A_pX_{t-p} + E_t$ is a pth-order model containing only one AR parameter of order p; i.e., what happens in the current period is related only to what happened p periods ago. This kind of relationship suggests the existence of seasonal behavior in the series. We will not, however, get into the question of seasonal models for seasonal series until Part 3.

If you recall the discussion on smoothing models in Chapter 2, you will probably recognize a similarity between autoregressive models and some of the smoothing models described there. In fact, the averaging model discussed in Chapter 2 has exactly the same form as a third-order autoregressive model. What, then, does a Box-Jenkins AR model provide that is any different from what a simple smoothing model provides? The answer is, so far, nothing. But recall that the Box-Jenkins methodology will provide the means for determining which AR model form to choose and what values to use for the parameters. Moreover, there are many more parameters that can go into a Box-Jenkins model—AR parameters are only the beginning.

*The origin of the term *autoregressive* is derived from the fact that the equation describing the autoregressive model looks exactly like a regression equation, where X_t plays the role of the dependent variable and X_{t-1}, X_{t-2}, etc. play the role of the independent variables. Since X_{t-1}, X_{t-2}, etc. are really the same data as X_t (but "offset" by one period, two periods, etc.), X_t is actually being regressed on *itself*—hence the term *auto*regressive.

6.3 MOVING-AVERAGE MODELS (MA MODELS)

The second type of basic model is built with parameters called *moving-average parameters* (MA parameters).

Although MA models closely resemble AR models in appearance, the concept behind the use of MA parameters is quite different. Moving-average parameters relate what happens in period t only to the random errors that occurred in past time periods, i.e., to E_{t-1}, E_{t-2}, . . . (as opposed to being related to the actual series values $X_{t-1}, X_{t-2}, \ldots$). For example, a Box-Jenkins model with one MA parameter is written as follows:

$$X_t = \overbrace{-B_1 E_{t-1}}^{F_t} + E_t$$

The term $-B_1E_{t-1}$ represents the fit to the series value X_t, and B_1 is called an *MA parameter of order 1*. (The use of the minus sign in front of B_1 is conventional only and has no other significance.)

The above model simply says that any given value X_t in the series is directly proportional only to the random error E_{t-1} from the previous period plus some current random error E_t.

Like AR models, MA models can be extended to include q MA parameters, as follows:

$$X_t = \overbrace{-B_1 E_{t-1} - B_2 E_{t-2} - \cdots - B_q E_{t-q}}^{F_t} + E_t$$

where B_1, B_2, . . . , B_q are the MA parameters of order 1, 2, . . . , q, respectively. The highest order q is referred to as the order of the MA model. In this generalized moving-average* model any series value X_t is expressed as an arithmetic combination of q past random errors plus some current random error. As in the case of AR parameters, a qth-order MA model may contain fewer than q MA parameters; i.e., lower-order terms may be excluded.

The moving-average concept is an interesting one, since it implies that any series value is really only a by-product of past random errors. This conclusion makes sense, however, if you think of the random error as small "shocks" that initially set the process in motion and continue to keep it in motion thereafter.

The use of past random errors in the formulation of model formulas is an important concept, one that sets Box-Jenkins models apart from most traditional models. In Box-Jenkins models, as we will see, the random error component plays a dominant role in determining the structure of the model.

*The origin of the term *moving average* is derived from the fact that the moving-average model formula is simply a weighted average of a fixed number of past random errors that "moves forward" in time as t increases.

6.4 MIXED AR AND MA MODELS

Both AR and MA parameters may be used in the same model. Models containing both types of parameters are called ARMA models and are written in the following form:

$$X_t = \overbrace{(A_1X_{t-1} + \cdots + A_{t-p}X_p) - (B_1E_{t-1} + \cdots + B_qE_{t-q})}^{F_t} + E_t$$

The order of an ARMA model is expressed in terms of both p and q.

6.5 REVIEW OF KEY CONCEPTS

In Part 2 only models for stationary series are discussed. A *stationary series* is one whose values vary more or less uniformly about some fixed level over time. The models for stationary series are called the basic Box-Jenkins models. Models for more complex series behavior (e.g., nonstationary and seasonal behavior) are built on these basic models and will be addressed in Part 3.

There are two types of basic Box-Jenkins models: autoregressive (AR) models and moving-average (MA) models. The AR and MA models may also be combined to form ARMA models. These models are written as follows:

1. *AR models.*

$$X_t = A_1X_{t-1} + \cdots + A_pX_{t-p} + E_t$$

where X_t is directly related to one or more past series values.

2. *MA models.*

$$X_t = -(B_1E_{t-1} + \cdots + B_qE_{t-q}) + E_t$$

where X_t is related to one or more past random errors.

3. *ARMA models.*

$$X_t = (A_1X_{t-1} \cdots + A_pX_{t-p}) - (B_1E_{t-1} + \cdots + B_qE_{t-q}) + E_t$$

where X_t is related to both past series values and past random errors.

The A_i are called *autoregressive parameters* and the B_i, *moving-average parameters*. The subscripts on the A's and B's are called the *orders* of the parameters. In an AR model p is the order of the model, and in an MA model q is the order of the model. The order of an ARMA model is expressed in terms of both p and q.

Given a stationary series, our task is to determine which model is the appropriate one for a given series, i.e., which AR and MA parameters should be included in the model. Chapters 7, 8, and 9 explain how this task is done.

CHAPTER 7

Tools for Identification of the Basic Models

Given the basic AR, MA, and ARMA models described in Chapter 6, how do you determine the appropriate one for your stationary series? Fortunately, the Box-Jenkins method provides some tools to help you select the correct number and type of parameters to include in a model.

In this chapter you will learn about two model identification tools:

- Autocorrelations.
- Partial-autocorrelations.

Autocorrelations and partial-autocorrelations are the most important tools you will encounter in the model-building process. They are used during the initial stages of model selection and later when checking the validity of a model. Autocorrelations and partial-autocorrelations can tell you a great deal about the historical behavior of a time series.

7.1 AUTOCORRELATIONS (AC PATTERNS)

Autocorrelations are statistical measures (numerical values) that indicate how a time series is related to itself over time. More precisely, an autocorrelation measures how strongly time series values at a specified number of periods apart are correlated to each other over time. The number of periods apart is usually called the *lag*. Thus an autocorrelation for lag 1 is a measure of how successive values (one period apart) are correlated to each other throughout the series. An autocorrelation for lag 2 measures how series values two periods away from each other are correlated throughout the series.

For example, if all higher-than-average series values tend to be followed n periods later by a higher-than-average value, and all lower-than-average values tend to be followed n periods later by a lower-than-average value, then there would be a large *positive autocorrelation* for lag n. On the other hand, if all higher-than-average (lower-than-average) values tend to be followed by a lower-than-average (higher-than-average) value n periods later, then there would be a large *negative autocorrelation* for lag n. If there is an inconsistency, or randomness, in the way values n periods apart behave with respect to each other, then there would be no (zero) autocorrelation for lag n.

An autocorrelation value is always computed as some value between -1 and $+1$. The value $+1$ means strong positive autocorrelation; -1 means strong opposite (or negative) autocorrelation; and 0 means no autocorrelation. A sample table of autocorrelations computed from an actual time series is given in Table 7.1.

The entries in the table show the autocorrelations of the series for lags 1 through 24. The first entry in the table indicates that there is a correlation of about .75 (high positive correlation) between series values one period apart. The last entry in the first row indicates there is a correlation of about $-.04$ between series values 12 periods apart. See Appendix A for details on how autocorrelations are actually computed.

How do you use the autocorrelations of a time series to help select an appropriate Box-Jenkins model for the series? It turns out that moving-average and autoregressive relationships in a stationary series produce recognizable *autocorrelation patterns* (AC patterns). If you learn to recognize these patterns, you can identify MA and AR relationships—and hence the appropriate MA and AR parameters. Since moving-average and autoregressive relationships account for a wide variety of stationary time series behavior, this identification allows you to obtain an optimum or near-optimum forecasting model for your series. Autocorrelation patterns for MA and AR relationships are described in Chapter 8.

7.2 GRAPHICAL DISPLAY OF AUTOCORRELATIONS

Autocorrelations are usually displayed either as a table of values, as shown in Table 7.1, or as a plot or graph of the correlation values, called a *correlogram*. In a correlogram the lags are printed on the horizontal axis and the correlation values on

Table 7.1 Autocorrelations

Lag	1	2	3	4	5	6	7	8	9	10	11	12
AC	.75	.65	.50	.41	.35	.28	.22	.18	.16	.06	.02	−.04
Lag	13	14	15	16	17	18	19	20	21	22	23	24
AC	−.03	−.07	−.11	−.18	−.18	−.20	−.23	−.22	−.16	−.12	−.00	−.01

the vertical axis from -1 to $+1$. The correlogram shown in Figure 7.1 corresponds to the table of autocorrelations in Table 7.1. Note that the magnitude of each autocorrelation is represented by a bar or spike. Correlograms are probably the easiest and most widely used method of displaying autocorrelations, since they allow you to visually determine the autocorrelation patterns.

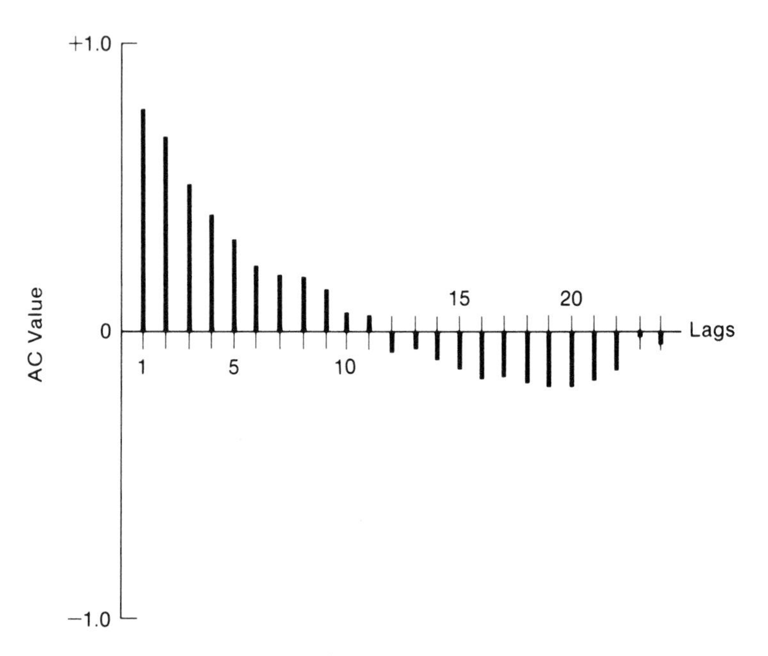

Figure 7.1 An Autocorrelation Correlogram

7.3 PARTIAL-AUTOCORRELATIONS (PAC PATTERNS)

Partial-autocorrelations are another set of statistical measures, similar to autocorrelations, that are used to evaluate relationships among time series values. As will be seen later, they are complementary to autocorrelations with respect to the patterns they produce for autoregressive and moving-average relationships. Partial-autocorrelations are very useful in some situations where the autocorrelation patterns are hard to determine. Partial-autocorrelation values also range from -1 to $+1$, and they may be displayed in a table or a correlogram in the same manner as autocorrelations. The partial autocorrelations shown in Table 7.2 and the correlogram of Figure 7.2 were computed from the same series that the autocorrelations in Table 7.1 and Figure 7.1 were computed from.

Table 7.2 Partial-Autocorrelations

Lag	1	2	3	4	5	6	7	8	9	10	11	12
PAC	.75	.20	−.07	−.01	.08	−.03	−.06	.02	.08	−.22	−.02	−.01
Lag	**13**	**14**	**15**	**16**	**17**	**18**	**19**	**20**	**21**	**22**	**23**	**24**
PAC	.08	−.11	−.06	−.10	.05	−.04	−.09	.05	.19	−.05	.20	−.12

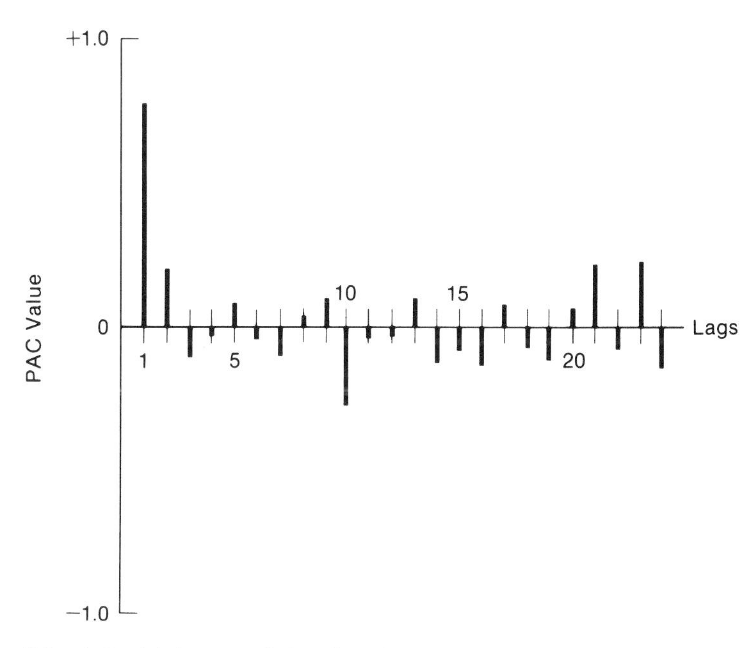

Figure 7.2 A Partial-Autocorrelation Correlogram

The definition and statistical meaning of partial-autocorrelations are somewhat more complicated than those of autocorrelations. It is not critical, however, to understand the meaning and origin of partial-autocorrelations. Rather, it is the interpretation of the patterns they produce that is important. For a more thorough description and definition of partial-autocorrelations, see Appendix B.

7.4 REVIEW OF KEY CONCEPTS

The main tools for identifying the correct parameters to include in a Box-Jenkins model are the autocorrelations and partial-autocorrelations of a series.

Autocorrelations are certain statistical measures computed from time series data; they have the following properties:

- An autocorrelation measures how strongly the series values at a specified number of periods apart are related to each other; the specified number of periods apart is called the *lag*.
- A set of autocorrelation values may be computed for a given series corresponding to lags of 1, 2, 3, etc.
- An autocorrelation value may range from -1 to $+1$, where a value close to $+1$ indicates a strong positive correlation, a value close to -1 indicates a strong negative (opposite) correlation, and a value close to 0 indicates no correlation.

Partial-autocorrelations are similar to autocorrelations in that they provide another, but different, measure of certain relationships among series values at each lag. They also range in value from -1 to $+1$. Partial-autocorrelations are useful since they can reveal relationships among series values that are sometimes difficult to detect from autocorrelations alone.

Correlograms are special graphs in which a set of autocorrelations or partial-autocorrelations for a given series is plotted as a series of bars or spikes. The length of each bar represents the magnitude of the autocorrelation at a given lag. Correlograms are extremely useful for visually demonstrating any patterns that may exist in the autocorrelations or partial-autocorrelations of a series.

It turns out that the autocorrelations (and partial-autocorrelations) of a series generated by a given MA, AR, or ARMA model will demonstrate particular patterns or sequences of behavior that are uniquely associated with the given model. If you learn to recognize the unique autocorrelation patterns associated with each model, you can then select the model whose autocorrelation pattern matches the autocorrelation pattern of the series you want to model.

CHAPTER **8**

Identifying the Basic Box-Jenkins Models (In Theory)

As mentioned previously, time series generated by MA, AR, or ARMA models have easily identifiable autocorrelation (and partial-autocorrelation) patterns. In this chapter you will see, *theoretically,* what these patterns look like so that you can recognize similar patterns in the autocorrelations of a real-life series.

Specifically, in this chapter you will learn:

- What the theoretical AC and PAC patterns are for MA models.
- What the theoretical AC and PAC patterns are for AR models.
- What the theoretical AC and PAC patterns are for mixed ARMA models.

By matching the computed autocorrelation patterns of your real-life series to the known theoretical autocorrelation patterns corresponding to the MA, AR, and ARMA models, you can identify the MA and AR parameters required. Real-life series, however, will usually not exhibit autocorrelation patterns as clean as the theoretical ones. In Chapter 9 you will learn why, and you will learn how you can compare autocorrelations computed from real-life series with those associated with the theoretical models.

8.1 THEORETICAL AC PATTERNS FOR MA MODELS

AC patterns for MA models are particularly easy to recognize, so we will begin our discussion with these models.

8.1.1. One MA Parameter of Order 1

Let's take a simple MA model containing only one MA parameter of order 1:

$$X_t = -B_1E_{t-1} + E_t$$

To find out what the autocorrelation pattern is for the series X_t, we need to determine the relationship between X_t and X_{t-1} for all t (i.e., the autocorrelation at lag 1), between X_t and X_{t-2} (i.e., the autocorrelation at lag 2), etc. In general, we need to determine the relationship between X_t and X_{t-n} (i.e., the autocorrelation at lag n) for all $n = 1, 2, \ldots, etc.$

Now X_t in the above model is directly related, by definition, to E_{t-1}. And since E_{t-1} is part of $X_{t-1} = -B_1E_{t-2} + E_{t-1}$, then X_t must also be related to X_{t-1} for all periods t. Thus X_t is autocorrelated for lag 1; i.e., the autocorrelation for lag 1 is nonzero.

What about the relationship of X_t to X_{t-2}? Well X_t contains two terms involving E_{t-1} and E_t. By definition, E_t and E_{t-1} are random errors that are not correlated to any other E_i; i.e., there is no relationship between E_t and any other E_i (except E_t itself) or between E_{t-1} and any other E_i. Thus since $X_{t-2} = -B_1E_{t-3} + E_{t-2}$ does not contain E_t or E_{t-1}, there can be no relationship between X_t and X_{t-2}. Hence the autocorrelation for lag 2 is zero. This same argument can be applied to show that the autocorrelation for any lag greater than 1 is zero.

The above analysis yields a very simple autocorrelation pattern for a model containing one MA parameter of order 1. The correlograms in Figure 8.1 illustrate this pattern for two different MA models, each containing one MA parameter of order 1. Both correlograms show only one large spike at lag 1. The direction of the spike (up or down) depends on whether the MA parameter is positive or negative, respectively.

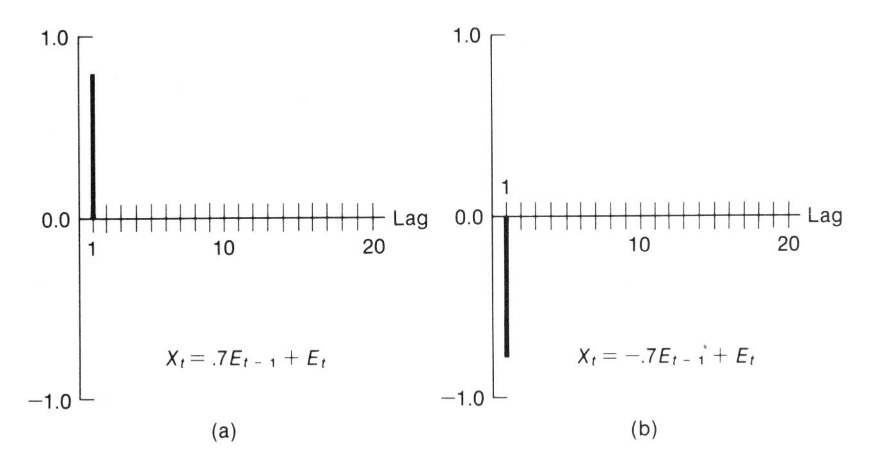

Figure 8.1 Theoretical AC Patterns for One MA Parameter of Order 1

8.1.2 Higher-Order MA Parameters

Now let's look at a two-parameter MA model:

$$X_t = -B_1E_{t-1} - B_2E_{t-2} + E_t$$

which contains the two elements E_{t-1} and E_{t-2}. Since E_{t-1} is an element of $X_{t-1} = -B_1E_{t-2} - B_2E_{t-3} + E_{t-1}$ and E_{t-2} is an element of $X_{t-2} = -B_1X_{t-3} - B_2X_{t-4} + E_{t-2}$, then X_t must be correlated to both X_{t-1} and X_{t-2}. On the other hand, with the same arguments used for the one-parameter model, you can see that X_t cannot be related to X_{t-3}, X_{t-4}, etc. Thus the autocorrelations are nonzero for lags 1 and 2 and zero for lags greater than 2.

The correlograms in Figure 8.2 show the autocorrelation patterns associated with two different models, each containing two MA parameters of orders 1 and 2. Here you see two large spikes at lags 1 and 2 and zero autocorrelations at all other lags.

In general, then, for an MA model of order q we have

$$X_t = -(B_1E_{t-1} + \cdots + B_qE_{t-q}) + E_t$$

and the autocorrelations associated with this model are nonzero for lags 1 through q and zero for lags greater than q.

As a final example of the AC pattern for a higher-order MA model, consider the following model containing only one MA parameter, but the order of the parameter is 3:

$$X_t = -B_3E_{t-3} + E_t$$

With the same arguments as before, you can see that X_t must be related to X_{t-3} since X_{t-3} contains E_{t-3}, but X_t is not related to X_{t-1}, X_{t-2}, or any X_{t-n} for n

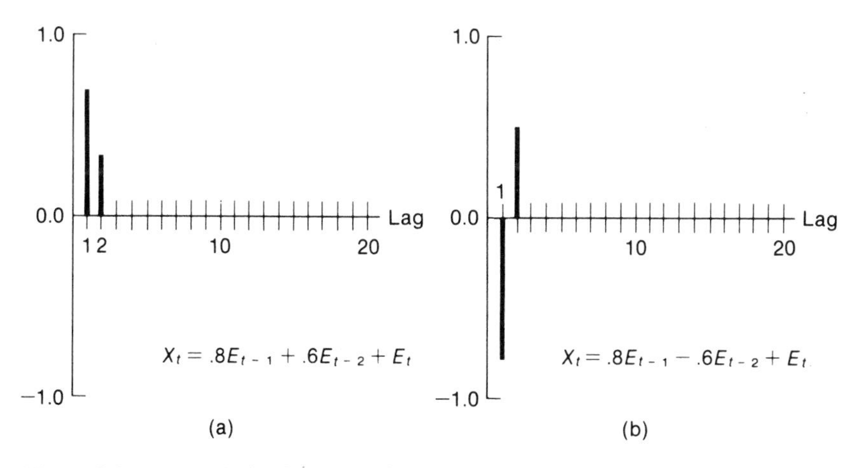

Figure 8.2 Theoretical AC Patterns for Two MA Parameters of Orders 1 and 2

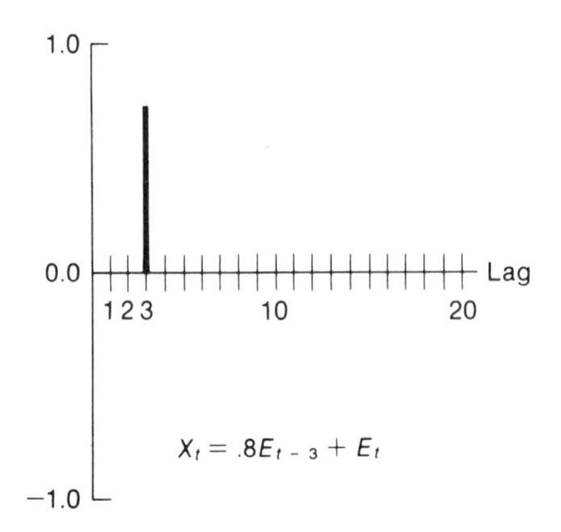

Figure 8.3 Theoretical AC Pattern for One MA Parameter of Order 3

greater than 3. Thus the autocorrelations associated with this model are zero for all lags except lag 3. This pattern is illustrated in Figure 8.3.

8.2 THEORETICAL AC PATTERNS FOR AR MODELS

The AC patterns for AR models are somewhat more complicated than those for MA models. So let's start with a simple AR model.

8.2.1 One AR Parameter of Order 1

Our first example is a model containing only one AR parameter of order 1:

$$X_t = A_1X_{t-1} + E_t$$

Here it is a clear that X_t is directly related to X_{t-1} for all periods, and so there is a nonzero autocorrelation for lag 1. What about the relationship of X_t to X_{t-2}? Note that $X_{t-1} = A_1X_{t-2} + E_{t-1}$; i.e., X_{t-1} is directly related to X_{t-2}. But then X_t is also indirectly related to X_{t-2} (since X_t is already related to X_{t-1}) and, in turn, indirectly related to X_{t-3}, etc. In other words, the current value X_t is actually related in some way to *all* past X_i, for i less than t.

It turns out, however, that this relationship on previous X_i decreases as i decreases; i.e., the further back in time you go, the weaker is the relationship. Thus the autocorrelations associated with such a model will be relatively large at lag 1 (positive or negative), somewhat smaller at lag 2, smaller still at lag 3, etc., until the autocorrelations trail out, or dampen out, to (almost) zero. Also, the autocorrelations should decrease fairly rapidly, usually within the first three to six lags. This rapid

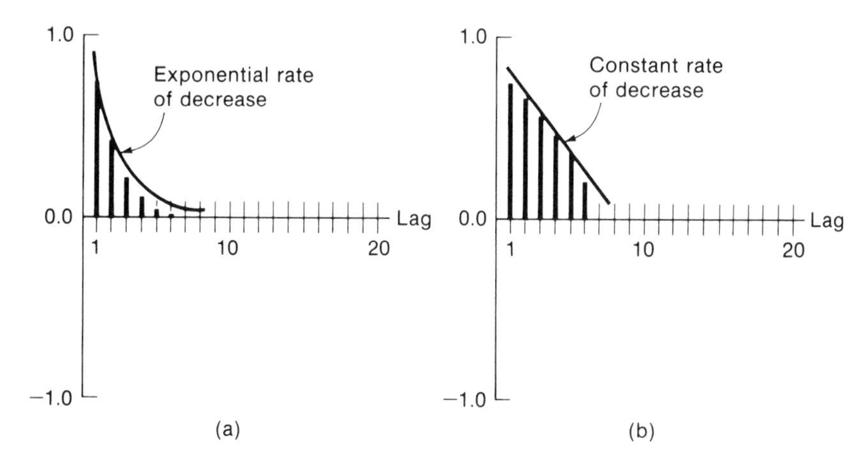

Figure 8.4 Exponential Versus Constant Rate of Decrease

rate of decrease is often referred to as an *exponential rate of decrease,* versus, for example, a constant rate of decrease, as illustrated in Figure 8.4. The AC pattern for an AR model must dampen out at an exponential rate.

The correlograms in Figure 8.5 illustrate the autocorrelation patterns for two different AR models, each containing one AR parameter of order 1. The correlogram for the model in part (a) shows the characteristic dampening out, where all the autocorrelations are positive. The correlogram for the model in part (b) shows the same dampening effect, except the autocorrelations alternate in sign. The difference in the patterns generated by these two similar models is due to the sign (positive or negative) of the AR parameter.

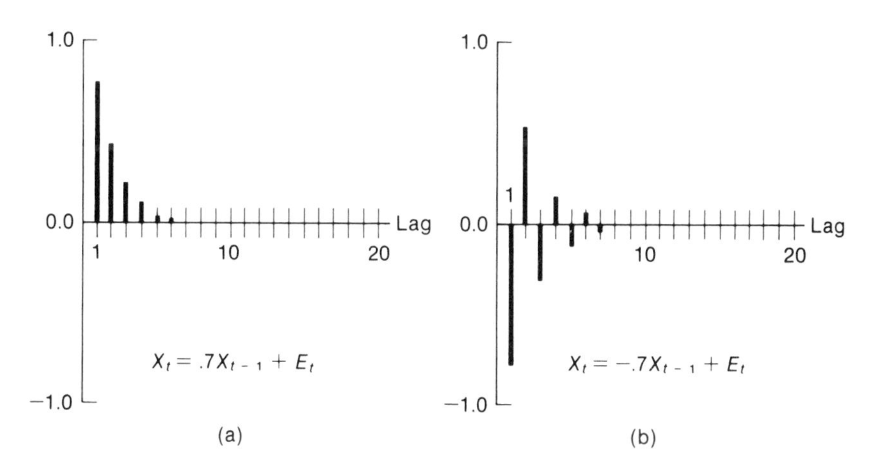

Figure 8.5 Theoretical AC Patterns for One AR Parameter of Order 1

8.2.2 Higher-Order AR Parameters

Using similar arguments to those used for a one-parameter AR model, you can convince yourself that AR models with higher-order AR parameters should produce similar types of patterns. Actually, a number of variations of the basic exponentially decreasing pattern are possible. For example, the autocorrelations may show an alternating positive-negative pattern (as shown in Figure 8.5b) or an alternating sine wave (humped) pattern about the lag axis. The overall pattern, however, will always show a rapid trailing off, or dampening out, as the lag increases.

The correlogram in Figure 8.6 illustrates the autocorrelation pattern for an AR model with two parameters of orders 1 and 2. This pattern shows a sine wave, or humped, pattern superimposed on an exponentially decreasing pattern. This pattern is typical, but not always the case, for AR models with more than one parameter.

As you can see, using autocorrelations to determine the correct number of AR parameters can be more difficult than it is for MA parameters. Fortunately, our second identification tool, the partial-autocorrelations, can be used to overcome this problem.

8.3 THEORETICAL PAC PATTERNS FOR MA AND AR MODELS

The partial-autocorrelations of a series, it turns out, produce patterns that are exactly the reverse of autocorrelation patterns with respect to AR and MA parameters. That is, partial-autocorrelation patterns for AR models look like autocorrelation patterns for MA models, and vice versa.

For example, a model with p AR parameters will generate a series whose PACs have p spikes at lags 1 through p and zeros at all remaining lags. The PAC pattern, therefore, can be extremely useful in identifying the presence and number of AR parameters. On the other hand, a model with q MA parameters will generate a series

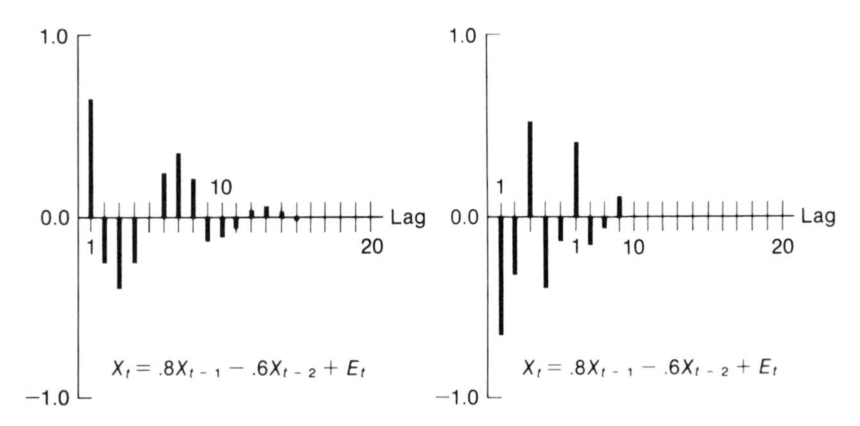

Figure 8.6 Theoretical AC Pattern for Two AR Parameters of Orders 1 and 2

whose PAC pattern is an exponentially decreasing pattern of some type. Figure 8.7 illustrates the comparison of AC and PAC patterns for a model with one AR parameter and a model with one MA parameter. Both parameters are of order 1.

8.4 THEORETICAL AC AND PAC PATTERNS FOR ARMA MODELS

The autocorrelation and partial-autocorrelation patterns of mixed relationships, involving both MA and AR parameters, are combinations of the individual patterns already discussed for each of these types of parameters. For example, Figure 8.8 shows the AC and PAC correlograms for a model containing one MA and one AR parameter of order 1. Note that the AC pattern, part (a), indicates the presence of the AR parameter, and the PAC pattern, part (b), indicates the presence of the MA parameter.

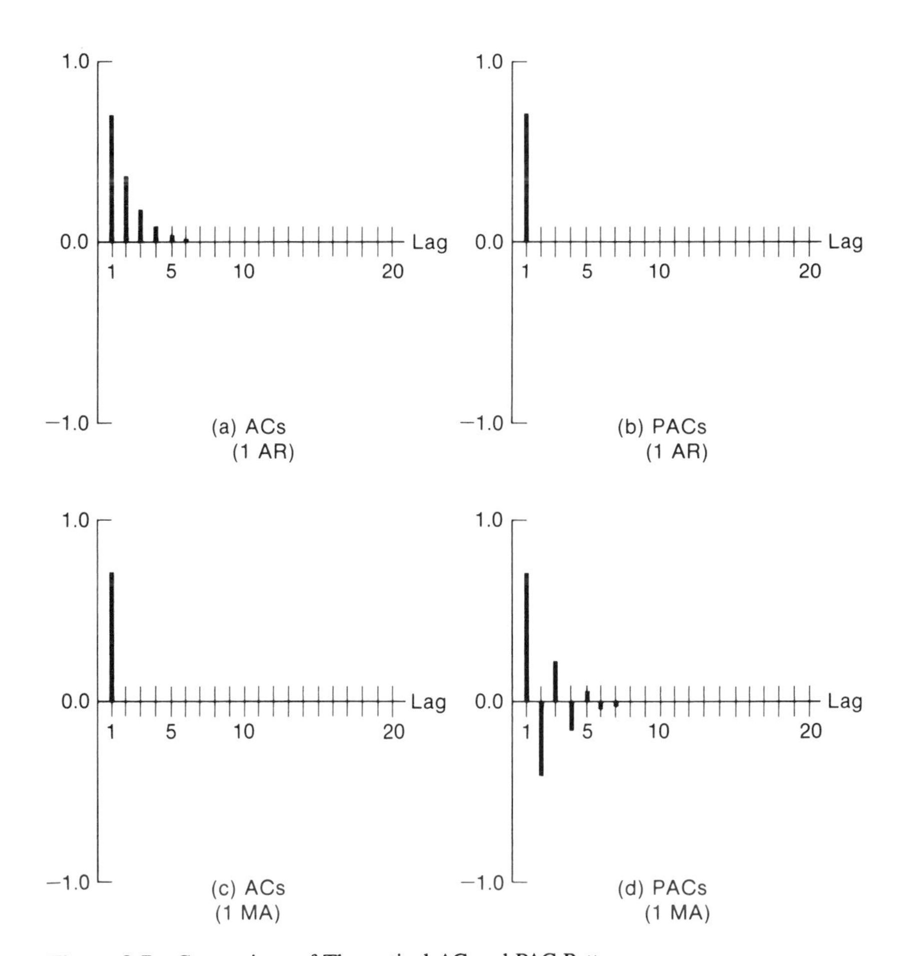

Figure 8.7 Comparison of Theoretical AC and PAC Patterns

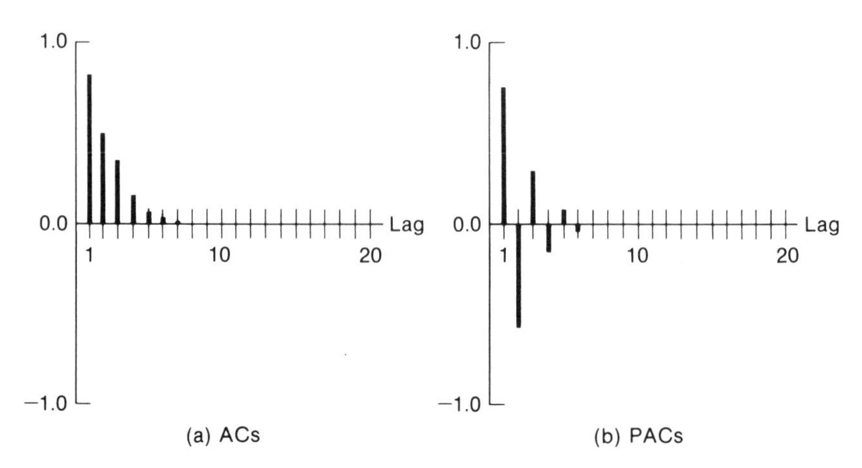

Figure 8.8 Theoretical AC and PAC Patterns for a Model with One MA and One AR Parameter of Order 1

Other exponentially decreasing patterns are, of course, possible for a one-MA and one-ARMA model, depending on the signs of the parameters.

In general, for a model containing q MA parameters and p AR parameters, the associated autocorrelation pattern will show an irregular pattern of spikes at lags 1 through q, followed by the normal exponentially decreasing pattern associated with the p AR parameters at lags greater than q. As might be expected, mixed MA and AR patterns are somewhat difficult to recognize; but with a little experience, and the use of both the ACs and the PACs, you can easily detect the correct patterns. In addition, you will learn in Chapter 10 how to use the estimation phase of Box-Jenkins as an aid in identifying the correct model. Fortunately, you will seldom, if ever, need to consider more than one or two of each type of parameter in a model for a stationary series.

8.5 SOME EXAMPLES TO TEST YOURSELF

The best way to gain proficiency in recognizing AC and PAC patterns is to study a variety of examples. In this section you can try your hand at identifying the AR, MA, or ARMA model that corresponds to the AC and PAC correlograms shown in each of the following examples. The correct answers are given in Appendix G.

Example 8.1

Identify the model that corresponds to the correlograms in Figure 8.9.

Example 8.2

Identify the model that corresponds to the correlograms in Figure 8.10.

Example 8.3

Identify the model that corresponds to the correlograms in Figure 8.11.

Example 8.4

Identify the model that corresponds to the correlograms in Figure 8.12.

Example 8.5

Identify the model that corresponds to the correlograms in Figure 8.13.

Example 8.6

Identify the model that corresponds to the correlograms in Figure 8.14.

Example 8.7

Identify the model that corresponds to the correlograms in Figure 8.15.

Example 8.8

Identify the model that corresponds to the correlograms in Figure 8.16.

Example 8.9

Identify the model that corresponds to the correlograms in Figure 8.17.

Example 8.10

Identify the model that corresponds to the correlograms in Figure 8.18.

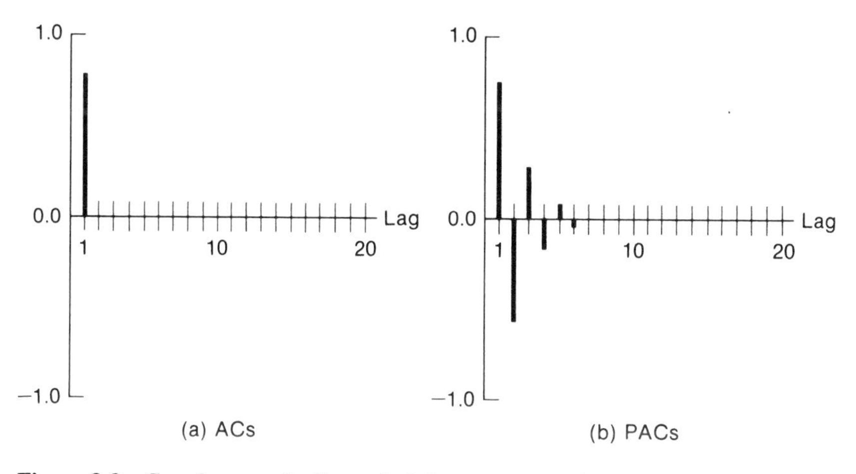

Figure 8.9 Correlograms for Example 8.1

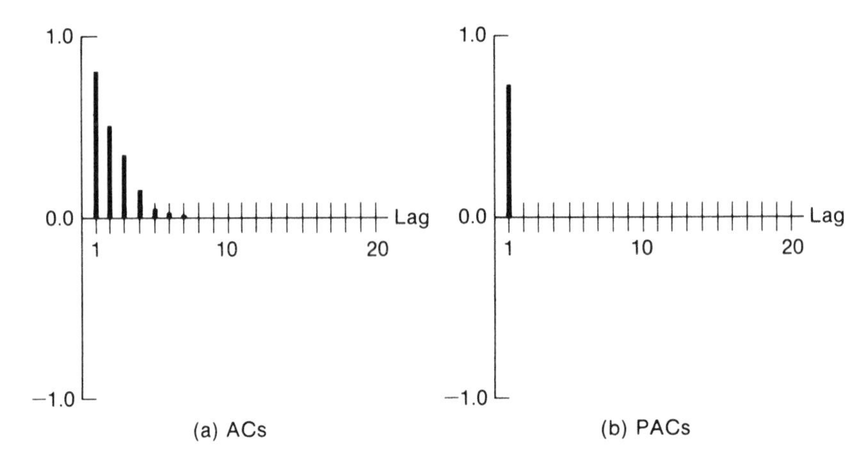

Figure 8.10 Correlograms for Example 8.2

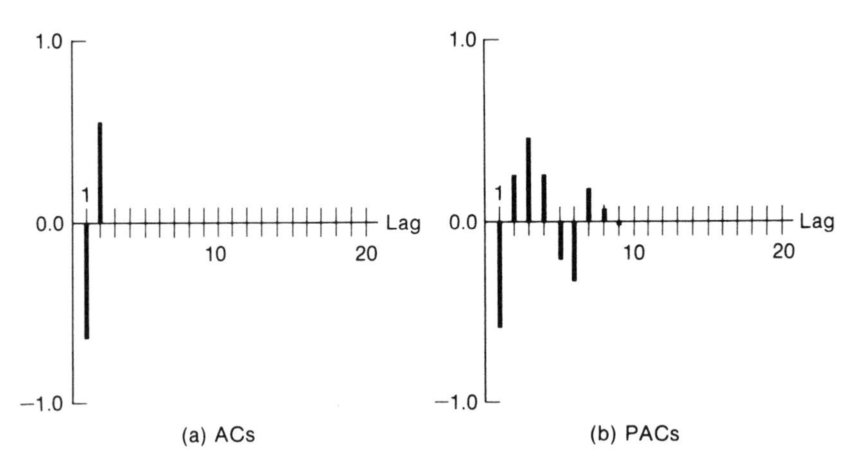

Figure 8.11 Correlograms for Example 8.3

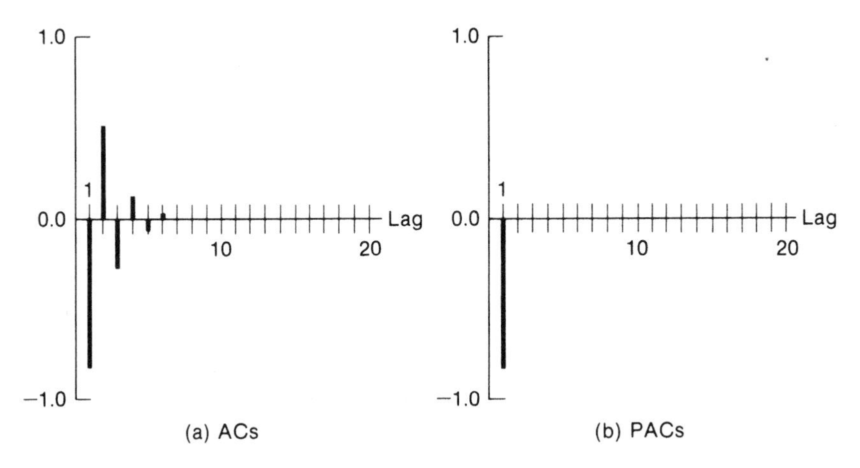

Figure 8.12 Correlograms for Example 8.4

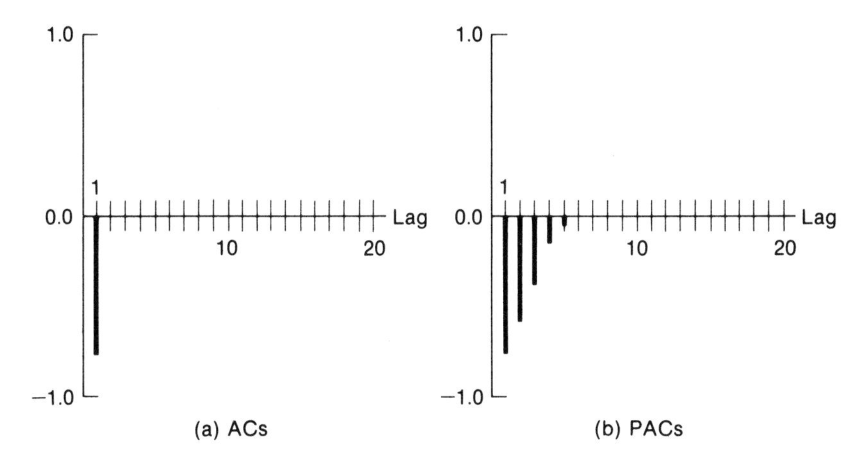

Figure 8.13 Correlograms for Example 8.5

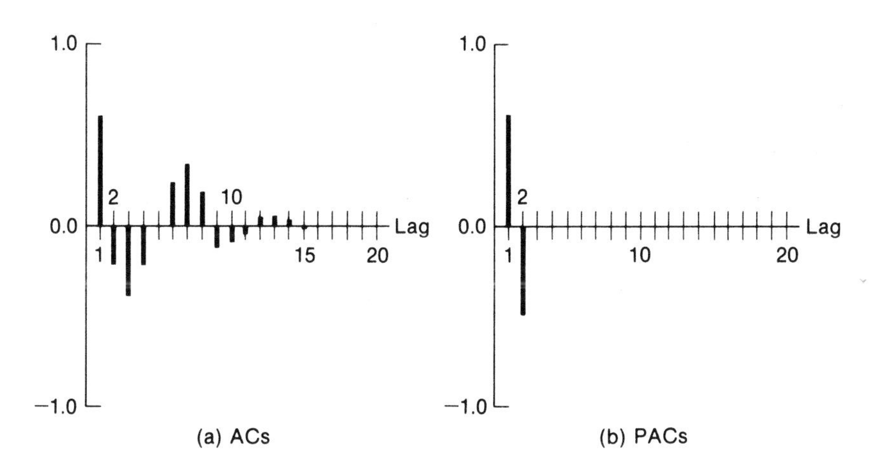

Figure 8.14 Correlograms for Example 8.6

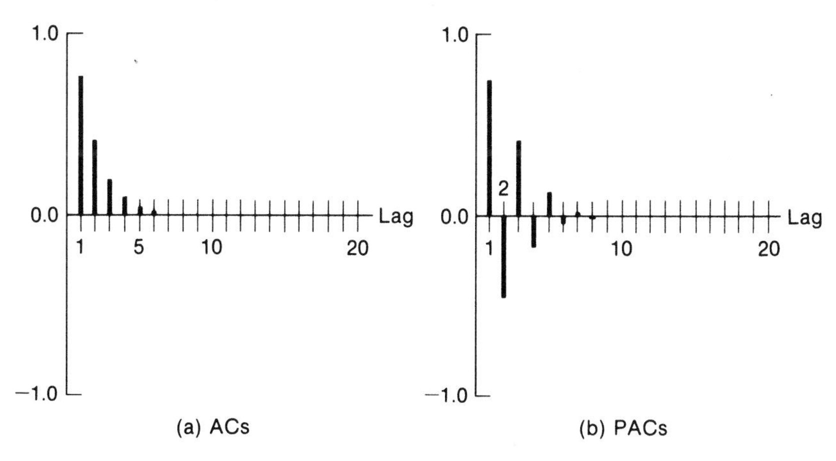

Figure 8.15 Correlograms for Example 8.7

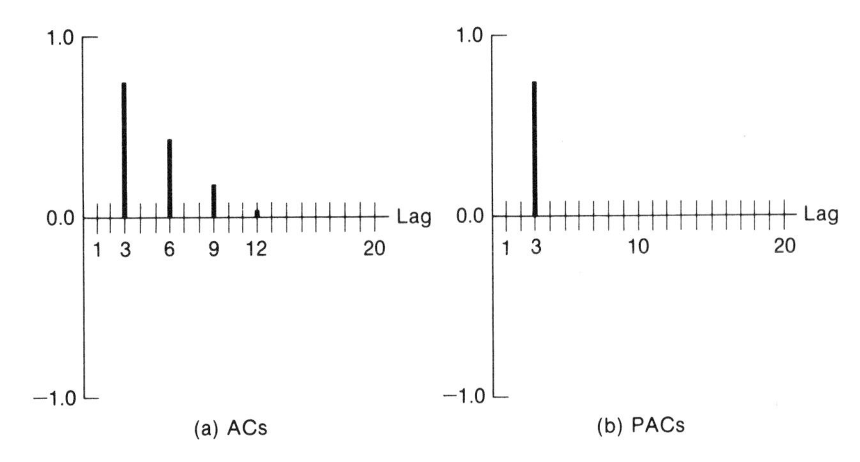

Figure 8.16 Correlograms for Example 8.8

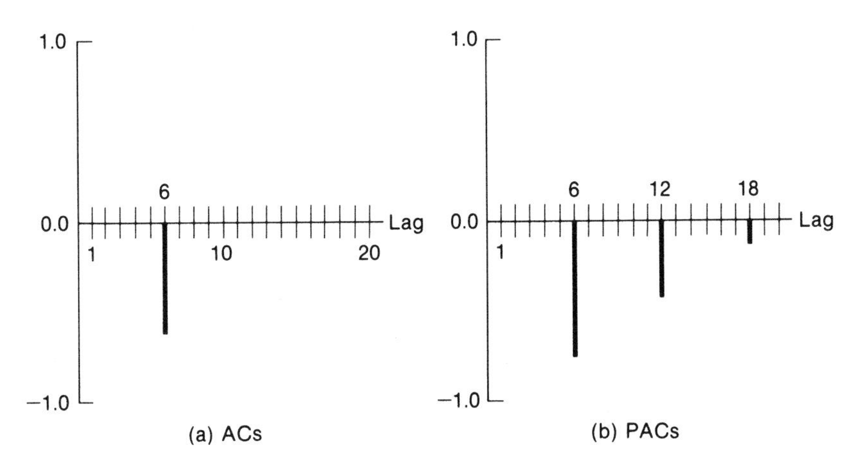

Figure 8.17 Correlograms for Example 8.9

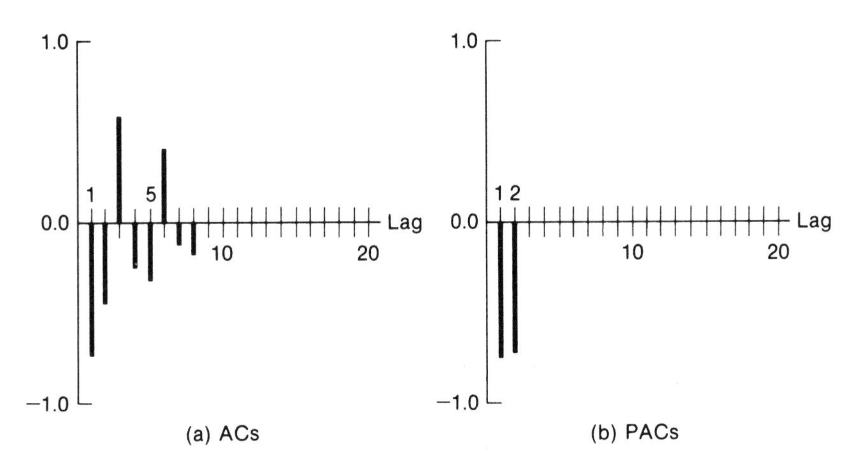

Figure 8.18 Correlograms for Example 8.10

8.6 REVIEW OF KEY CONCEPTS

The basic AR, MA, and ARMA models give rise to some easily recognizable theoretical autocorrelation and partial-autocorrelation patterns. To identify the correct Box-Jenkins model for a series, we compute the AC and PAC from a given series and compare them with known theoretical patterns. When a match is found, the appropriate model for the series can then be identified. How these comparisons are made is addressed in the next chapter.

A summary of the theoretical AC and PAC patterns associated with AR, MA, and ARMA models is given in Table 8.1. Keep in mind, however, that you will probably never have to worry about identifying more than one or two of each type of parameter.

Table 8.1 Summary of AC and PAC Patterns for AR, MA, and ARMA Models

Type of Model	Number and Order of Parameters	AC Pattern	PAC Pattern
MA	q parameters of order $1, 2, \ldots, q$	Spikes at lags 1 through q; zeros elsewhere	Spikes decreasing exponentially, beginning at lag 1; for q greater than 1, damped sine waves are superimposed on the pattern
AR	p parameters of order $1, 2, \ldots, p$	Spikes decreasing exponentially, beginning at lag 1; for p greater than 1, damped sine waves are superimposed on the pattern	Spikes at lags 1 through p; zeros elsewhere
ARMA	q MA and p AR parameters of orders $1, \ldots, q$ and $1, \ldots, p$ respectively	Irregular pattern of spikes at lags 1 through q; remaining pattern as in AR ACs	Irregular pattern of spikes at lags 1 through p; remaining pattern as in MA PACs

CHAPTER 9

Identifying the Basic Box-Jenkins Models (In Practice)

The derivation of the theoretical AC and PAC patterns discussed in the previous chapter was based on the mathematical and statistical definition of a given model and the assumptions made about certain elements of the model (e.g., the assumption that the correlation of any random error E_i to any other E_j is equal to zero when i is not equal to j. This derivation was a theoretical one—no autocorrelation values were actually computed from real data.

In this chapter you will learn:

- What the AC and PAC patterns computed from a real-life series look like.
- How to compare the computed AC and PAC patterns of a real-life series with the theoretical patterns discussed in Chapter 8.

9.1 SAMPLE AUTOCORRELATIONS VERSUS THEORETICAL AUTOCORRELATIONS

A mathematical or statistical model, of course, is only an idealization of some process or activity in real life. The model tells us how we can expect the activity or process to most likely behave. But if we were to examine any *sample occurrence* of this activity in real life, it would seldom turn out to be precisely what the model said it would be—probably close, but not the same. For example, flipping a fair coin 100 times will seldom produce exactly 50 tails and 50 heads, although that is what is expected to happen. The model in this case states that half the time a head should appear and half the time a tail should appear, if the coin is fair. We would say that such a model was probably correct, even if it turned out that 48 heads and 52 tails occurred, but we would suspect something was fishy if 20 heads and 80 tails occurred; either the model was wrong or the coin wasn't fair.

In a similar manner, a real-life series is only *one sample* (probably the only sample you'll ever get) of the process or activity that is represented by a given Box-Jenkins model. It follows that the autocorrelations computed from such a series will never precisely match the theoretical autocorrelations associated with the correct model for the series. They will, however, be close to, or approximate, the theoretical autocorrelations. Autocorrelations computed from real-life series are called *sample autocorrelations* to distinguish them from the theoretical autocorrelations associated with a given model.

9.2 INTERPRETING SAMPLE AUTOCORRELATIONS AND PARTIAL-AUTOCORRELATIONS

Sample autocorrelations (and partial-autocorrelations)—i.e., autocorrelations computed from actual time series data—are the only autocorrelations you have to look at in trying to determine what AC and PAC patterns are representative of the process you are trying to model. As we noted in the previous section, however, sample autocorrelations are only approximations of the "true" autocorrelations associated with the correct model for the series. In particular, some nonzero sample autocorrelations will be computed whose corresponding theoretical autocorrelations are actually zero. Thus if you were to look for sample autocorrelation patterns that looked precisely like the theoretical ones discussed previously, you would probably never find a precise match. What we need, then, is some means by which we can tell when a sample autocorrelation is close enough to zero to be interpeted as zero. Or to put it another way, we need to know when a sample autocorrelation can really be considered to be a significant one (i.e., nonzero). We can then compare the pattern of significant autocorrelations to the theoretical AC and PAC patterns.

It turns out that the degree to which a sample autocorrelation may vary from the "true" autocorrelation it represents can be measured, in some sense, by a computed quantity called the *standard error of the sample autocorrelation*. Table 9.1 lists some sample autocorrelations along with their standard errors. In particular, the standard error of an autocorrelation can tell us whether a computed nonzero sample autocorrelation is really significantly different from zero, or whether it is just within the "noise level" of its possible variation from zero. Statistically, a sample autocorrelation is regarded as being significantly different from zero (with roughly 95% confidence) if it is larger in magnitude than twice its standard error. The ± 2 times the standard error values are often called the *confidence limits* for the sample autocorrelation.

Correlograms printed out by most Box-Jenkins computer programs will plot these confidence limits along with the autocorrelation values. The correlogram in

Table 9.1 Sample Autocorrelations and Their Standard Errors

Lag	1	2	3	4	5	6	7	8	9	10	11	12
AC	−.61	.45	−.07	.11	−.06	.07	.03	−.08	.16	−.20	.22	−.19
Sd. Error	.08	.11	.12	.12	.12	.12	.12	.12	.12	.12	.13	.13
Lag	13	14	15	16	17	18	19	20	21	22	23	24
AC	.11	−.01	−.04	.07	−.05	.05	−.09	.09	−.11	.10	−.12	.12
Sd. Error	.13	.13	.13	.13	.13	.13	.13	.13	.13	.13	.13	.13

Figure 9.1(a), for example, illustrates a typical program-generated correlogram that was printed on a standard terminal printer device. The columns of asterisks are the bars representing the autocorrelation values, and the dashed lines are confidence bands representing the confidence limits. Where the autocorrelation bars and the confidence bands intersect, an X is printed.

Generally, for stationary, nonseasonal series, you will only need to look at the first 10–20 autocorrelations in order to get a good feel for what their pattern is. For example, the sample autocorrelations for lags 1 and 2 in Figure 9.1 can be assumed to be nonzero (with 95% confidence), while those for lags greater than 2 are assumed to be zero (more precisely, we don't reject the possibility that they are zero). With this kind of analysis, then, the theoretical and sample autocorrelation patterns shown in Figure 9.1 are a "perfect" match. The sample autocorrelations thus indicate a model with two MA parameters of orders 1 and 2.

9.3 ADDITIONAL POINTERS FOR INTERPRETING THE SAMPLE ACs AND PACs

As stressed in the previous section, the sample ACs and PACs obtained from real-life series will never precisely match the theoretical AC and PAC patterns associated with AR and MA models. The problem, then, is in deciding which nonzero sample autocorrelations should be ignored and which should be counted as significant. Confidence limits for the autocorrelations, of course, can be used as a rough guide in making these decisions, but they should not be strictly relied on at all times. Recall also that in the use of the confidence limits there is always some chance that an autocorrelation may appear significant when it really isn't, and vice versa. Thus

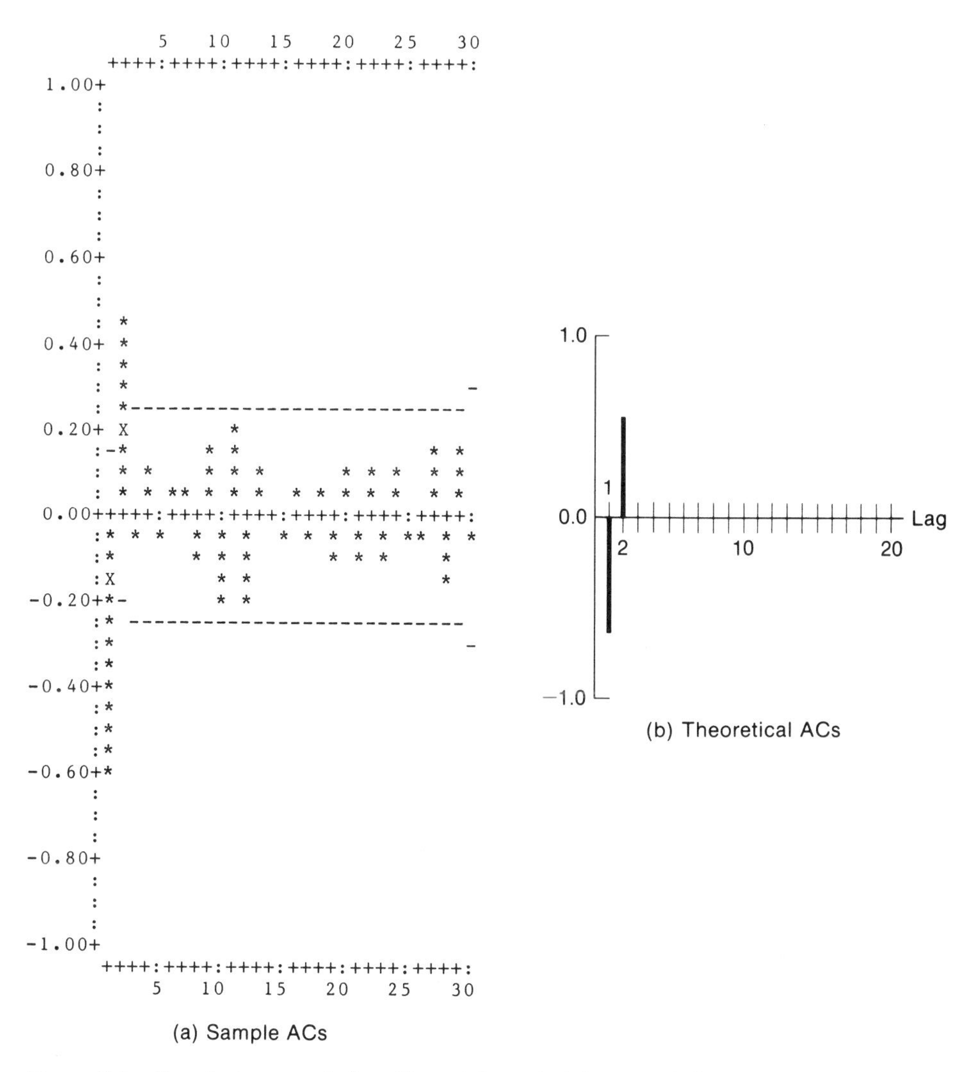

Figure 9.1 Sample Autocorrelations Versus Theoretical Autocorrelations

a fair amount of judgment is always required in interpreting the sample AC and PAC patterns.

For example, an AC pattern representing a model with one AR parameter is exponentially decreasing and trails off to zero rapidly, usually within the first 3–6 lags. All of these autocorrelations are significant in terms of being part of an identifiable pattern, but most often all but the first one or two autocorrelations will fall between the confidence limits.

Also keep in mind that each series that can be described by the same Box-Jenkins model will produce a slightly different set of sample autocorrelations. Figure 9.2, for instance, shows five different sample AC patterns produced from five different series whose Box-Jenkins model is the same in each case, namely, a model with one AR parameter. You therefore need to develop flexibility in interpreting such patterns.

Knowledge of what the data represents can also be helpful in making judgments about sample autocorrelations. For example, if you have a stationary, monthly time series representing unit sales, and a large autocorrelation appears at lag 7, there might be reason to be suspicious of its significance since there is probably no physical or economic reason to believe that a relationship exists between unit sales volumes seven months apart. On the other hand, for the same seasonal time series, if the autocorrelation at lag 12 is almost, but not quite, significant, there may still be reason to suspect that it really is significant because of the strong possibility of seasonal relationships.

9.4 SOME EXAMPLES TO TEST YOURSELF

Experience in interpreting sample AC and PAC patterns is one of the most important prerequisites for successful model construction. With this in mind, try to identify the model suggested by the correlograms in the following examples. The correct answers are given in Appendix G.

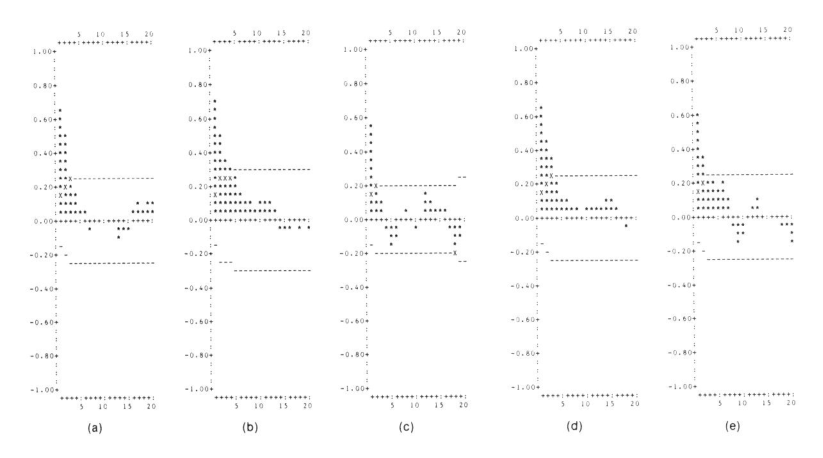

Figure 9.2 Five Sample AC Patterns Indicating a Model with One Positive AR Parameter of Order 1

Example 9.1

Identify the model that corresponds to the correlograms in Figure 9.3.

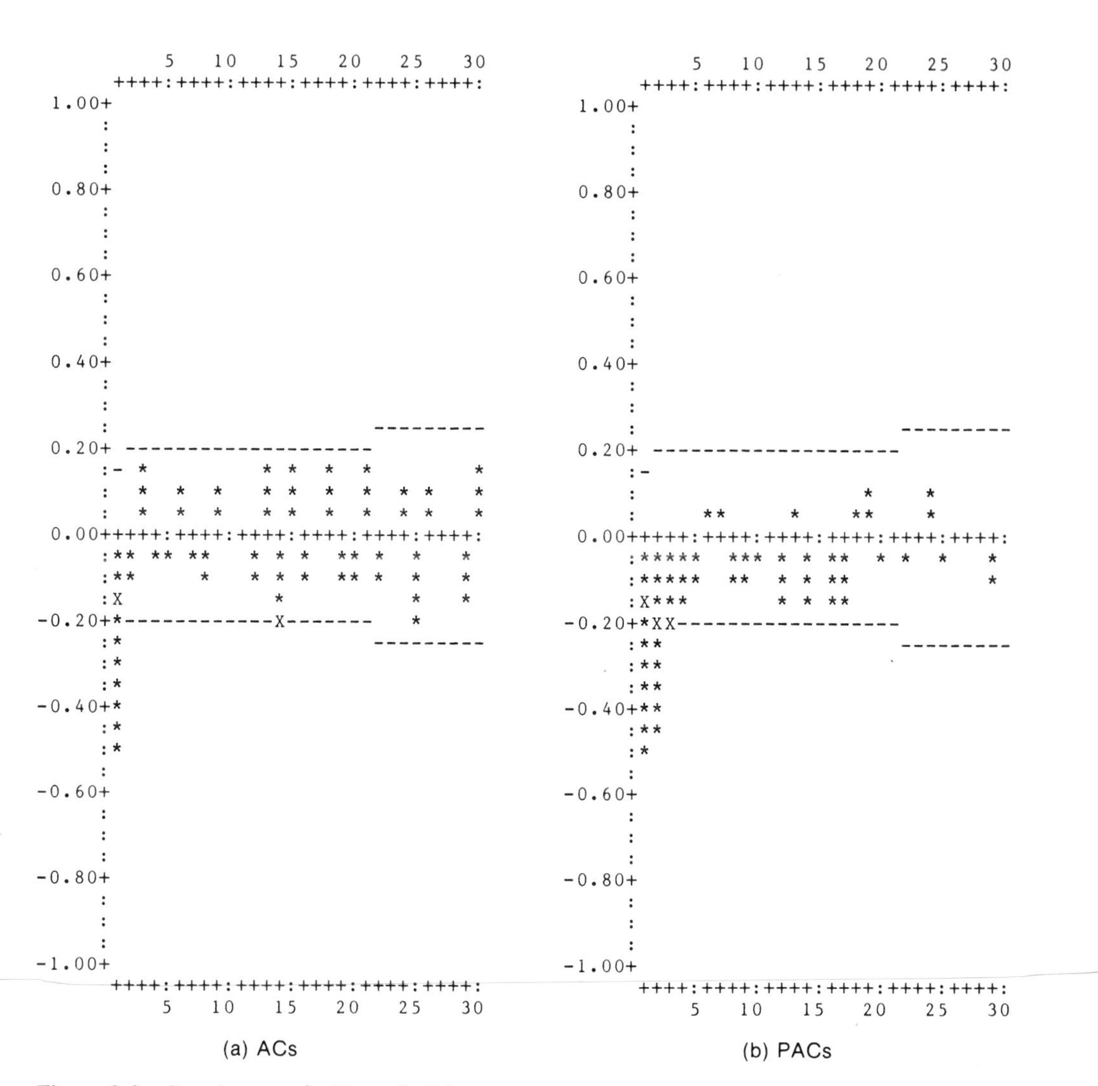

Figure 9.3 Correlograms for Example 9.1

Example 9.2

Identify the model that corresponds to the correlograms in Figure 9.4.

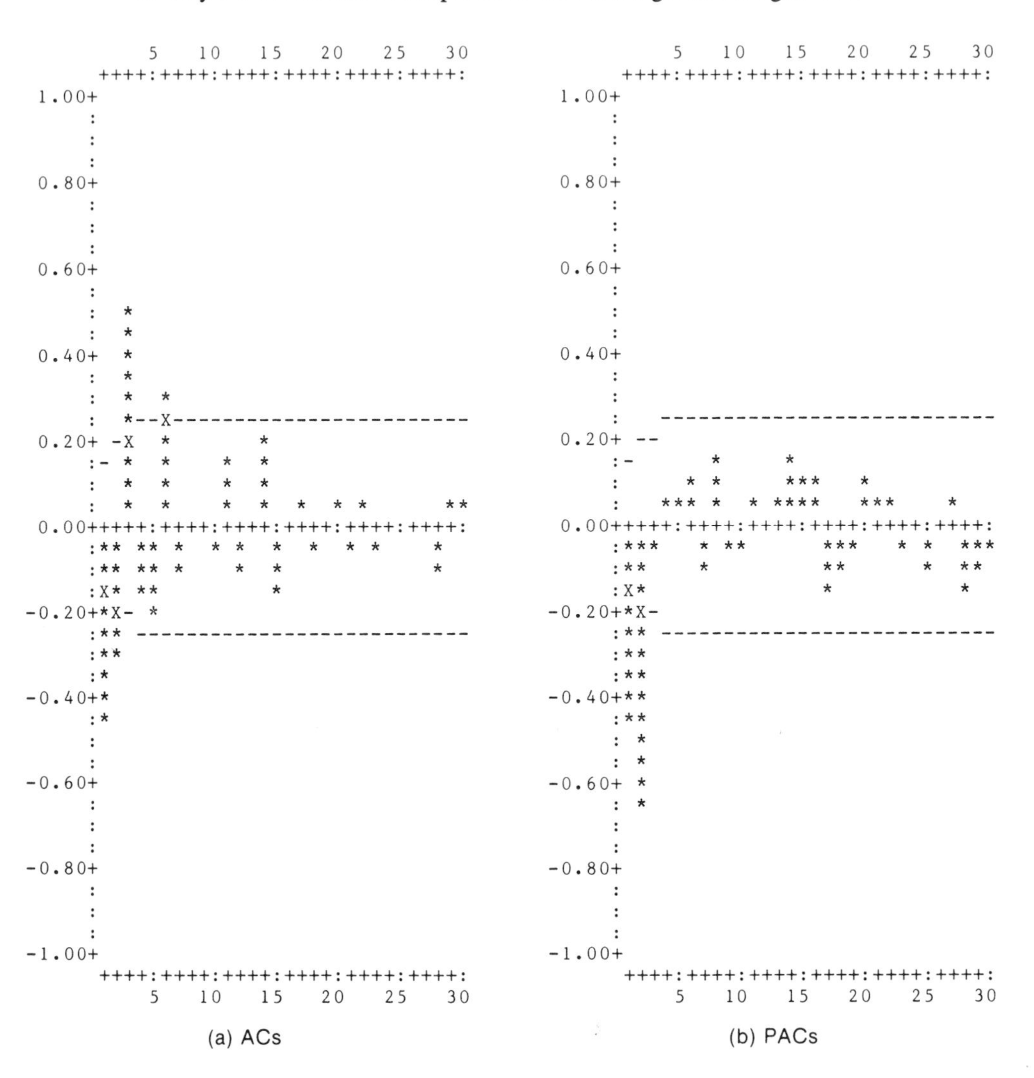

Figure 9.4 Correlograms for Example 9.2

Example 9.3

Identify the model that corresponds to the correlograms in Figure 9.5.

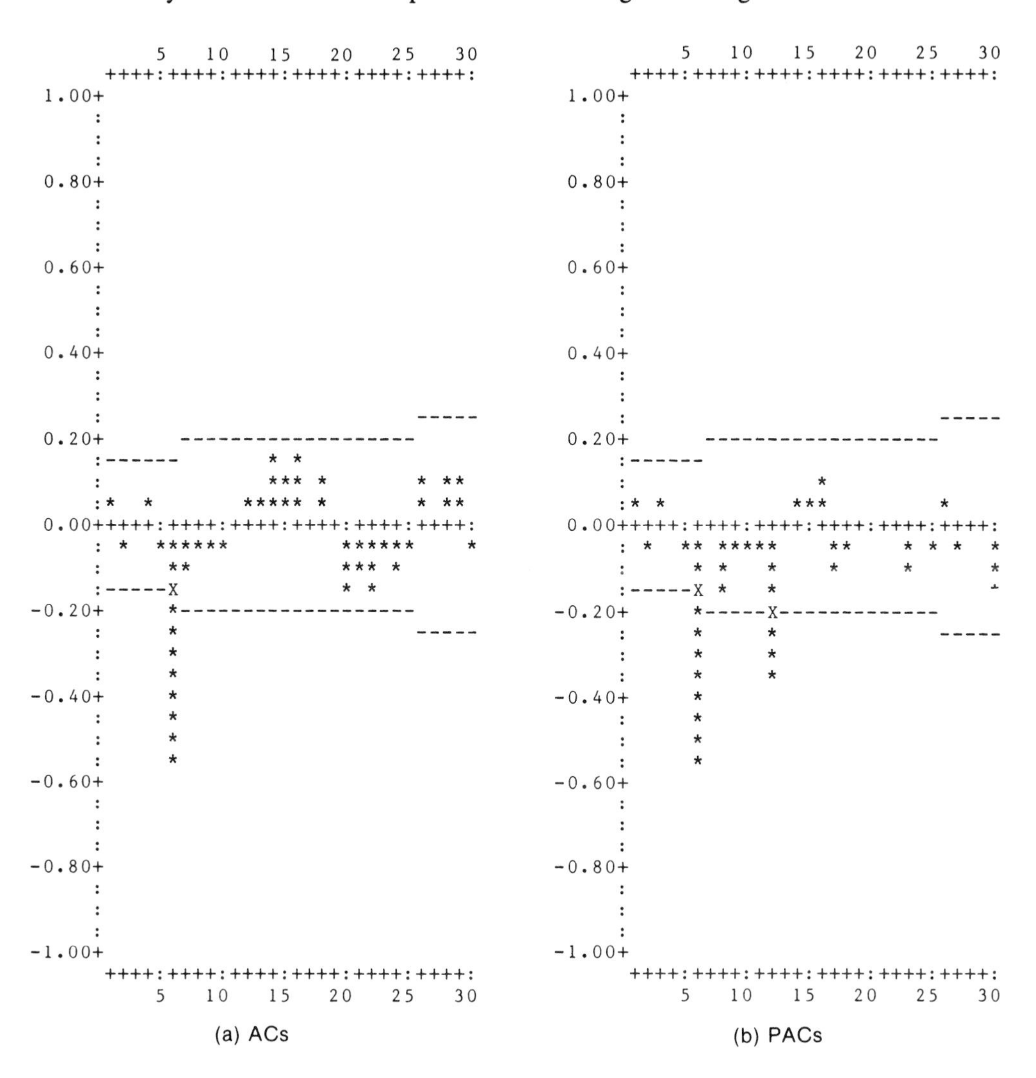

Figure 9.5 Correlograms for Example 9.3

Example 9.4

Identify the model that corresponds to the correlograms in Figure 9.6.

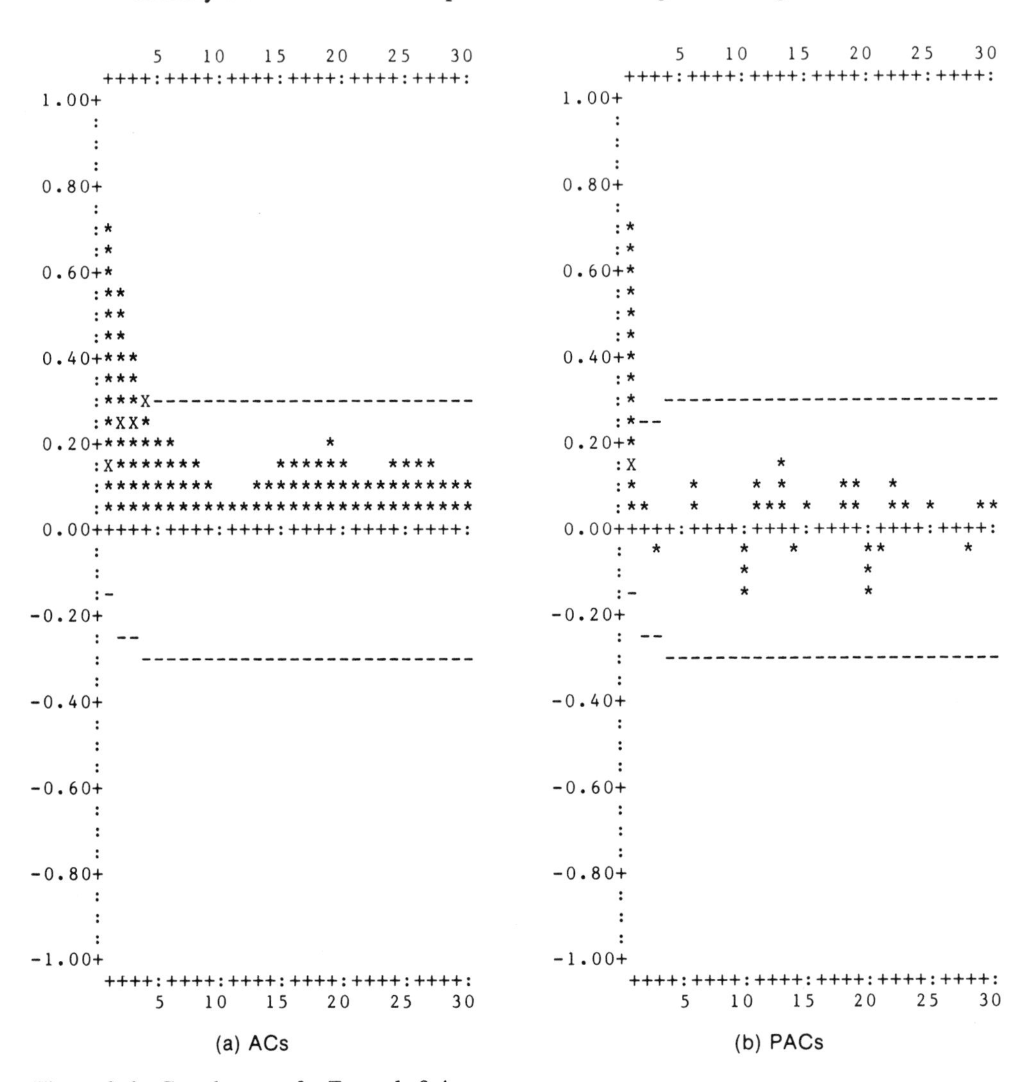

(a) ACs

(b) PACs

Figure 9.6 Correlograms for Example 9.4

Example 9.5

Identify the model that corresponds to the correlograms in Figure 9.7.

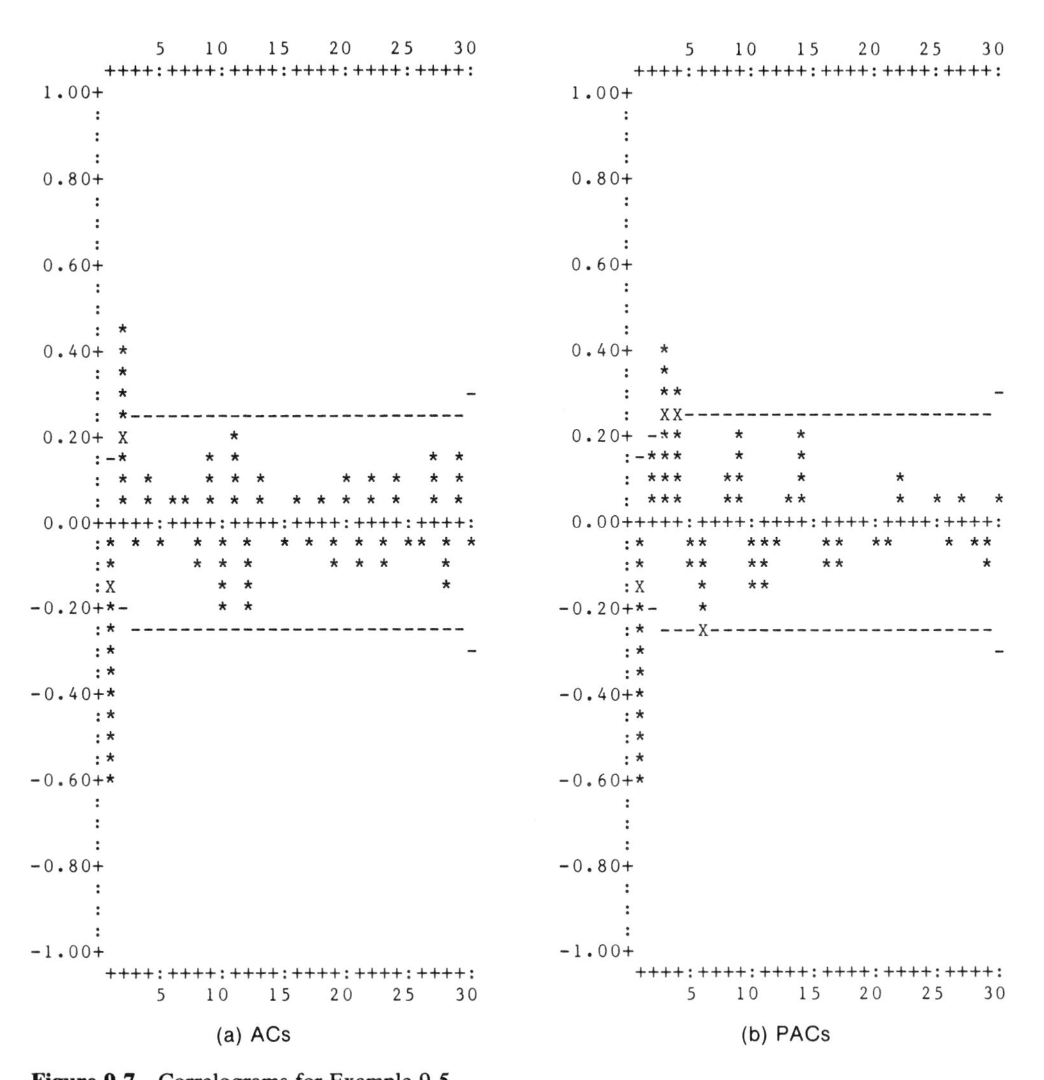

Figure 9.7 Correlograms for Example 9.5

Example 9.6

Identify the model that corresponds to the correlograms in Figure 9.8.

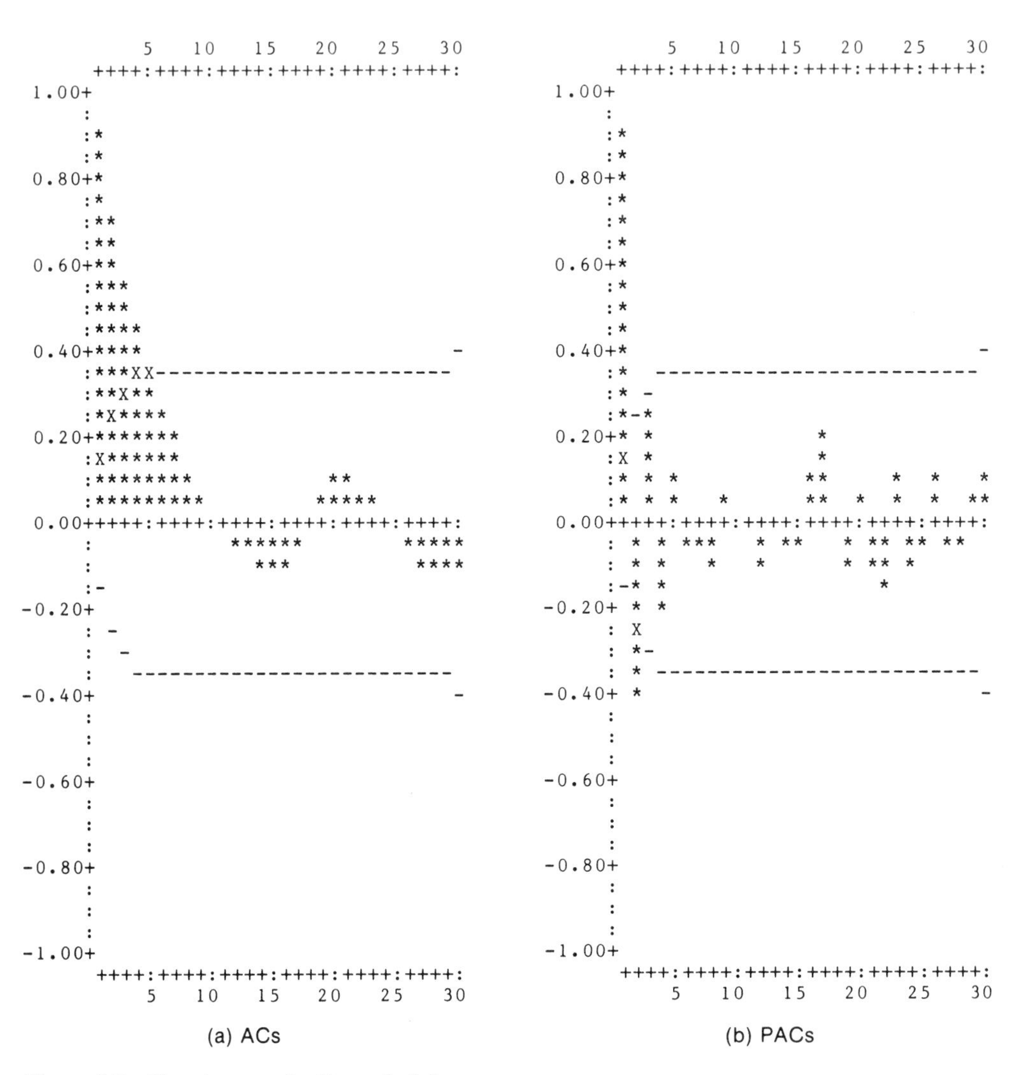

Figure 9.8 Correlograms for Example 9.6

Example 9.7

Identify the model that corresponds to the correlograms in Figure 9.9.

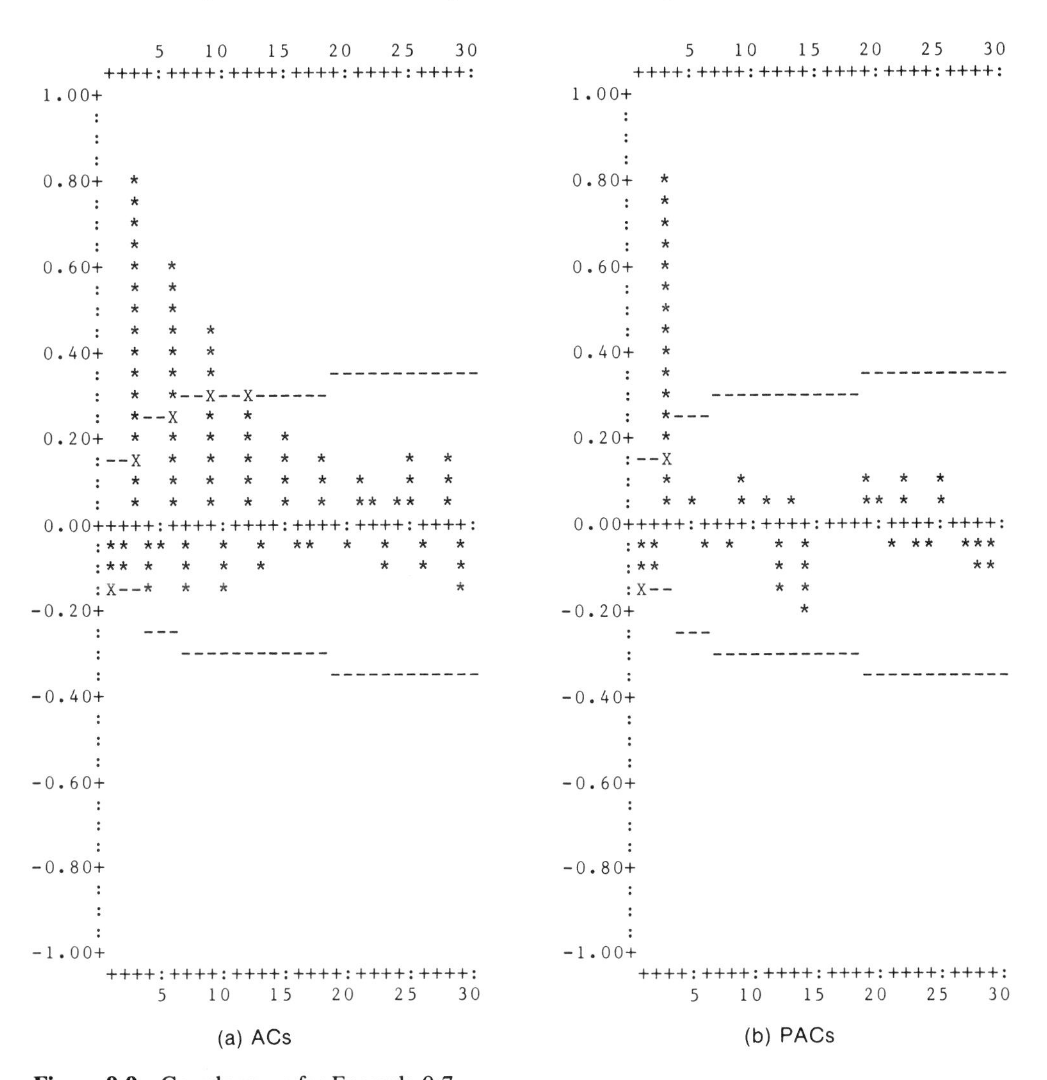

(a) ACs

(b) PACs

Figure 9.9 Correlograms for Example 9.7

Example 9.8

Identify the model that corresponds to the correlograms in Figure 9.10.

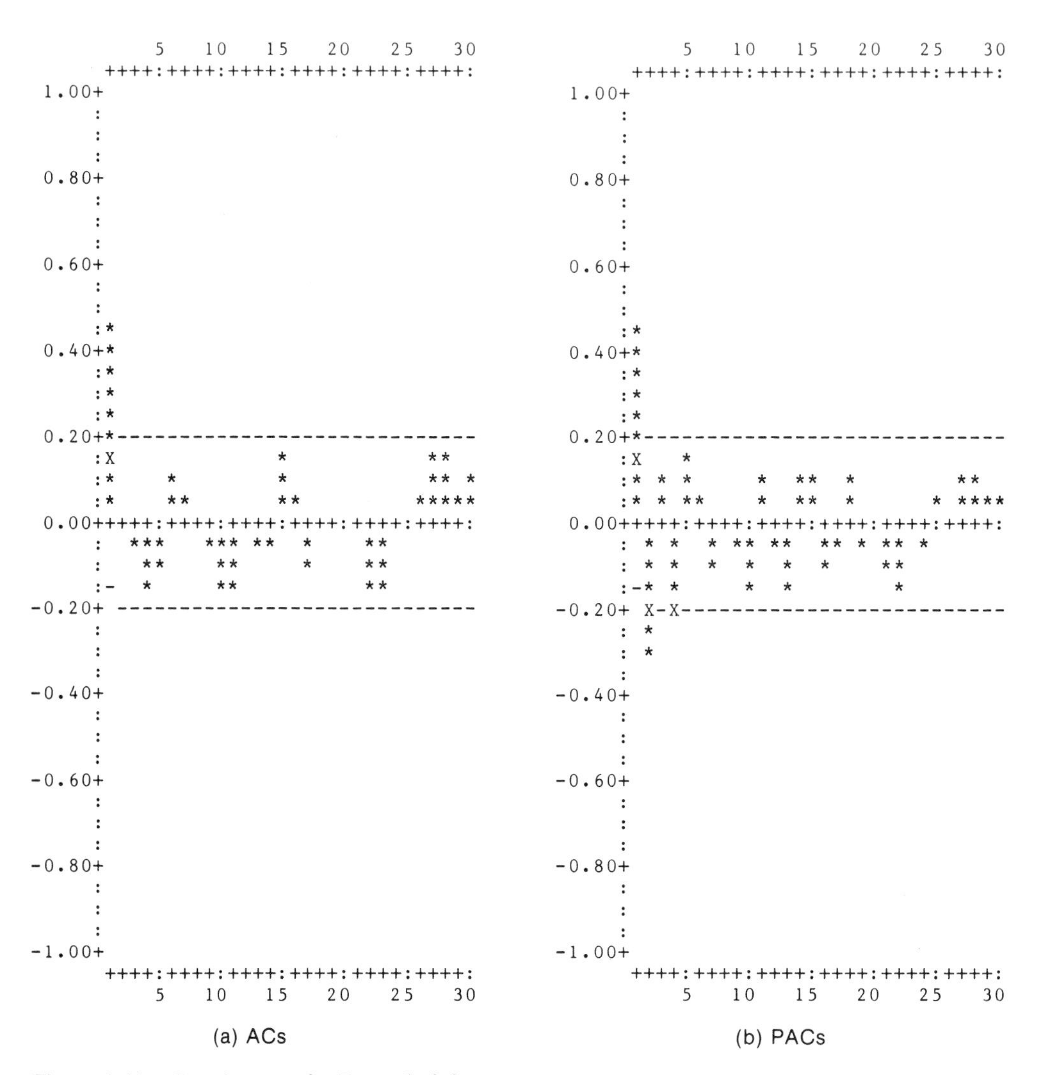

(a) ACs

(b) PACs

Figure 9.10 Correlograms for Example 9.8

Example 9.9

Identify the model that corresponds to the correlograms in Figure 9.11.

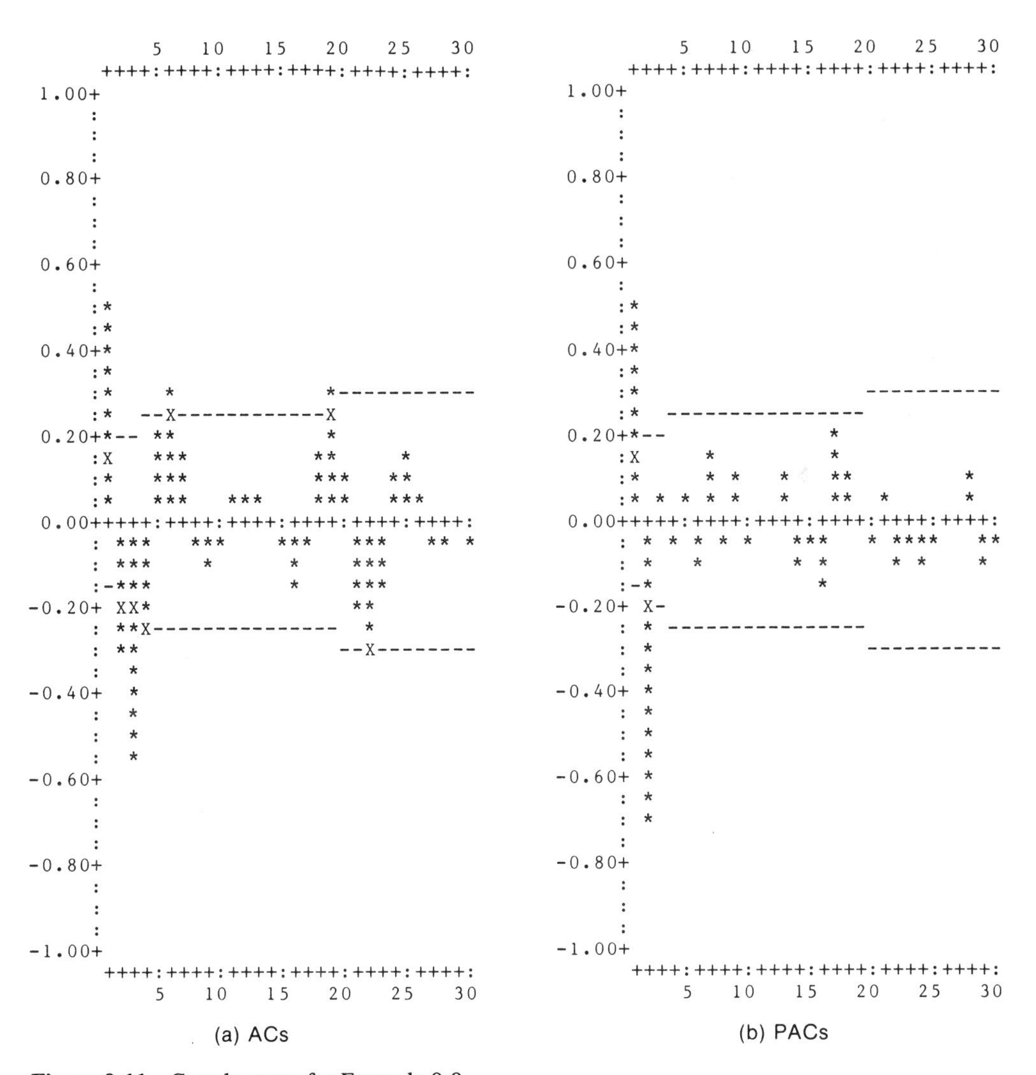

Figure 9.11 Correlograms for Example 9.9

Example 9.10

Identify the model that corresponds to the correlograms in Figure 9.12.

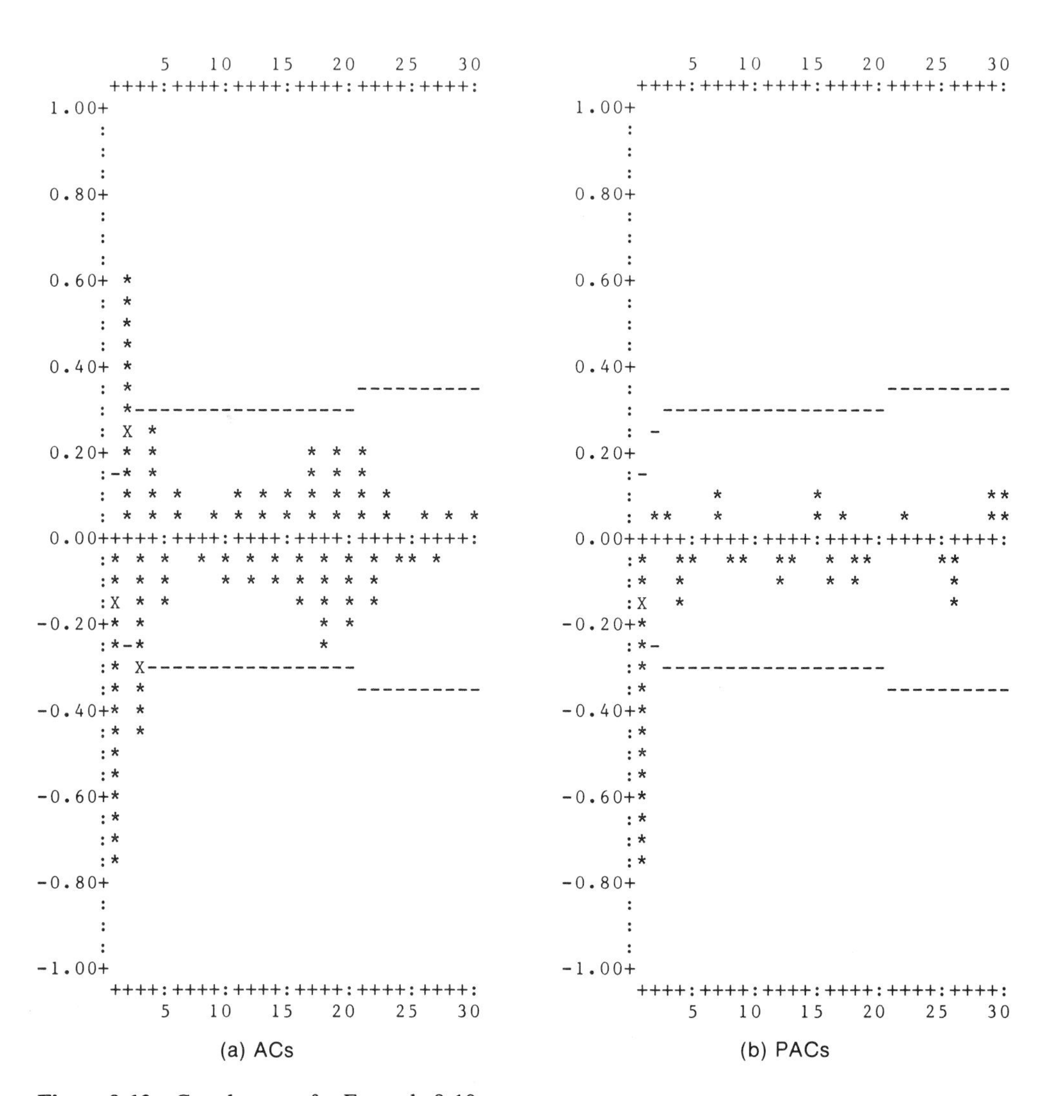

Figure 9.12 Correlograms for Example 9.10

9.5 REVIEW OF KEY CONCEPTS

Sample autocorrelations (and partial-autocorrelations) are the autocorrelation values actually computed from real-life series. Sample autocorrelations are only approximations to the theoretical autocorrelations associated with the correct model for the series, because the time series itself is only a *sample* of the true process represented by the data.

The *standard error of an autocorrelation* can also be computed from the data. It measures the potential magnitude of the autocorrelation's variation from the "true" or expected value of the autocorrelation.

The *autocorrelation confidence limits* are values located a distance of two times the standard error on either side of zero. If a sample autocorrelation falls between ± 2 times its standard error, that autocorrelation can be considered to be insignificant (with 95% confidence), i.e., essentially zero. In this way only the significant sample autocorrelations need to be compared with the theoretical patterns to find a match.

You should take care, however, not to be too rigid in the interpretation of the sample AC and PAC. A fair amount of latitude and judgment should always be exercised, for the following reasons:

- The autocorrelation confidence limits are only guidelines; there is always a chance that what appears to be significant really isn't, and vice versa.
- Knowledge of what the series actually represents can have a bearing on the proper interpretation of the ACs and PACs; a significant autocorrelation at lag 7 for a monthly series, for example, probably should be viewed with suspicion.
- You will always have the opportunity later to test your judgments and suspicions during the estimation phase of Box-Jenkins.

Examples of model identification and estimation in Chapters 11, 13, and 16 will serve to illustrate the above concepts.

CHAPTER **10**

Tools for Computing and Validating the Basic Models

In this chapter you will learn:

- What the objectives are in estimating a model.
- How a Box-Jenkins model is estimated.
- What tools are available for checking the validity of an estimated model.
- How to use the model validation tools.

In Chapter 11 you will see how these tools are applied when estimating models for some actual stationary series.

After you have selected an appropriate set of model parameters for your series, the next step is to compute specific values for these parameters from the series data. Since your time series is only one sample of the process it represents, you can only compute approximations, or estimates, of the true parameter values. Hence this step is generally referred to as *model estimation*. Any Box-Jenkins computer program will compute the estimated parameter values for you automatically. All you have to do is indicate to the program which parameters you want to include in your model and what your time series data is.

After the model has been estimated (i.e., the estimated parameter values, fitted values, and residuals have been computed), you must then check to see whether the model is an appropriate one for your series. This step is called *model validation*. To check the validity of a model, you must first understand what the objectives are in estimating a model and learn how to check whether these objectives have been met.

10.1 OBJECTIVES IN ESTIMATING A MODEL

There are three objectives in determining a specific Box-Jenkins model for a given series:

1. Obtain fitted values F_t that come as close as possible to the original series values X_t (what is meant by "as close as possible" is described below).
2. Obtain residuals, $E_t = X_t - F_t$, that are not correlated to one another (i.e., E_i is not correlated to E_j for i not equal to j).
3. Use as few parameters as necessary to obtain an adequate model; i.e., make sure the model is not overspecified.

The *first objective*—obtaining fitted values that come as close as possible to the original series value—is met by determining optimum numerical values for the selected AR and/or MA parameters. These values are optimum in the sense that the fitted values generated by the model using these parameter values produce the smallest possible residuals. Specifically, the parameter values are picked so that the *sum of the squared residuals is minimized* (i.e., the sum of the E_t^2 for all t is minimized). This is what is meant by "as close as possible." From a practical perspective it is not important to know how this minimization is done, since any Box-Jenkins computer program will do these computations automatically. Suffice it to say that the parameter values that minimize the sum of the squared residuals are called the *estimated parameters,* and the procedure for computing them is called *least squares estimation.*

The *second objective*—obtaining residuals that are not correlated to each other—is met if the parameters included in the model are really the correct ones for your series. In general, if needed parameters are left out, then the residuals will be correlated and hence will not behave the way the random error component is supposed to. In other words, the model will not capture all of the pattern component in the original series, and part of the pattern component will remain in the residuals. One purpose of this chapter is to describe some tools that will help you decide whether the resulting residuals from an estimated model are uncorrelated and, if they are not, suggest some ways to improve the model. These tools are called *residual diagnostics.*

The *third objective*—using as few parameters as necessary—is one of the guiding principles in developing any time series forecasting model. Using too many parameters in a model only complicates matters unnecessarily and can lead to erroneous judgments when comparing the results of different models. Another purpose of this chapter is to describe tools that will help you decide whether there are parameters included in your model that are not needed or are incorrect. These tools are called *parameter diagnostics.*

Residual diagnostics and parameter diagnostics are certain statistics that are automatically computed when the estimated parameters are computed. These diagnostics enable you to take the following steps:

- Decide whether your model adequately explains the behavior of your time series.
- Decide whether your model is overspecified, i.e., has unnecessary parameters in it.
- Compare alternative models.
- Determine how the current model might be improved.

If the diagnostics indicate that improvements can be made, other models can be estimated and their residual and parameter diagnostics interpreted and compared. In practice, a number of prospective models are usually estimated before the correct model is finally obtained. It is therefore important to learn how to interpret the residual and parameter diagnostics so that the best possible model can be constructed. Examples of model estimation and validation for actual stationary series are given in Chapter 11.

10.2 CHECKING THE RESIDUALS—RESIDUAL DIAGNOSTICS

Residual diagnostics include the following items:

- *Residual mean* and *mean percent error,* which indicate the presence of bias in the residuals.
- *Autocorrelations of the residuals,* which measure the correlation among the residuals.
- *Q-statistics for the residual autocorrelations,* which indicate whether the autocorrelations of the residual series, *as a whole,* are significantly nonzero; i.e., whether the residuals show any significant correlation to one another.
- *Closeness-of-fit statistics,* which measure how close the fit is to the original series and allow comparison of alternative models.

We will discuss each diagnostic in turn in the following subsections.

The first three diagnostics will help you decide whether your model has accounted for all patternable behavior in the series. If it hasn't, the computed residuals will still contain some patterns; i.e., they will be correlated in some way. Some of these diagnostics also provide help in determining how to improve your model.

Once you are satisfied that you have a valid model for your series, the closeness-of-fit statistics will tell you how big the random error component in the series is. Some of these statistics will be important later in measuring the uncertainty in forecasts produced by the model. They are also used for comparing and selecting a model from among two or more valid alternatives.

10.2.1 Residual Mean and Mean Percent Error

The *residual mean* (or mean error) is simply the average of all the computed residuals, i.e., the sum of all the E_t divided by the number of residuals.

We are looking for a residual mean that is zero or close to zero (i.e., not significantly nonzero). For if the residual mean is significantly nonzero, then either the fitted values are consistently higher or lower than the original series values, or the residuals of one sign (positive or negative) are consistently larger than the residuals of the other sign. In either case the fit is *biased,* so you can expect a biased forecast.

In addition, the statistical assumptions made about the random error component E_t in a theoretical Box-Jenkins model imply that the residual mean should be zero, and the computation of other residual diagnostics is based on this assumption. If the residual mean is significantly nonzero, then the other diagnostics are of questionable validity.

Many Box-Jenkins programs will tell you whether the residual mean is significantly nonzero. But if your program does not tell you, you can easily check for significance by comparing the magnitude of the residual mean (i.e., disregarding any minus sign) with the quantity 2 times the residual standard error divided by $\sqrt{N}$, where N is the number of residuals (see Section 10.2.4 for a description of the residual standard error). If the residual mean is greater than this quantity, then you can conclude that it is significantly greater than zero. For example, if there are 100 residuals, the residual mean is .8, and the residual standard error is 3, then $2 \times 3 \div \sqrt{100} = .6$, which is smaller than .8; so the residual mean is significantly nonzero.

The *mean percent error* is similar to the residual mean, except that the residuals are first divided by the magnitude of the original series values. Specifically, it is the average of the percent differences between the fitted and actual series values, i.e., the sum of the ratios E_t/X_t divided by the number of residuals, times 100. The mean percent error just gives you another way to judge whether the residual mean is significantly large. For example, if the residual mean is 100 and the original series values have magnitudes on the order of 1,000,000, then the mean percent error will only be about .01%, which indicates a very small residual mean relative to the magnitude of the series values.

10.2.2 Autocorrelations of the Residuals (Residual ACs)

Just as sample autocorrelations of the original series can be computed, so can sample autocorrelations of the series of residuals. It is important to know whether the residuals are correlated, since one of the objectives stated at the beginning of this chapter is to *obtain residuals that are not correlated.* If all the sample autocorrelations of the residuals are not significant and show no discernible patterns, then you may safely conclude that the residuals are uncorrelated. On the other hand, if the residual ACs and PACs show significantly large spikes at certain lags, then the

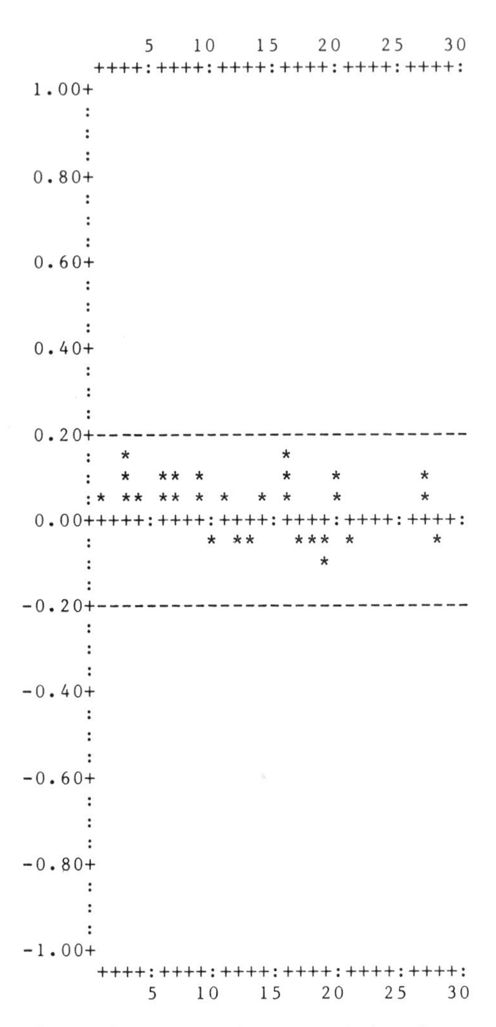

Figure 10.1 Residual Autocorrelations from an Adequate Model

residuals are probably correlated in some way, and the model is inadequate—i.e., is *underspecified*—or doesn't contain the correct parameters.

The AC pattern of the residuals can also be used to indicate how the model might be improved. For example, the residual AC correlogram in Figure 10.1 indicates that the sample residual autocorrelations are not significant, and hence the model is adequate. Figure 10.2, on the other hand, illustrates the resulting residual correlogram when another model was specified but an MA parameter was omitted. Here the large (greater than the confidence bands) spike at lag 2 reflects the fact that there is still a need for one MA parameter of order 2.

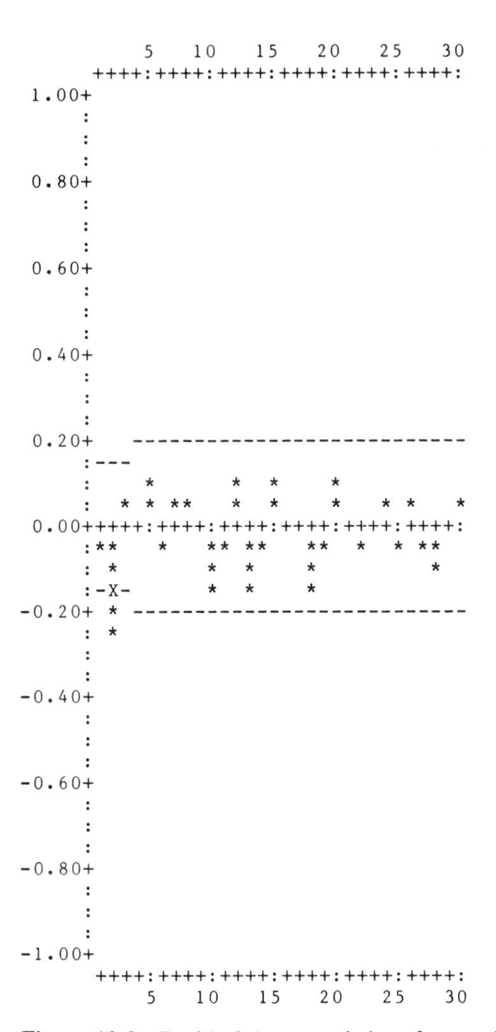

Figure 10.2 Residual Autocorrelations from an Inadequate Model

10.2.3 The Q-Statistic for the Residual Autocorrelations

The *Q-statistic* is a certain statistical measure computed from the sample autocorrelations of the residual series. It can be used to test whether a set of residual autocorrelations, *as a whole,* are statistically significant. The objective, of course, is to produce residual autocorrelations that are *not* statistically significant. Usually, you will want to look at the Q-statistic corresponding to a small set of autocorrelations (e.g., lags 1 to 12) and then a larger set (e.g., lags 1 to 24), etc., to make sure that no subset of the residual autocorrelations is significant.

The Q-statistic thus provides you with an additional test for determining whether the residuals display any correlation among themselves. The test is simply made by comparing the Q-statistic with a critical test value (called a chi-square value). If the Q-statistic is larger than the critical test value, then you conclude (usually with 95% confidence, or a 5% risk of being wrong) that the residual autocorrelations being tested, as a whole, are significant. If the Q-statistic is less than the critical value, then you cannot conclude that they are significant, and you assume that they probably are not. For example, the Q-statistics shown in Table 10.1 indicate that the residual autocorrelations for lags 1 to 12 have a Q-statistic of 7.94 with a corresponding critical test value (chi-square value at 5% risk) of 18.3. You can conclude from this result that these 12 residual autocorrelations, as a whole, are probably not significant.

Table 10.1 Q-Statistics

Lags	Q-Statistic	Chi-square at 5%
1–12	7.94	18.30
1–20	20.41	28.87

Some Box-Jenkins programs display the appropriate critical test value along with each Q-statistic. If your program does not, the test value can be looked up in a chi-square distribution table found in most statistics texts. To find the right test value in the table, use a 95% confidence level (5% risk level) and a degree of freedom equal to the number of autocorrelations being tested minus the number of parameters included in the model.

The Q-statistics only give you an indication of an inadequate model. They do not provide any information about how the current model might be improved. However, you can get this kind of information by looking at the sample autocorrelations of the residuals themselves or by reexamining the sample AC pattern for the original stationary series.

10.2.4 Closeness-of-Fit Statistics

Some or all of the following statistics are also computed and displayed by most computer programs:

- Average absolute error
- Residual standard error
- Average absolute percent error
- Index of determination

These statistics are primarily descriptive. That is, after you are satisfied that you have a valid model, you can the use these statistics to measure how big the random error (residual) component is and get a feeling for how much uncertainty you will have to live with in the forecasts produced by your model. They are also useful for comparing different models that all pass the validation tests. The model with the better-fit statistics, in this case, would be the natural one to choose. Examples of how these statistics are computed are given at the end of this section.

The average absolute error and the residual standard error provide absolute measures of how close the fit is to the original data, i.e., how big the residuals are, on the average. The *average absolute error* is simply the average of all residuals without regard to the sign of each residual (i.e., all residuals are treated as positive values). The *residual standard error* is a little more complicated: It is the square root of the average of the squared residuals (the average of the squared residuals is often called the *residual mean square*). In symbolic terms the average absolute error and the residual standard error are computed as follows, where N is the number of residuals:*

$$\text{Average absolute error} = \frac{1}{N}(\text{Sum of } |E_t|)$$

$$\text{Residual standard error} = \sqrt{\frac{1}{N}(\text{Sum of } E_t^2)}$$

The symbol $|E_t|$ means that any negative E_t is changed to a positive value.

These measures are useful in the following way: If the residuals are truly random errors, then the average absolute error and residual standard error tell you how big the error component of the original series is, on the average. This result, in turn, provides you with some idea of how big you can expect the error to be in forecasts generated by your model.

The residual standard error is generally used more often than the average absolute error in statistical work. In particular, it is used in the computation of other statistical measures such as forecast confidence limits. Forecast confidence limits provide a measure of the uncertainty in a forecast and are discussed in Chapter 17. Recall also that the estimated parameter values are computed so that the sum of the squared residuals is minimized. Since the residual standard error is the square root of the average of this sum, the residual standard error is also minimized.

The *average absolute percent error* is similar to the average absolute error (in the same way that the residual mean is similar to the mean percent error), except that the residuals are compared with the magnitude of the original series values. Specifically, it is the sum of the ratios $|E_t/X_t|$ divided by the number of residuals, times 100. This statistic allows you to judge the magnitude of the residuals relative to the magnitude of the original series values.

*Actually, the residual standard error is computed by using a value for N that is equal to the number of residuals minus the number of parameters in the model. This value is called the *degrees of freedom*.

The *index of determination** is a useful measure for quickly comparing alternative valid models and for understanding how much variation in the original series has been accounted for by the fit. Specifically, it is a ratio that compares the amount of variation present in the original series values with the amount of that variation that has been accounted for by the fit. The term *variation* is used here to mean the variation of the data from its mean value. Thus if a fit explains all the variation (i.e., a perfect fit), the ratio will be 1. If the fit is no better than just using the average, or mean, of the series as the fit, then the ratio will be 0. If a fit explains 75% of the variation, then the ratio will be .75.

If the residuals are truly uncorrelated, then the fit will have explained all variation, except for the random error. It would be nice, of course, if the random error in the series data were only a small percentage of the total variation. In this case a correct model will produce an index of determination close to 1. Keep in mind, however, that a high index of determination does not necessarily mean you have a good model. You should make sure that your model passes all the validation tests before you look at the index of determination. On the other hand, a small index of determination doesn't necessarily mean that you have a bad model. In this case, if the model is valid, it simply means that the random error component in the series is large.

Typically, the index of determination is computed by finding the ratio of the variation in the original series due to random error compared to the total variation in the original series, and subtracting this ratio from 1. The variation due to random error is computed as the sum of the squared residuals, and the total variation in the original series is computed as the sum of the squared deviations of the original series from its mean value.

To illustrate how all the closeness-of-fit statistics are computed, we give the following simple example, where M is the mean of the original series, $X_t - M$ are the deviations of the original series from the mean, and there are five periods of data. The data are shown in Table 10.2.

Mean of series	$= 100$
Sum of residuals	$= -1$
Sum of residuals (without minus signs)	$= 9$
Sum of ratios	$= -.014$
Sum of ratios (without minus signs)	$= .09$
Sum of squared residuals	$= 27$
Sum of squared deviations	$= 66$
Residual mean	$= \frac{-1}{5} = -.2$

*In regression this statistic is usually called the R^2 statistic.

Table 10.2 Data for Example

t	1	2	3	4	5
Series, X_t	94	98	104	101	103
Fit, F_t	97	98	100	102	104
Residuals, E_t	−3	0	4	−1	−1
Deviations, $X_t - M$	−6	−2	4	1	3
Ratios, E_t/X_t	−.032	0	.038	−.01	−.01

Mean percent error $= 100\left(\frac{-.014}{5}\right) = -.3\%$

Average absolute error $= \frac{9}{5} = 1.8$

Residual standard error $= \frac{27}{5} = 2.3$

Average absolute percent error $= 100\left(\frac{.09}{5}\right) = 1.8\%$

Index of determination $= 1 - \frac{27}{66} = .591$

10.3 CHECKING THE COMPUTED PARAMETERS—PARAMETER DIAGNOSTICS

Parameter diagnostics include the following items:

- *Correlations between parameters,* which measure the degree of correlation between any two parameters in the model.
- *Parameter confidence limits,* which test the significance of each parameter included in the model.

These diagnostics will help you decide whether you have unneeded or incorrect parameters in your model. We discuss each diagnostic in the following subsections.

10.3.1 Correlation Between Parameters

When an unneeded parameter is included in a model and estimated along with the correct parameters, its estimation will sometimes tend to be influenced significantly by the existence of one or more of the other valid parameters (since it has no relationship of its own to model). It turns out that, statistically, this degree of influence, or correlation, can be measured and computed. In fact, the correlation between any two estimated parameters can be obtained. If two estimated parameters are highly correlated, then one of them can probably be eliminated without affecting the adequacy of the model. Thus examining the measure of correlation between parameters can be helpful in determining overspecification.

The measure of parameter correlation, like any other type of correlation measure, is computed as some value between -1 and $+1$, where 0 indicates no correlation and a value close to -1 or $+1$ indicates high correlation. Usually, these correlation measures are arranged in the form of a triangular matrix, as shown in Table 10.3. The correlation between parameter 1 (PAR 1) and parameter 2, in the table, for example, is only .012, while the correlation between parameter 2 and parameter 3 is quite high, at .856. This result suggests that either parameter 2 or parameter 3 should be removed from the model. Since parameter 1 is more highly correlated to 3 than it is to 2, parameter 3 is the most likely candidate for removal.

Table 10.3 Parameter Correlation Matrix

	PAR 1	PAR 2	PAR 3
PAR 1	—		
PAR 2	.012	—	
PAR 3	.235	.856	—

10.3.2 Parameter Confidence Limits

Since the estimated parameter values are computed from sample data, there is an associated degree of uncertainty in the values obtained, relative to the theoretical true values of the parameters (in the same sense that sample ACs are only approximations of the theoretical ACs). Because of this uncertainty, confidence limits for a parameter (just like confidence limits for the autocorrelations) can be computed that indicate a range of values in which the true or theoretical parameter value can be expected to lie with some degree of confidence (usually 95% confidence). The estimated value of the parameter is located at the midpoint of this range.

Thus we can judge the significance or nonsignificance of a parameter by noting whether the confidence range includes zero (i.e., the lower limit is less than zero and the upper limit is greater than zero). If the range includes zero, then there is a strong possibility that the true value of the parameter is in fact zero (i.e., the parameter is not significant), and we can probably eliminate the parameter from the model without affecting its adequacy. For example, Table 10.4 shows the estimation results for a three-parameter model. In this case the confidence range for parameter 3 brackets zero, and hence it is a candidate for removal.

Table 10.4 Parameter Confidence Limits

	Estimated Value	95% Limits	
		Upper	Lower
PAR 1	.532	.222	.742
PAR 2	.710	.560	.860
PAR 3	.151	−.160	.464

10.4 REVIEW OF KEY CONCEPTS

There are three objectives in estimating, or computing, the parameters in a model:

1. Obtain fitted values as close as possible to the original series values; i.e., minimize the sum of the squared residuals. The parameter values that accomplish this minimization can be computed by any Box-Jenkins program that uses a technique called least squares estimation.
2. Obtain residuals that are not correlated to each other. This objective is accomplished if you have included the correct (and sufficient) parameters in your model. A set of tools called residual diagnostics will help you decide whether the residuals are uncorrelated and will suggest ways you might improve your model.
3. Use as few parameters as possible in the model. A set of tools called parameter diagnostics will help you determine whether you have overspecified your model and will suggest parameters you might eliminate.

Residual diagnostics and parameter diagnostics comprise the tools available to you for determining whether a selected model is valid.

Residual diagnostics are helpful in determining whether the residuals are, in fact, uncorrelated and whether the model is underspecified, i.e., whether additional parameters are needed to explain the behavior of the series. These diagnostics include the following:

- The *residual mean* and the *mean percent error,* which should not be significantly nonzero.
- The *residual autocorrelations,* which should not show any significant autocorrelations.
- The *Q-statistic* for the residual autocorrelations, which should be less than its corresponding critical test value (chi-square at 5% risk).

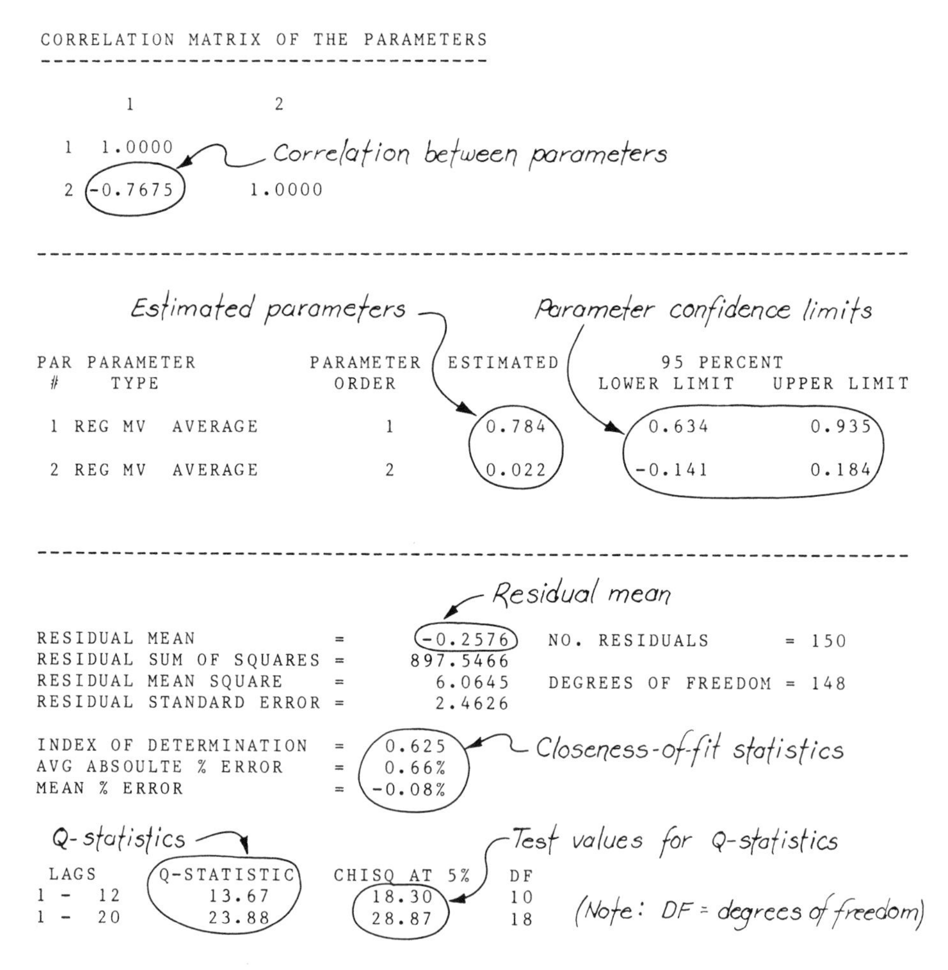

```
CORRELATION MATRIX OF THE PARAMETERS
------------------------------------

          1          2

  1    1.0000
  2   -0.7675      1.0000

--------------------------------------------------------------------

PAR PARAMETER          PARAMETER  ESTIMATED        95 PERCENT
 #     TYPE              ORDER               LOWER LIMIT   UPPER LIMIT

 1 REG MV   AVERAGE        1        0.784      0.634          0.935

 2 REG MV   AVERAGE        2        0.022     -0.141          0.184

--------------------------------------------------------------------

RESIDUAL MEAN             =      -0.2576   NO. RESIDUALS        = 150
RESIDUAL SUM OF SQUARES   =     897.5466
RESIDUAL MEAN SQUARE      =       6.0645   DEGREES OF FREEDOM   = 148
RESIDUAL STANDARD ERROR   =       2.4626

INDEX OF DETERMINATION    =    0.625
AVG ABSOULTE % ERROR      =    0.66%
MEAN % ERROR              =   -0.08%

 LAGS       Q-STATISTIC     CHISQ AT 5%    DF
1 -  12         13.67          18.30       10
1 -  20         23.88          28.87       18
```


Figure 10.3 Typical Residual and Parameter Diagnostics from a Computer Program Indicating an Overspecified Model

If the model is not adequate, the residual AC can be used to help determine how you can improve the model. *Closeness-of-fit statistics* are additional residual diagnostics that are used to describe the magnitude of the random error in the original series. These statistics allow you to compare the results of alternative models and to measure the degree of uncertainty to be expected in forecasts produced by the model (see Chapter 17).

Parameter diagnostics are helpful in determining whether a model is overspecified, i.e. whether the model contains superfluous or incorrect parameters. These diagnostics include the following:

- *Correlations between parameters,* which should not be significantly large.
- *Parameter confidence limits,* which should not bracket the value zero.

A parameter that is highly correlated to other parameters, or has confidence limits that bracket zero, is a candidate for removal from the model.

Figure 10.3 shows a typical set of results displayed after running a Box-Jenkins computer program. Two MA parameters were specified for the model in this example (in the computer output an MA parameter is labeled as a "REG MV AVERAGE," meaning a "regular moving average"). The correlation between the two parameters is approximately −.768, which is quite high. This result indicates that one of the parameters is probably not needed. Looking for further verification, we find that the upper and lower confidence limits for the second parameter are approximately −.141 and .184, respectively, which bracket zero. This result indicates that the second parameter is not significantly different from zero and hence can be eliminated. On

```
PAR PARAMETER           PARAMETER  ESTIMATED         95 PERCENT
 #     TYPE               ORDER                  LOWER LIMIT   UPPER LIMIT

 1 REG MV   AVERAGE          1         0.801         0.704          0.897


---------------------------------------------------------------------------


RESIDUAL MEAN              =        -0.2497    NO. RESIDUALS        = 150
RESIDUAL SUM OF SQUARES =        895.2495
RESIDUAL MEAN SQUARE       =          6.0084    DEGREES OF FREEDOM = 149
RESIDUAL STANDARD ERROR =          2.4512

INDEX OF DETERMINATION    =     0.626
AVG ABSOULTE % ERROR      =     0.65%
MEAN % ERROR              =    -0.08%


 LAGS       Q-STATISTIC    CHISQ AT 5%    DF
1 -   12        13.67          19.67       11
1 -   20        23.77          30.14       19
```

Figure 10.4 Residual and Parameter Diagnostics Indicating a Valid Model

the other hand, the Q-statistics are both less than their test value, indicating that the first 12 and first 20 residual autocorrelations, as a whole, are not significant; i.e., the residuals are uncorrelated. This result indicates that probably no needed parameters have been left out.

Thus it would appear that a model with only one MA parameter might be the appropriate model in this case. The estimation results for a one-MA-parameter model are shown in Figure 10.4.

The results in Figure 10.4 now indicate that the estimated model is a valid one: The parameter confidence limits for the MA parameter do not bracket zero and the Q-statistics are less than their test values (note that no parameter correlation matrix was displayed since there is only one parameter in the model. In examining the fit statistics, we find that the average absolute percent error and the mean percent error are both relatively small, indicating a close fit on a percentage basis. The index of determination, on the other hand, indicates that over 37% of the variation in the original series is attributable to random error.*

*Note that two other residual statistics are printed in the estimation results in both Figures 10.3 and 10.4 that were not previously explained: the residual sum of squares and the residual mean square. These statistics do not provide any additional important information since they are directly related to the residual standard error. In particular, the residual mean square is the square of the residual standard error, and the residual sum of squares is the residual mean square times the degree of freedom, printed directly to the right (degrees of freedom equals the number of residuals minus the total number of parameters in the model).

CHAPTER **11**

Computing and Validating Some Basic Box-Jenkins Models

We are now ready to apply the tools and techniques described in Chapters 7 through 10 to identify, estimate, and validate some basic Box-Jenkins models for some actual stationary series.

In this chapter we will:

- Review the Box-Jenkins model-building process in terms of the material given in Chapters 7 through 10.
- Show four examples of identifying, estimating, and validating models for stationary series.

11.1 OBTAINING THE BEST MODEL IN PRACTICE—THE PROCESS REVISITED

As outlined in Chapter 5 and illustrated in Figure 5.1, the general process for obtaining a Box-Jenkins model for a series is to first identify the model and then estimate the optimum values of the parameters in the model. In practice, this process may involve a number of tentative model identifications and parameter estimations before the best model is found. This iterative process is usually necessary since the AC and PAC patterns of a series often suggest more than one possible model, and some AC and PAC patterns are easy to overlook. The best model can then be selected by using the model validation tools described in Chapter 10.

In light of the above comments and the new knowledge we have of the identification, estimation, and validation tools available to us, we can now revise the general model-building process of Figure 5.1. Figure 11.1 illustrates this revised model-building process for stationary series.

Before we apply this process to some actual stationary series, however, we need to digress briefly to reexamine an assumption about the mean value of the series that was made in Chapter 6.

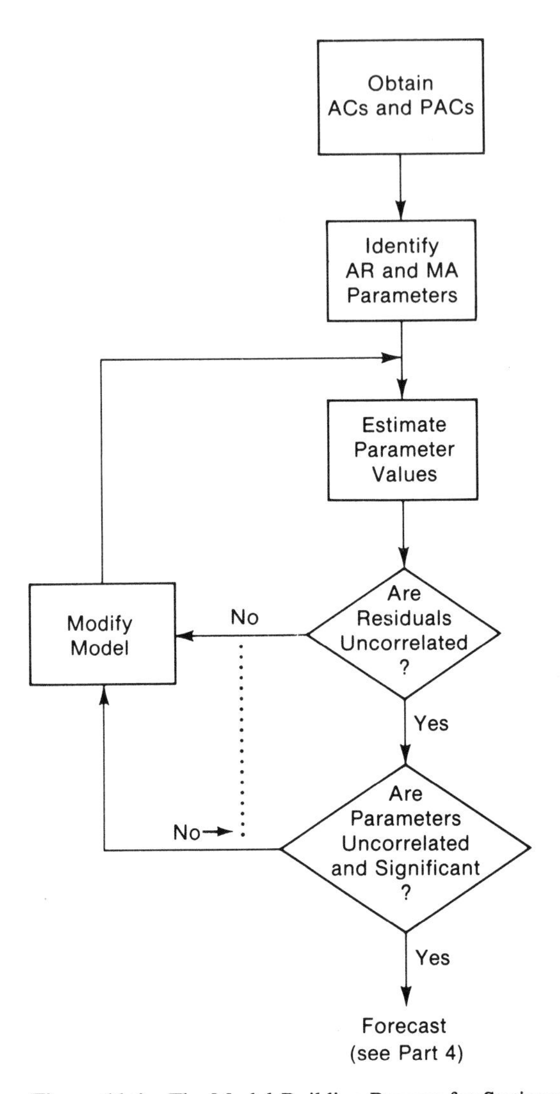

Figure 11.1 The Model-Building Process for Stationary Series

11.2 A BRIEF DIGRESSION—THE MEAN PARAMETER

In Chapter 6 the basic models were described as being applicable only to stationary series. It was also assumed that the average, or mean value, of the stationary series was zero. This assumption was made only for notational convenience. Thus the basic Box-Jenkins models can also be used to model stationary series whose mean is not zero. Most stationary time series, of course, will have a nonzero average value.

If it is assumed that a series has a nonzero mean value, then the basic models require a simple modification: namely, everywhere an X_i term appears in a written model, replace X_i by $X_i - M$, where M represents the mean value of all the X_i. For example, a model with one AR parameter of order 1 would be written as follows:

$$X_t - M = B_1(X_{t-1} - M) + E_t$$

or

$$\dot{X}_t = B_1\dot{X}_{t-1} + E_t$$

where $\dot{X}_t = X_t - M$ and $\dot{X}_{t-1} = X_{t-1} - M$

Now we might simply compute M from our real-life series, subtract it from every X_i, and proceed to model the series $\dot{X}_i$. Theoretically, however, M is supposed to be the mean value of the process or activity represented by the sample series. But the mean value of the real-life (sample) series is only a *sample mean* and not the "true" mean of the process. Therefore, M is included as a parameter in a basic Box-Jenkins model for a stationary series; i.e., it is estimated along with all the other AR and MA parameters that are included in a model.

Most Box-Jenkins programs automatically include the mean parameter in your model for you when your series is stationary (how the program knows your series is stationary is explained in Part 3). Therefore, you need not be concerned about "identifying" the mean parameter. You will, however, see that it has been included in your model when the estimation phase results are displayed. The examples of stationary series models in the following section all contain the mean parameter.

11.3 EXAMPLES OF MODEL IDENTIFICATION, ESTIMATION, AND VALIDATION FOR STATIONARY SERIES

In each of the following examples we will begin by looking at the ACs and PACs of the series. One or more tentative models will then be selected and estimated. From the parameter and residual diagnostics from each estimation, we will then determine the appropriate model for the series or see if the diagnostics suggest a better alternative.

Example 11.1

A graph of the stationary series for this example is shown in Figure 11.2, along with its AC and PAC correlograms.

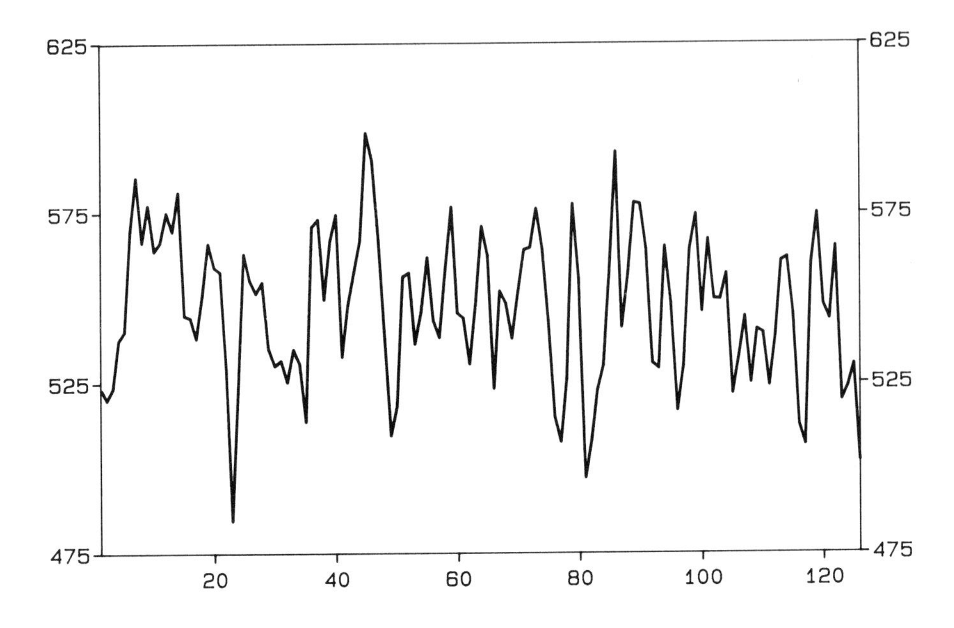

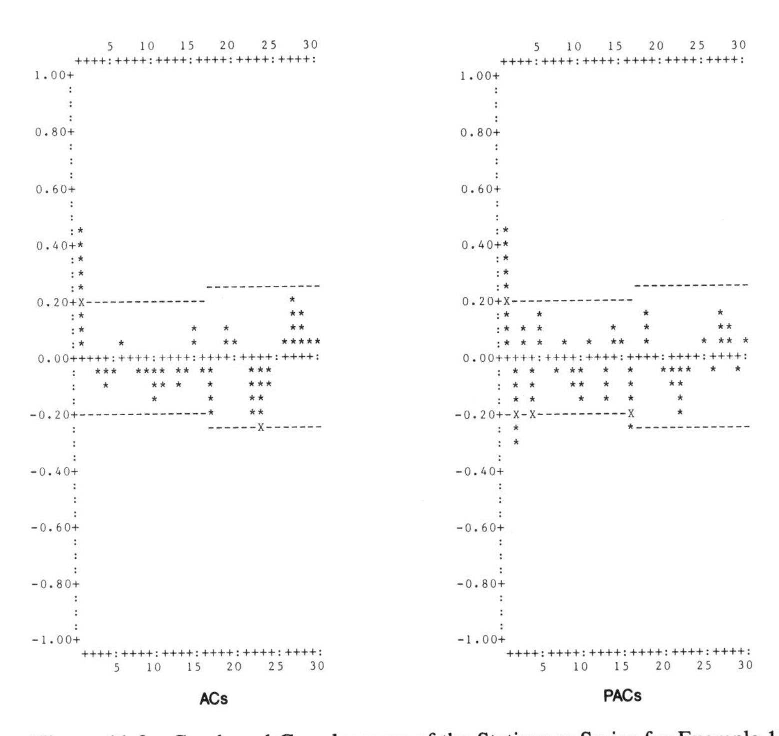

Figure 11.2 Graph and Correlograms of the Stationary Series for Example 11.1

The ACs with a large spike at lag 1 and the PACs with alternating patterns of rapidly decaying spikes clearly indicate a model with only one MA parameter of order 1. When this model and series data are entered into the estimation phase of a Box-Jenkins program, the estimation results of Figure 11.3 are obtained.

From Figure 11.3 the mean parameter and MA parameter appear to be significant (confidence limits exclude zero) and uncorrelated. The Q-statistics for the residual ACs are well below their test values, and the residual mean is not significantly greater than zero (this program would print a message if it were). Also, the residual AC correlogram shows no large spikes (there is no practical reason to suspect that the autocorrelations at lag 15 and 17 are significant, even though they touch the confidence bands). The one-MA model thus appears to be the correct one. In equation form this model is written as follows:

$$X_t = 546.748 + .676E_{t-1} + E_t$$

Note that the index of determination is quite low (.327), indicating that most of the variation in the original series is due to the random error component.

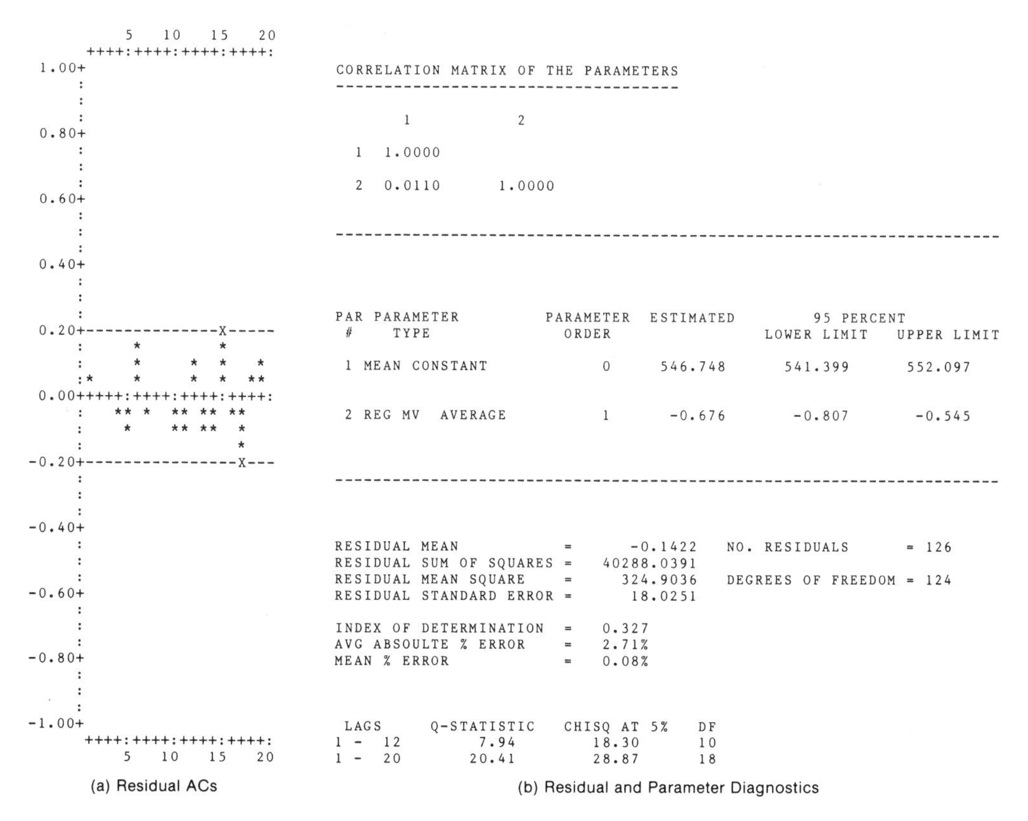

(a) Residual ACs (b) Residual and Parameter Diagnostics

Figure 11.3 Estimation Results for a Model with One MA of Order 1

Example 11.2

A graph of the stationary series for this example is shown in Figure 11.4, along with its AC and PAC correlograms.

The AC correlogram indicates at least one AR parameter, and the one large PAC spike at lag 1 clearly indicates a model with only one AR parameter of order 1. When this model and series data are entered into the estimation phase of a Box-Jenkins program, the estimation results of Figure 11.5 are obtained. The results for this model are similar to those for Example 11.1; i.e., all diagnostics point to this model being the correct one.

In equation form this model is written as follows:

$$X_t = 52.538 + .689(X_t - 52.538) + E_t$$

Note that again, as in Example 11.1, the random error component dominates the behavior of the series since the index of determination is only .475.

Example 11.3

A graph of the stationary series for this example is shown in Figure 11.6, along with its AC and PAC correlograms.

The AC correlogram indicates a model with at least one MA parameter of order 1 (large spike at lag 1) and possibly a second MA parameter of order 2 (large spike at lag 2). The PAC correlogram shows a dampened sine wave, or humped, pattern, which suggests the presence of two MA parameters.

In any event, we will estimate two models, one containing one MA parameter and the second, two MA parameters. The parameter diagnostics and residual diagnostics should tell us which is the correct model. The estimation results for the one-MA-parameter model are shown in Figure 11.7.

The Q-statistics for the one-*MA*-parameter model indicate conclusively that the residual autocorrelations are significant (both Q-statistics are much larger than their test value), and the residual AC correlogram shows large spikes at lags 2 and 3. The spike at lag 2 verifies the need for the second MA parameter. The spike at lag 3 might indicate the need for a third MA parameter of order 3, but this condition was not evident in the AC correlogram of the original series. We will therefore estimate the model with two MA parameters and see if they are sufficient. If not, we can add the third parameter later. The estimation results for the two-MA-parameter model are shown in Figure 11.8.

The diagnostics in Figure 11.8 now indicate an adequate model. Note that there is no large spike at lag 3 in the residual AC correlogram, as there was with the one-MA-parameter model. A third MA parameter is therefore not necessary. The correlation of −.39 between the two MA parameters (parameters numbered 2 and 3 in the printout) is not quite big enough to be alarming. Besides, the parameter confidence limits for each parameter indicate that they are both significant. Note also the

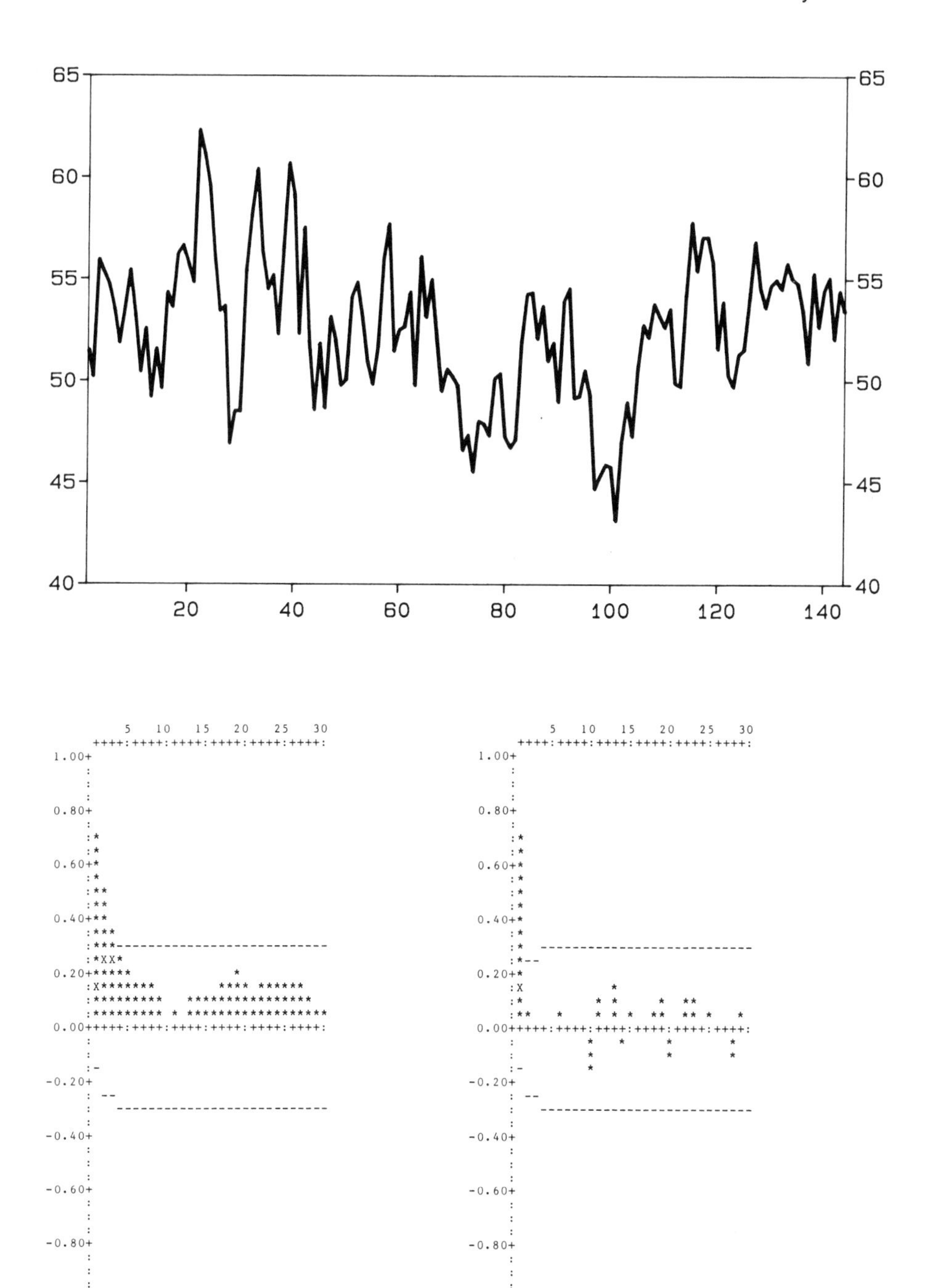

Figure 11.4 Graph and Correlograms of the Stationary Series for Example 11.2

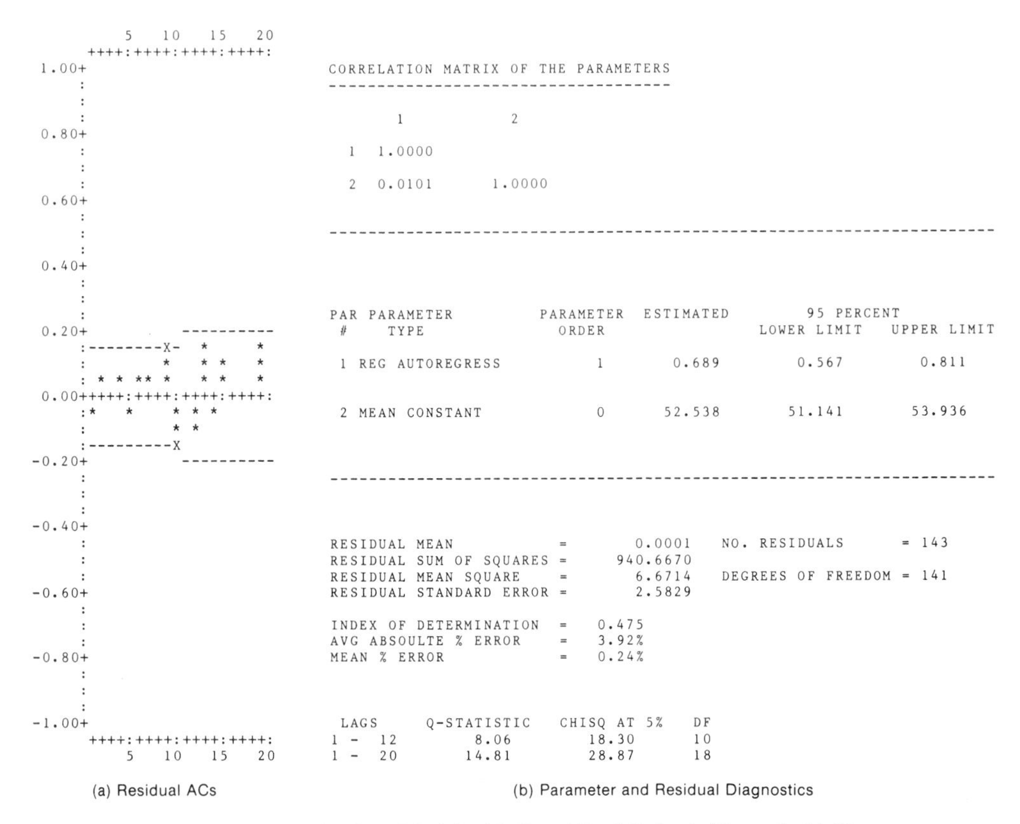

CORRELATION MATRIX OF THE PARAMETERS

	1	2
1	1.0000	
2	0.0101	1.0000

PAR #	PARAMETER TYPE	PARAMETER ORDER	ESTIMATED	95 PERCENT LOWER LIMIT	UPPER LIMIT
1	REG AUTOREGRESS	1	0.689	0.567	0.811
2	MEAN CONSTANT	0	52.538	51.141	53.936

RESIDUAL MEAN = 0.0001 NO. RESIDUALS = 143
RESIDUAL SUM OF SQUARES = 940.6670
RESIDUAL MEAN SQUARE = 6.6714 DEGREES OF FREEDOM = 141
RESIDUAL STANDARD ERROR = 2.5829

INDEX OF DETERMINATION = 0.475
AVG ABSOULTE % ERROR = 3.92%
MEAN % ERROR = 0.24%

LAGS	Q-STATISTIC	CHISQ AT 5%	DF
1 - 12	8.06	18.30	10
1 - 20	14.81	28.87	18

(a) Residual ACs (b) Parameter and Residual Diagnostics

Figure 11.5 Estimation Results for a Model with One AR of Order 1 (Example 11.2)

dramatic improvement of the index of determination, from .134 to .516, when the second MA parameter was added. However, almost 50% of the variation in the original series is still attributable to random error.

In equation form this model is written as follows:

$$X_t = 231.771 - .666E_{t-1} + .865E_{t-2} + E_t$$

Example 11.4

A graph of the stationary series for this example is shown in Figure 11.9, along with its AC and PAC correlograms.

The AC correlogram clearly indicates a model with at least one AR parameter. The PAC correlogram substantiates this result with the spike at lag 1. An additional large PAC spike is also seen at lag 2, which could indicate a model with two AR parameters. On the other hand, the first four PAC spikes might be part of an expo-

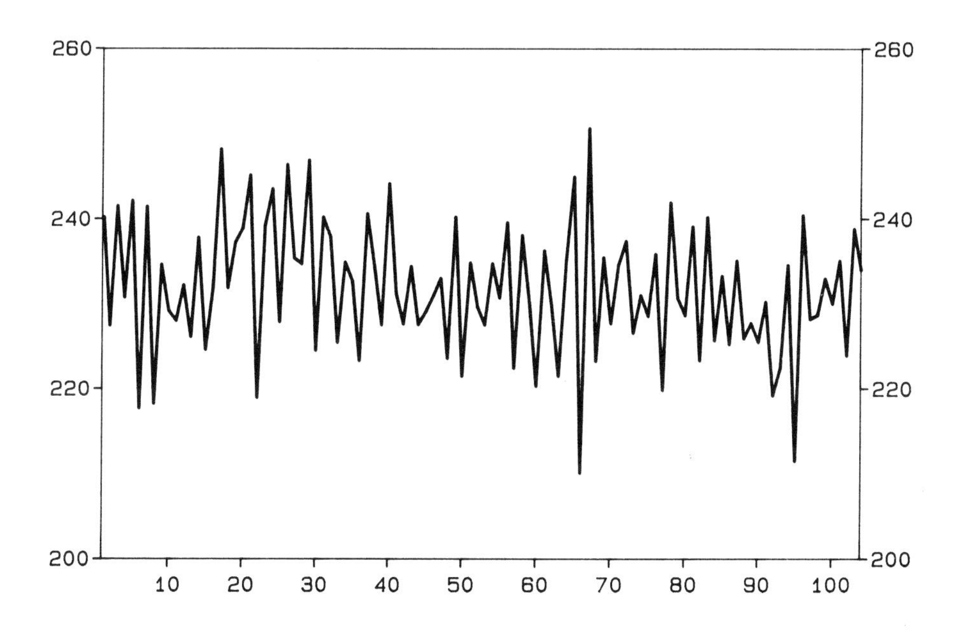

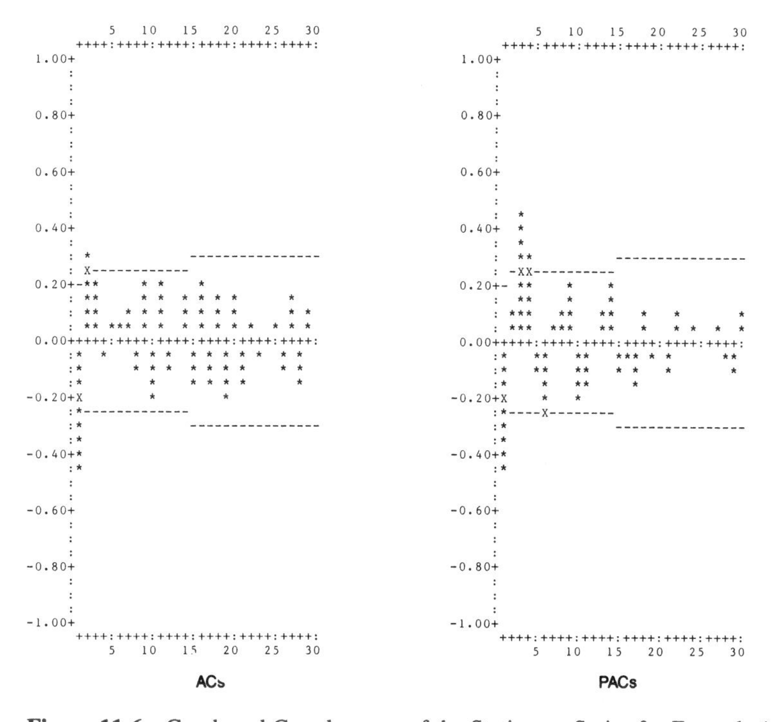

Figure 11.6 Graph and Correlograms of the Stationary Series for Example 11.3

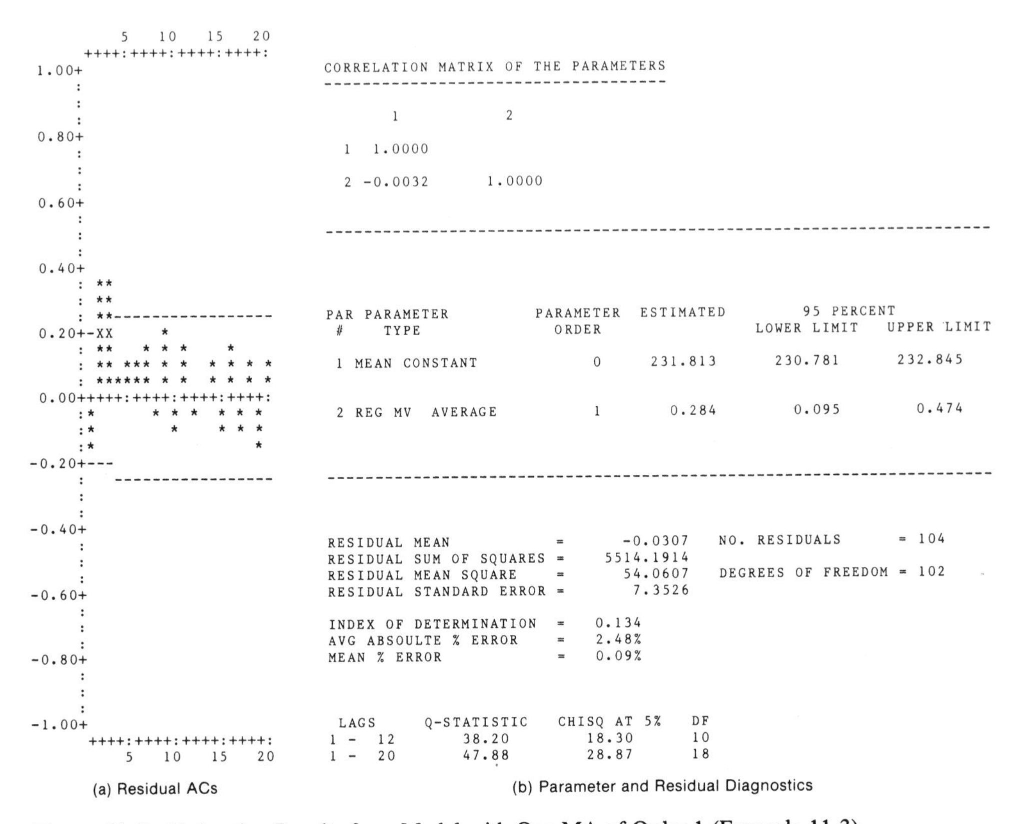

```
CORRELATION MATRIX OF THE PARAMETERS
-----------------------------------

        1          2

 1   1.0000

 2  -0.0032     1.0000

-------------------------------------------------------------------------

PAR PARAMETER          PARAMETER  ESTIMATED        95 PERCENT
 #    TYPE               ORDER                 LOWER LIMIT  UPPER LIMIT

1 MEAN CONSTANT            0       231.813       230.781      232.845

2 REG MV  AVERAGE          1         0.284         0.095        0.474

-------------------------------------------------------------------------

RESIDUAL MEAN            =      -0.0307   NO. RESIDUALS        = 104
RESIDUAL SUM OF SQUARES =    5514.1914
RESIDUAL MEAN SQUARE     =      54.0607   DEGREES OF FREEDOM = 102
RESIDUAL STANDARD ERROR =       7.3526

INDEX OF DETERMINATION   =   0.134
AVG ABSOULTE % ERROR     =   2.48%
MEAN % ERROR             =   0.09%

LAGS     Q-STATISTIC    CHISQ AT 5%    DF
1 - 12      38.20          18.30       10
1 - 20      47.88          28.87       18
```

(a) Residual ACs

(b) Parameter and Residual Diagnostics

Figure 11.7 Estimation Results for a Model with One MA of Order 1 (Example 11.3)

nentially decaying pattern, which would indicate a model with one MA parameter and one AR parameter, both of order 1. We thus have three possible models:

Model 1: One AR of order 1.

Model 2: Two AR of orders 1 and 2.

Model 3: One AR of order 1 and one MA of order 1.

The last two models, however, seem the most likely. Their estimation results are shown in Figures 11.10 and 11.11.

The model with two AR parameters (Figure 11.10) definitely has some problems. First, the two AR parameters are highly correlated (correlation = −.848), so one of them should be removed. Unfortunately, we don't get any guidance from the parameter confidence limits about which one should be removed since both sets of limits exclude zero. Logically, however, the second one should be removed. Second, the Q-statistic for the first 12 lags of the residual ACs is larger than its test value, indicating the lack of a needed parameter (as well as the inclusion of an incorrect

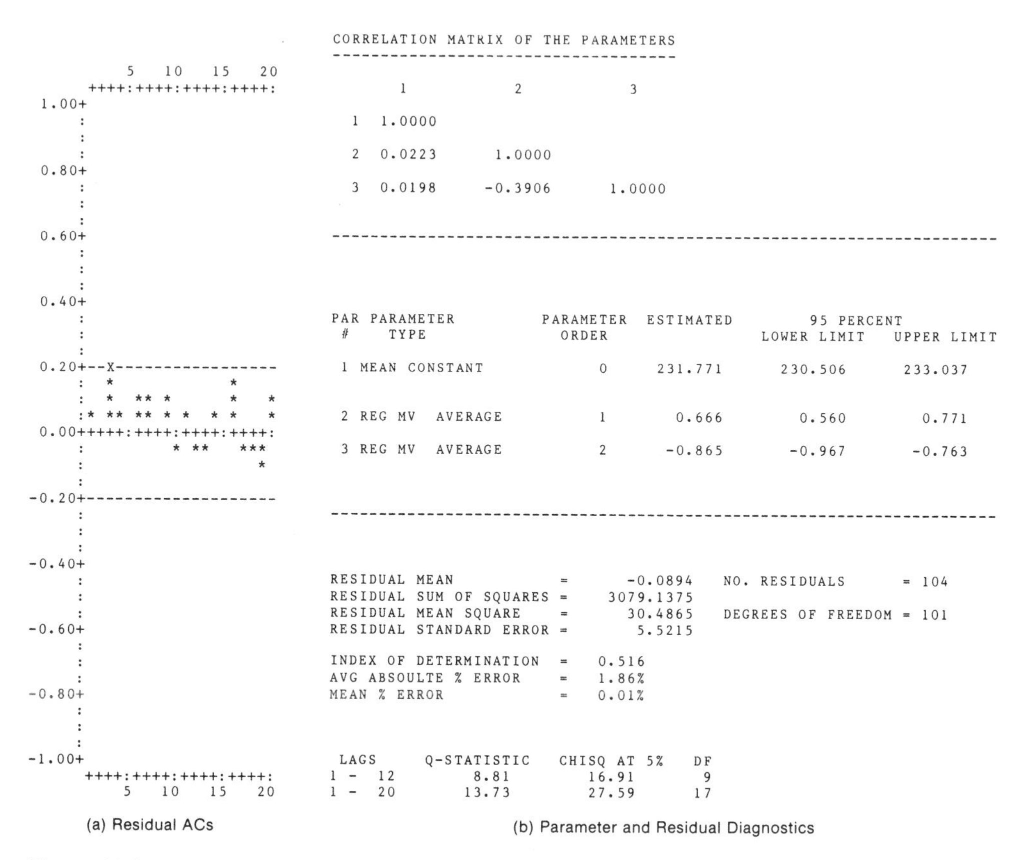

CORRELATION MATRIX OF THE PARAMETERS

	1	2	3
1	1.0000		
2	0.0223	1.0000	
3	0.0198	-0.3906	1.0000

PAR #	PARAMETER TYPE	PARAMETER ORDER	ESTIMATED	95 PERCENT LOWER LIMIT	UPPER LIMIT
1	MEAN CONSTANT	0	231.771	230.506	233.037
2	REG MV AVERAGE	1	0.666	0.560	0.771
3	REG MV AVERAGE	2	-0.865	-0.967	-0.763

RESIDUAL MEAN = -0.0894 NO. RESIDUALS = 104
RESIDUAL SUM OF SQUARES = 3079.1375
RESIDUAL MEAN SQUARE = 30.4865 DEGREES OF FREEDOM = 101
RESIDUAL STANDARD ERROR = 5.5215

INDEX OF DETERMINATION = 0.516
AVG ABSOULTE % ERROR = 1.86%
MEAN % ERROR = 0.01%

LAGS	Q-STATISTIC	CHISQ AT 5%	DF
1 - 12	8.81	16.91	9
1 - 20	13.73	27.59	17

(a) Residual ACs (b) Parameter and Residual Diagnostics

Figure 11.8 Estimation Results for a Model with Two MAs of Orders 1 and 2 (Example 11.3)

parameter, in this case). The residual AC indicates the possibility of the need for an MA parameter of order 2 due to the significant spike at lag 2. However, the residual AC can sometimes be misleading as an identification tool when incorrect parameters are currently in the model. We should therefore return to the alternative models identified from the original AC and PAC before trying to pick up on this kind of lead.

The model with one AR and one MA parameter (Figure 11.11), however, appears to be the correct one for the series. In this case all validation tests indicate that the model is adequate and not overspecified. In equation form this model is written as follows:

$$X_t = 3632.734 + .716(X_{t-1} - 3632.734) + .862E_{t-1} + E_t$$

Note that the index of determination is .831, so that less than 17% of the variation in the original series is attributable to random error.

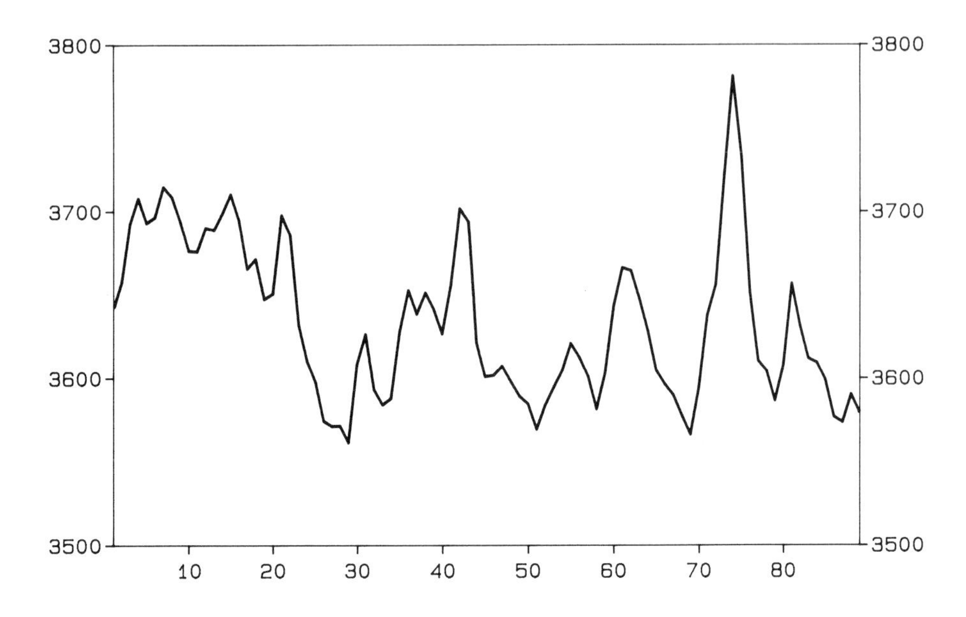

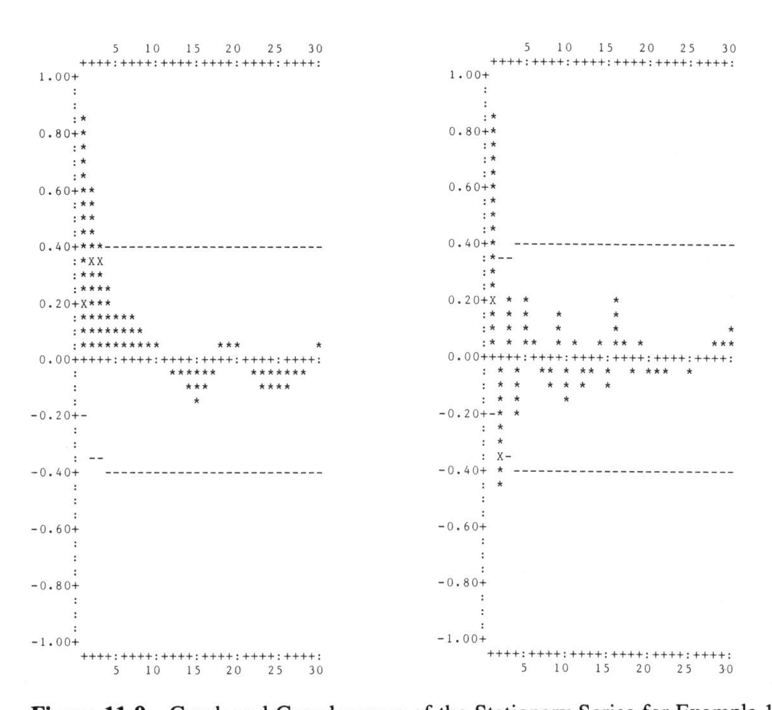

Figure 11.9 Graph and Correlograms of the Stationary Series for Example 11.4

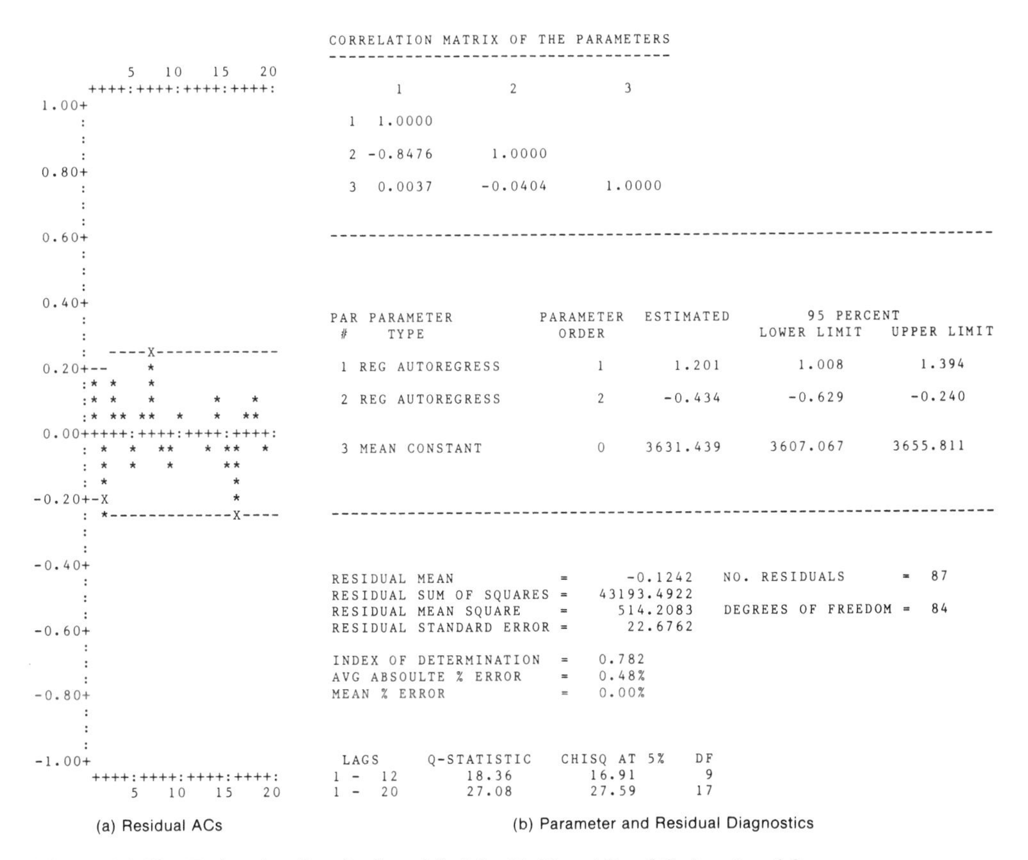

CORRELATION MATRIX OF THE PARAMETERS

	1	2	3
1	1.0000		
2	-0.8476	1.0000	
3	0.0037	-0.0404	1.0000

PAR #	PARAMETER TYPE	PARAMETER ORDER	ESTIMATED	95 PERCENT LOWER LIMIT	UPPER LIMIT
1	REG AUTOREGRESS	1	1.201	1.008	1.394
2	REG AUTOREGRESS	2	-0.434	-0.629	-0.240
3	MEAN CONSTANT	0	3631.439	3607.067	3655.811

RESIDUAL MEAN = -0.1242 NO. RESIDUALS = 87
RESIDUAL SUM OF SQUARES = 43193.4922
RESIDUAL MEAN SQUARE = 514.2083 DEGREES OF FREEDOM = 84
RESIDUAL STANDARD ERROR = 22.6762

INDEX OF DETERMINATION = 0.782
AVG ABSOULTE % ERROR = 0.48%
MEAN % ERROR = 0.00%

LAGS	Q-STATISTIC	CHISQ AT 5%	DF
1 - 12	18.36	16.91	9
1 - 20	27.08	27.59	17

(a) Residual ACs (b) Parameter and Residual Diagnostics

Figure 11.10 Estimation Results for a Model with Two ARs of Orders 1 and 2

11.4 REVIEW OF KEY CONCEPTS

We now have all the tools and model-building blocks necessary to identify, estimate, and validate models for stationary series. The process for doing so is summarized in the following steps:

1. Obtain ACs and PACs of the original series.
2. Identify one or more possible models, i.e., one or more possible combinations of AR and MA parameters.
3. Estimate the best parameter values for a given model.
4. Determine whether the residuals are uncorrelated. (Check the Q-statistics and residual AC.) If they are not, modify the model and go back to step 3.

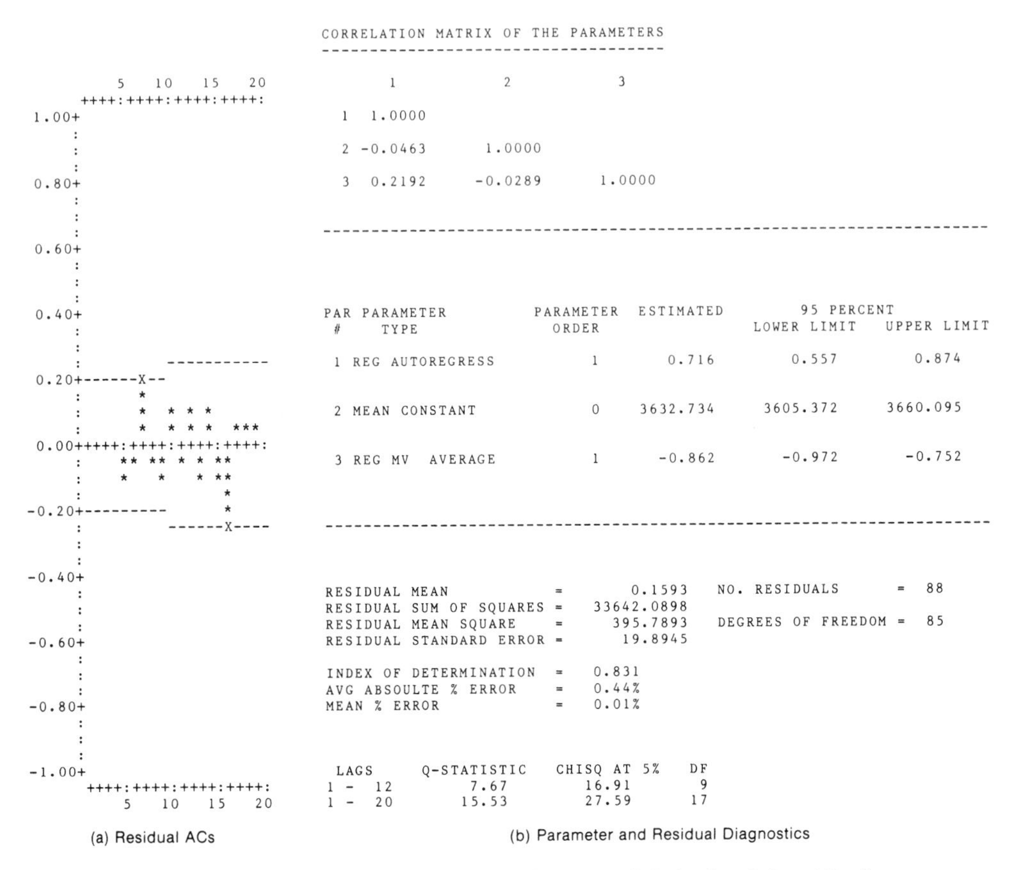

```
           5    10    15    20
        ++++:++++:++++:++++:
 1.00+
     :
     :
     :
 0.80+
     :
     :
     :
 0.60+
     :
     :
     :
 0.40+
     :
     :
     :               -----------
 0.20+------X--
     :      *
     :      *  *  *  *
     :      *  *  *  *    ***
 0.00++++:++++:++++:++++:
     :    **  **  *  *  **
     :    *    *     *  **
     :                   *
-0.20+---------          *
     :               ------X----
     :
     :
-0.40+
     :
     :
     :
-0.60+
     :
     :
     :
-0.80+
     :
     :
     :
-1.00+
        ++++:++++:++++:++++:
           5    10    15    20
```

(a) Residual ACs

CORRELATION MATRIX OF THE PARAMETERS

	1	2	3
1	1.0000		
2	-0.0463	1.0000	
3	0.2192	-0.0289	1.0000

PAR #	PARAMETER TYPE	PARAMETER ORDER	ESTIMATED	95 PERCENT LOWER LIMIT	UPPER LIMIT
1	REG AUTOREGRESS	1	0.716	0.557	0.874
2	MEAN CONSTANT	0	3632.734	3605.372	3660.095
3	REG MV AVERAGE	1	-0.862	-0.972	-0.752

```
RESIDUAL MEAN             =        0.1593   NO. RESIDUALS        =  88
RESIDUAL SUM OF SQUARES  =    33642.0898
RESIDUAL MEAN SQUARE      =      395.7893   DEGREES OF FREEDOM =  85
RESIDUAL STANDARD ERROR  =       19.8945

INDEX OF DETERMINATION    =   0.831
AVG ABSOULTE % ERROR      =   0.44%
MEAN % ERROR              =   0.01%
```

LAGS	Q-STATISTIC	CHISQ AT 5%	DF
1 - 12	7.67	16.91	9
1 - 20	15.53	27.59	17

(b) Parameter and Residual Diagnostics

Figure 11.11 Estimation Results for a Model with One MA of Order 1 and One AR of Order 1

5. Determine whether the parameters are uncorrelated and significant. (Check the parameter correlation matrix and parameter confidence limits.) If they are not, modify the model and go back to step 3.
6. Continue to the forecasting phase.

When you are modeling stationary series, an "automatic" parameter, called the mean parameter, is always included in the model you specify for the series. You do not have to be concerned about "identifying" the mean parameter because your Box-Jenkins program will automatically include it in your model for you. The mean parameter is estimated just like any other parameter during the estimation phase.

Part 3

Building on the Foundation

The foundation on which you will eventually be able to construct all univariate Box-Jenkins models was provided in Part 2. This foundation consisted of two model-building blocks called autoregressive parameters (AR parameters) and moving-average parameters (MA parameters). With these two building blocks you can construct the basic AR, MA, and ARMA Box-Jenkins models.

The basic models, however, are only applicable to a special type of time series: namely, time series that are stationary (a stationary series is one that only fluctuates consistently about some fixed level over time). Most economic and business time series, however, are rarely stationary. These series usually contain trend patterns, seasonal patterns, and other types of patterns that are nonstationary. Therefore to construct models for most real-life time series, you will need to have some additional model-building blocks at your disposal.

In Part 3, you will learn about:

- Two new building blocks for modeling nonstationary, nonseasonal series (Chapter 12).
- Three new building blocks for modeling seasonal series (Chapter 14).
- How to select the appropriate model-building blocks for nonstationary and/or seasonal series (Chapters 13, 15, and 16).

In Part 3, then, you will learn about five new model-building blocks. These building blocks, along with the AR and MA building blocks you are already familiar with, will allow you to build models for almost any time series you will ever encounter. These seven building blocks represent the complete set of building blocks for univariate Box-Jenkins models.

Fortunately, to identify models that require the new building blocks, you will still be able to use the same model identification tools (autocorrelations and partial-autocorrelations) that were used in Part 2 for identifying the basic models. All you need to do is learn to recognize some new AC and PAC patterns associated with the new building blocks. Also, the process of computing and validating a model containing the new building blocks is exactly the same as for the basic models; you do not have to learn about any new techniques or diagnostics.

CHAPTER **12**

Models for Nonstationary Series

This chapter describes two new building blocks that are used in constructing models for nonstationary series. In this chapter and Chapter 13, we will also assume that the series are nonseasonal. Models for seasonal series are addressed in Chapters 14, 15, and 16.

In this chapter you will learn:

- What a nonstationary series is.
- How to convert a nonstationary series into a stationary one.
- How models for nonstationary series are constructed.

The identification of models for nonstationary series is covered in Chapter 13.

12.1 NONSTATIONARY SERIES

In Chapter 6 you learned that a series is stationary if it varies more or less uniformly over time, about a constant, fixed level. On the other hand, a series is *nonstationary* if it appears to have no fixed level. The simplest example of a nonstationary series is one that exhibits an overall trend, as shown in Figure 12.1. In this case the "level" of the series is continuously shifting upward; i.e., there is no overall fixed level of the series.

Nonstationary behavior can also take other forms. For example, a series that exhibits random changes in *established levels* or random changes in *established levels and slopes* would also be nonstationary, as illustrated in Figures 12.2 and 12.3.

Most series of interest are almost always nonstationary. Thus the basic models described in Part 2 are not, by themselves, sufficient to model most series, since they are only applicable to stationary series. Fortunately, the Box-Jenkins method

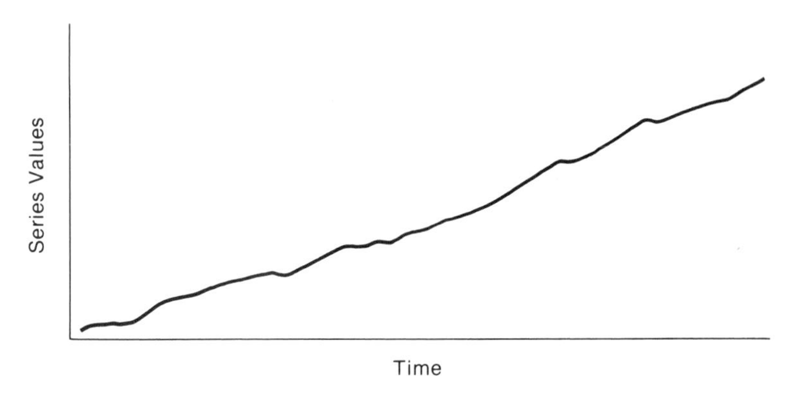

Figure 12.1 A Nonstationary Series: Overall Trend

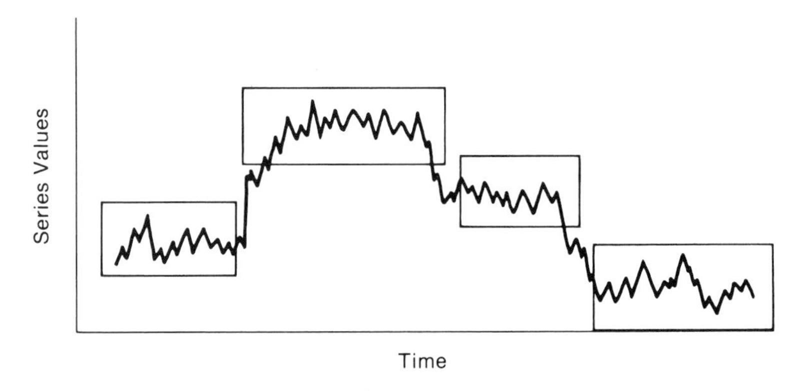

Figure 12.2 A Nonstationary Series: Random Changes in Level

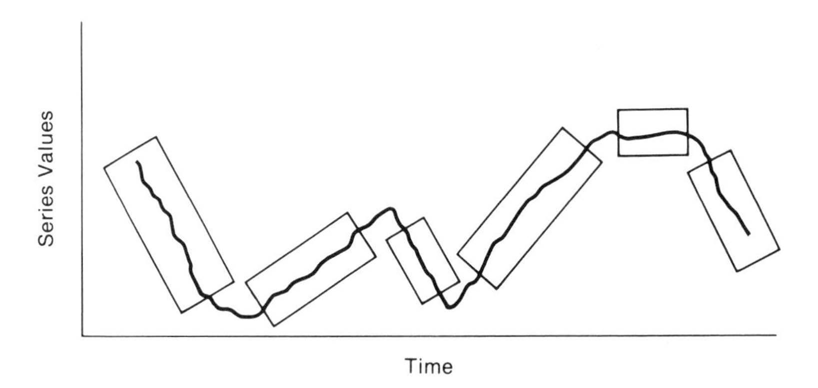

Figure 12.3 A Nonstationary Series: Random Changes in Both Level and Slope

provides another model-building block whose purpose is to convert most nonstationary behavior into stationary behavior. Once this conversion is accomplished, a basic model can then be constructed for the converted stationary series. This new building block, together with the AR and MA parameters identified for the converted stationary series, provides a model for the nonstationary series.

The new model-building block, however, is not a parameter in the same sense as an AR or MA parameter. Instead, it represents a computational process that is performed on the original series to convert the series into a stationary one. This computational process is called regular differencing.

12.2 REGULAR DIFFERENCING (THE RD BUILDING BLOCK)

Regular differencing is simply the process of computing the difference between every two successive values in a series. The resulting series of differences is called the *differenced series*.

For example, if we have a series $X_1, X_2, X_3, \ldots$, then the differenced series would consist of the values $X_2 - X_1$, $X_3 - X_2$, $X_4 - X_3$, etc. In general, the differenced series may be written in equation form as follows:

$$Z_t = X_t - X_{t-1}, \qquad \text{for } t = 2, 3, \ldots, N$$

The results of applying this differencing process to a sample series are illustrated in Table 12.1.

Table 12.1 Differencing a Time Series

Period, t	Original Series, X_t	Differenced Series, Z_t
1	110	—
2	115	5
3	90	−25
4	130	40
5	140	10
6	150	10
7	120	−30
8	115	−5
9	130	15
10	135	5

Table 12.2 Differencing a Nonstationary, Straight-Line Trend Series

Period, t	Original Series, X_t	Differenced Series, Z_t
1	23	—
2	33	10
3	43	10
4	53	10
5	63	10
6	73	10
7	83	10
8	93	10
9	103	10
10	113	10

In Table 12.1 the differenced value for the second period is $X_2 - X_1 = 115 - 110 = 5$ and for the third period is $X_3 - X_2 = 90 - 115 = -25$, etc. Note that no differenced value can be computed for the first period, since no time series value precedes X_1.

Since these differences are computed by subtracting each series value from the series value *one* period ahead of it, the result is called a *difference of order 1*, or a *regular difference*. From now on we will use the term *regular difference*, or the acronym RD, to refer to a difference of order 1. (In Chapter 14, we will consider differences of orders greater than 1 when modeling seasonal series.)

The reason differencing can be used to convert a nonstationary series into a stationary series is that differencing tends to remove the short- and long-term trends in a series, such as those illustrated in Figures 12.2 and 12.3. This fact can be clearly demonstrated by applying a regular difference to the hypothetical series shown in Table 12.2. In this example the series X_t follows a perfect straight-line trend (and is therefore nonstationary), but the differenced series Z_t, which has the same value for every period, falls on a horizonal line (and is therefore stationary).* In other words, applying one regular difference to X_t has "removed" the straight-line trend. This is exactly what we want to happen if we are to obtain a stationary series.

As another example, consider a hypothetical series with the kind of nonstationary behavior illustrated in Figure 12.2; i.e., the series exhibits random changes in *established levels*. The result of applying one regular difference to such a series is illus-

*The process of differencing a time series is analogous to the operation of differentiating a continuous function in the calculus. For example, if the function for a straight line is differentiated once, the resulting function (derivative) is a constant, or a horizontal line. Differencing achieves the same effect when applied to a time series that follows a straight-line trend.

trated in Table 12.3. In this example the differenced series is essentially a constant (equal to zero) with occasional random "blips" at the points in time where the original series changes levels. This differenced series is therefore also stationary.

The two graphs in Figure 12.4 illustrate an actual nonstationary series (part a) and the stationary series obtained after applying one regular difference (part b).

Sometimes, the application of one regular difference is not enough to obtain stationarity; i.e., the differenced series is still nonstationary. This is the case, for example, if the overall trend of the series is not a straight-line trend but a curved trend (parabolic trend). In this case you may obtain stationarity by applying two successive regular differences to the original series. Thus the differenced series obtained by applying one regular difference would, itself, be differenced by one regular difference (i.e., the differences $Z_t - Z_{t-1}$ would be computed). Table 12.4 illustrates the results of applying two regular differences to a time series. Note that

Table 12.3 Differencing a Nonstationary Series with Random Changes in Established Levels

Period, t	Original Series, X_t	Differenced Series, Z_t
1	100	—
2	100	0
3	100	0
4	100	0
5	50	−50
6	50	0
7	50	0
8	50	0
9	50	0
10	125	75
11	125	0
12	125	0
13	125	0
14	130	5
15	130	0
16	130	0
17	130	0
18	130	0
19	130	0
20	130	0
⋮	⋮	⋮

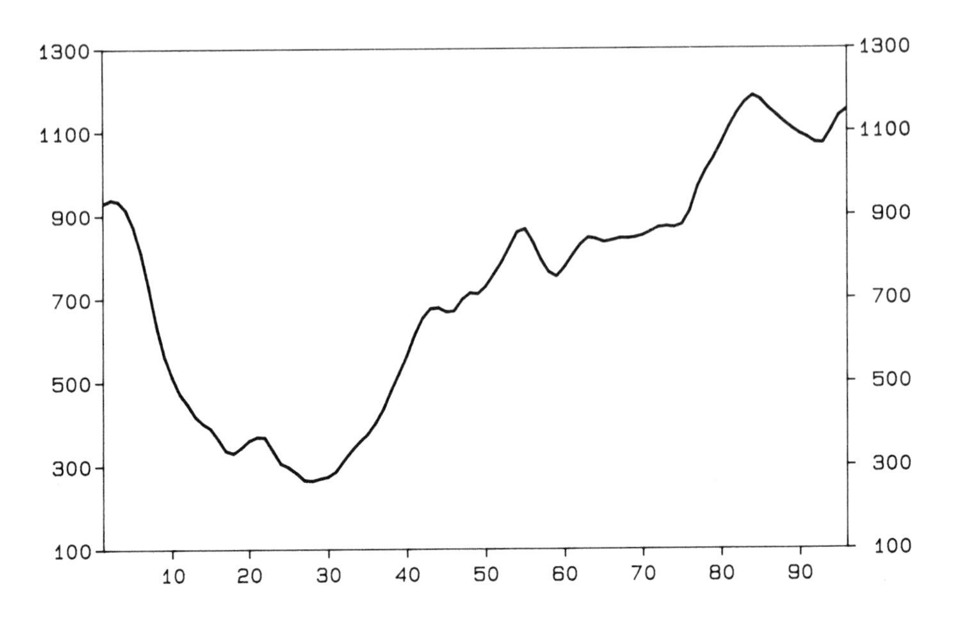

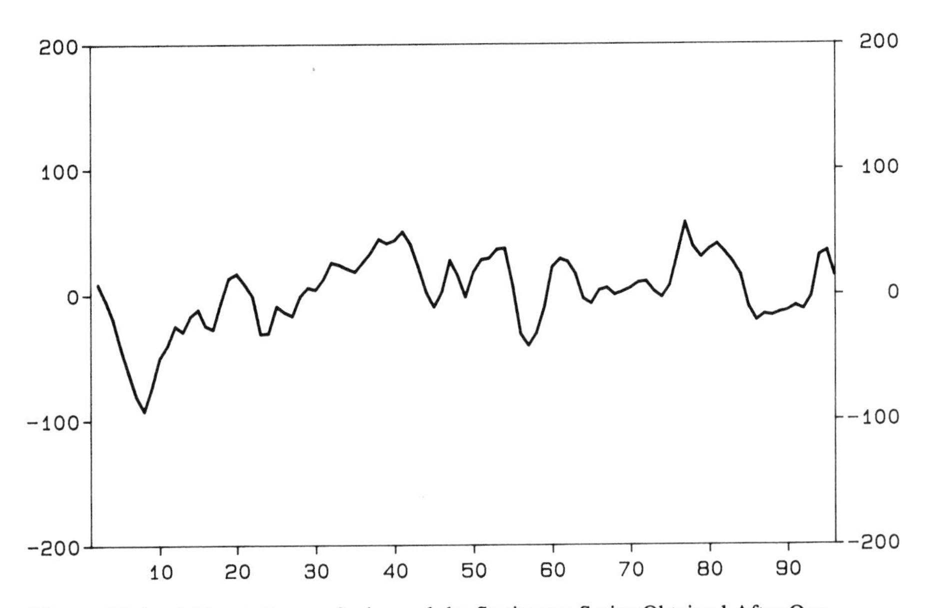

Figure 12.4 A Nonstationary Series and the Stationary Series Obtained After One Regular Difference

Table 12.4 Differencing a Series with Two Regular Differences

Period, *t*	Original Series	Series After One RD	Series After Two RDs
1	110	—	—
2	115	−5	—
3	90	−25	−30
4	130	40	65
5	140	10	−30
6	150	10	0
7	120	−30	−40
8	115	−5	25
9	130	15	20
10	135	5	−10

after the application of two regular differences, no values are available for the first two periods.

For most series it is unlikely that more than two regular differences would ever be needed to achieve stationarity. But in Chapter 13 you will learn how to determine how many regular differences are needed to convert a nonstationary series into a stationary one.

There are other types of nonstationary behavior, however, where differencing, by itself, is not sufficient to achieve stationarity. Such is the case, for example, when the series follows a certain type of overall trend, such as an exponential trend, or when the variation of the series is changing. In such cases you may have to apply a special mathematical transformation on the series prior to computing any differences (e.g., by taking the logarithm or the square root of every series value). See Appendix C for more information on the use of prior transformations.

12.3 NONSTATIONARY MODELS (ARIMA MODELS)

Once you have achieved stationarity by applying regular differences to your original series, you are ready to proceed with the modeling process as described in Part 2. Thus you will need to identify a basic AR, MA, or ARMA model for the converted stationary series. These models are written in the same way as the basic models, except that the differenced (stationary) series Z_t is substituted for the original series X_t, as shown below:

$$Z_t = A_1Z_{t-1} + \ldots + A_pZ_{t-p} + E_t \qquad \text{(AR model for } Z_t\text{)}$$

$$Z_t = -(B_1E_{e-1} + \ldots + B_qE_{t-q}) + E_t \qquad \text{(MA) model for } Z_t\text{)}$$

$$Z_t = (A_1Z_{t-1} + \ldots + A_pZ_{t-p}) - (B_1E_{t-1} + \ldots + B_qE_{t-q}) + E_t \qquad \text{(ARMA model for } Z_t\text{)}$$

In terms of the original series such models are called *integrated models* and are often denoted by the acronyms ARI, IMA, and ARIMA. Thus we can talk about *A*uto*R*egressive *I*ntegrated (ARI) models, *I*ntegrated *M*oving-*A*verage (IMA) models, and *A*uto*R*egressive, *I*ntegrated, *M*oving-*A*verage (ARIMA) models. The term *integrated,* which is a synonym for *summed,* is used because the differencing process can be reversed to obtain the original series values by *summing* successive values of the differenced series.* This reverse process, or "integration," is exactly what is done by a Box-Jenkins computer program to obtain the fitted values to the original nonstationary series from the fitted values computed for the differenced (stationary) series during the estimation process.

12.4 ACCOUNTING FOR OVERALL TRENDS (THE TREND PARAMETER)

Recall from Chapter 11 that a Box-Jenkins model for a stationary series must contain a parameter, called the mean parameter, when the stationary series has a significantly large mean value. You do not need to worry about specifying this mean parameter or deciding whether you need it—your Box-Jenkins program will always automatically include it in the model if your series is stationary (whether or not the mean value of the series is significantly large). The program assumes your series is stationary if you do not specify any differencing.

What happens, then, when a nonstationary series is sufficiently differenced to obtain a stationary series, and the mean value of the differenced series is significantly large? It turns out that in this case *you* must include a new type of parameter in your model, called the *trend parameter.* Most Box-Jenkins programs will *not* include this parameter for you automatically.

You might expect the trend parameter to represent the mean value of the differenced series in the same way that the mean parameter represents the mean of an original stationary series. In general, this is only true if there are no AR parameters

*In the calculus the operation of integration is the opposite, or reverse, operation to differentiation. Since differentiation is analogous to differncing and since summing is the reverse operation of differencing, the term *integrated* is used as a synonym for *summed*.

in the rest of the model; otherwise, the trend parameter is related to, but does not represent, the mean of the differenced series (it could be computed from the estimated mean of the differenced series and the estimated values of the AR parameters, if they were all available).

In a model equation the trend parameter appears as a constant parameter that is simply added to the other AR and MA terms in the model; i.e.,

$$Z_t = B_0 + (\text{AR and MA terms}) + E_t$$

where B_0 is the trend parameter.

The reason that this new parameter is called a trend parameter can be best illustrated in an example. Consider the hypothetical straight-line trend series in Table 12.2. Here the differenced series obtained after applying one regular difference has the same value, 10, for every period. The mean value of the differenced series is therefore 10. The value 10 is also the slope of the original trend line. Since there are no random errors in this hypothetical case, a trivial model for the original series would simply be

$$Z_t = B_0 \qquad (\text{or } X_t = 10 + X_{t-1})$$

where $B_0 = 10$. Thus the only parameter in the model is a constant parameter, which, as noted above, represents the slope of the trend line—hence the name *trend parameter.*

Many Box-Jenkins programs will automatically tell you whether the mean value of the differenced series is significantly nonzero. If the program does not, you may make this determination by using the same procedure described in Chapter 10 for testing the significance of the residual mean. Remember, in most Box-Jenkins programs if the mean value of the differenced series is significantly nonzero, you are responsible for including the trend parameter in your model. In most real-life applications, however, the mean value of a differenced series will not be significantly large, so you will seldom need to worry about using a trend parameter. The only time a regularly differenced series will have a significantly nonzero mean value is when the *only* demonstrable pattern in the original series is an overall trend (versus random changes in level and slope, and/or seasonal behavior).

12.5 REVIEW OF KEY CONCEPTS

The basic models described in Part 2 can only be used to model stationary series. If a series is nonstationary, you will have to use additional model-building blocks to construct a model for your series.

A *nonstationary series* is one that appears to have no fixed level. Examples of nonstationary series include the following:

- A simple overall trend.
- A series that randomly changes from one established level to another.
- A series that randomly changes from one established trend to another.

A *regular difference* (RD) is a new Box-Jenkins building block that is used in modeling nonstationary series. A regular difference is not a parameter in the same sense as an AR or MA parameter; instead, it is used to indicate that a certain computational process, called regular differencing, is to be applied to the series.

Regular differencing is the process of computing the difference between every two successive values in a series. The resultant series of differenced values is called the differenced series. Regular differencing tends to remove any overall trending behavior and other forms of nonstationary behavior from the original series; it converts nonstationary behavior into stationary behavior.

To build a Box-Jenkins model for a nonstationary (nonseasonal) series, you must first convert the series into a stationary one by using the RD building block. The following are points to remember about the use of the RD building block:

- Usually, only one RD is required to convert a nonstationary series into a stationary one.
- If stationarity is not achieved after one regular difference, additional regular differences may be applied (to the newly differenced series) until stationarity is achieved.
- You will seldom, if ever, need more than two RDs.
- You do not have to difference the series yourself; all you need to do is indicate to the Box-Jenkins program how many regular differences are to be applied and the program will compute the final differenced series automatically.
- You will learn in Chapter 13 how to determine when stationarity has been achieved.

Once a stationary series has been obtained through differencing, you must then identify a basic AR, MA, or ARMA model for the differenced (stationary) series as described in Part 2.

Integrated models are models for nonstationary series. The basic AR, MA, or ARMA model for the differenced (stationary) series, together with the specified RD, constitutes the integrated model for the original nonstationary series. Integrated models are called ARI, IMA, and ARIMA models, where the I stands for the term *integrated.*

The *trend parameter* is another parameter that may have to be included in a model for a nonstationary series. The trend parameter must be included in a model if the differenced (stationary) series has a significantly large mean value. Some Box-Jenkins programs will tell you whether the mean value of the differenced series is significantly nonzero; if your program doesn't, you can employ the technique described in Chapter 10 for testing the significance of the residual mean value. You will seldom need to use the trend parameter.

The general Box-Jenkins integrated model for nonstationary (nonseasonal) series is written as follows:

$$Z_t = B_0 + A_1 Z_{t-1} + \ldots + A_p Z_{t-p} - (B_1 E_{t-1} + \ldots + B_q E_{t-q}) + E_t$$

where Z_t represents the differenced (stationary) series and B_0 represents the trend parameter.

CHAPTER 13

Identifying, Computing, and Validating Nonstationary Models

Now that we know about the building blocks necessary for construcing models for nonstationary (nonseasonal) series, we will need to know how to select the correct combination of these building blocks for a particular nonstationary series.

In this chapter you will learn:

- How to identify the need for regular differencing.
- How to select the correct amount of regular differencing.
- What the model-building process for nonstationary series is.
- How to compute and validate an integrated model for a nonstationary (nonseasonal) series.

Examples of computing and validating some integrated models for nonstationary series are given at the end of this chapter.

13.1 SELECTING THE CORRECT AMOUNT OF DIFFERENCING

As indicated in Chapter 12, a nonstationary (nonseasonal) series must first be transformed into a stationary one by applying one or more regular differences. After this conversion is done, the AC and PAC of the differenced (stationary) series can be used to identify the correct AR and MA parameters for a given model.

Most series you will encounter will be nonstationary. Hence there is almost always a need for differencing. The question is, how much differencing is required; i.e., how many successive differences should be applied? To determine the correct amount of differencing, you can again make use of the autocorrelations of the series.

In particular, the need for regular differencing is indicated when the autocorrelations of the series tend to remain large for many successive lags, beginning at lag 1. Although the autocorrelations may be decreasing, they generally will decrease slowly (e.g., at a constant rate), rather than very fast (e.g., at an exponential rate). Visually, the correlogram for a series that requires regular differencing will often (but not always) have an overall heavy, solid appearance or have large alternating heavy clumps of autocorrelation spikes, as shown in Figure 13.1.

The reason these patterns indicate nonstationarity can be seen by considering a nonstationary series that follows a constantly rising straight-line trend. In this case every series value will always be greater than the immediately preceding value by approximately the same amount, since the trend grows at a constant rate. This relationship is a rather strong one that indicates a high autocorrelation at lag 1. Likewise, every series value will always be greater than the series value two periods back by approximately the same amount, which means high autocorrelation at lag 2. The same argument applies for any lag. Thus the autocorrelations for such a series should all tend to remain large for all lags; i.e., none of them will be zero and they won't decrease rapidly in magnitude.

You will know when you have applied the correct number of differences when you are able to identify the typical AC patterns associated with the basic AR and MA parameters in the ACs of the differenced series. At this point you can then proceed to identify the model for the differenced series. Note that the Box-Jenkins program will automatically compute the differenced series and the ACs associated with the differenced series. All you have to do is indicate the number of differences to be applied.

For example, Figure 13.2 illustrates the progression of AC correlograms for a series that requires two regular differences to achieve stationarity. In the correlogram in part (c), which corresponds to the differenced series after applying two regular differences, the autocorrelations are all insignificant except at lag 1. This result, of course, indicates one MA parameter of order 1. The model for this series would therefore consist of two RDs and one MA.

Sometimes, the decision about whether stationarity has been achieved is not so clear-cut. You may already have noted that there is a thin line between some AC patterns indicating nonstationary behavior (see, for example, the AC correlogram in Figure 13.1d) and some AC patterns indicating the presence of AR parameters. It's just a question of how fast the autocorrelations decrease. This is no accident. In fact, an AR model can represent a relationship that is almost equivalent to differencing. For example, the AR model

$$X_t = A_1 X_{t-1} + E_t$$

where X_t is stationary, can be rewritten as

$$X_t - A_1 X_{t-1} = E_t$$

But if $A_1 = 1$, then the above equation becomes

$$X_t - X_{t-1} = E_t$$

and the left side of the above equation now represents one regular difference.

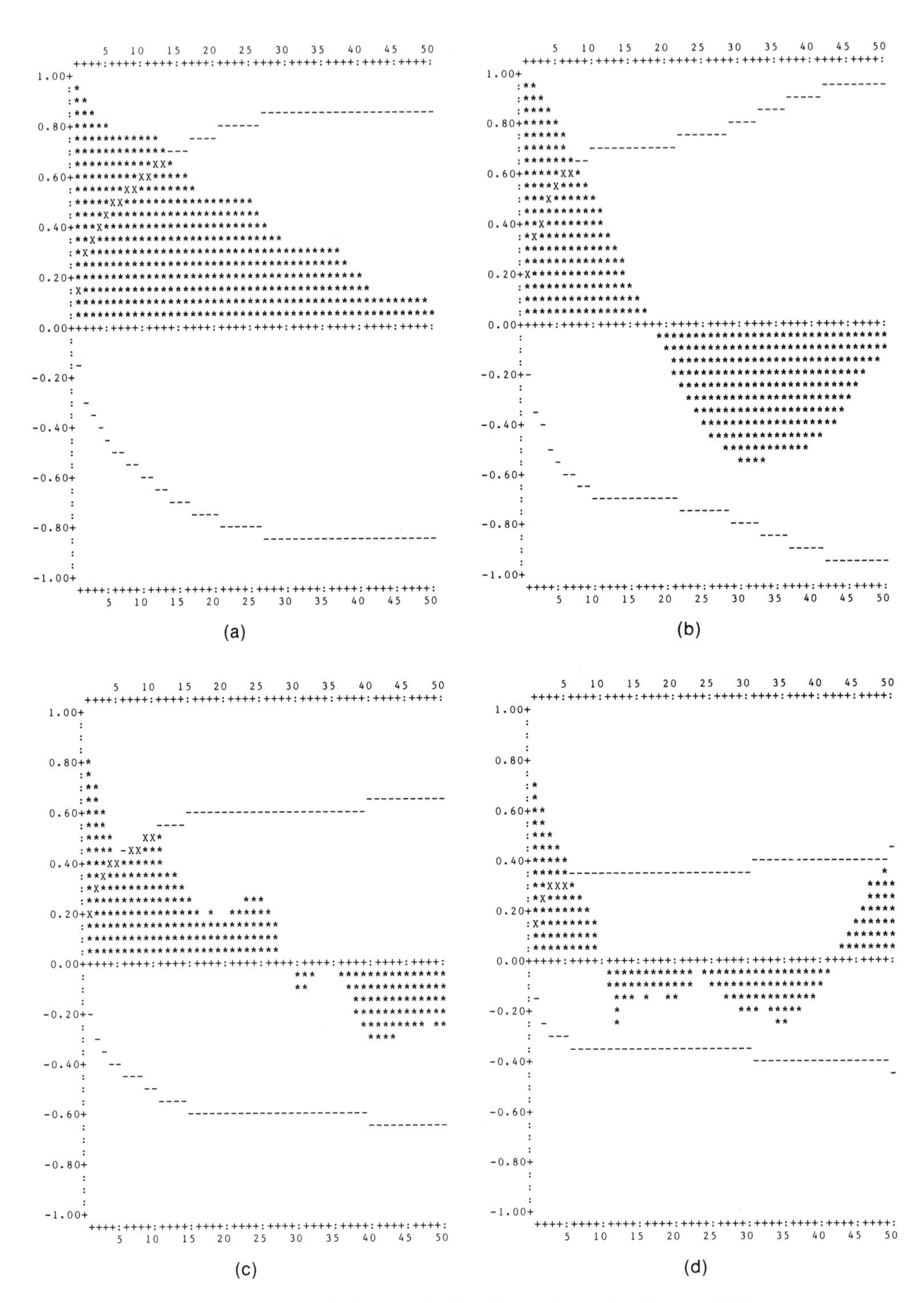

Figure 13.1 Sample AC Patterns Indicating the Need for at Least One Regular Difference

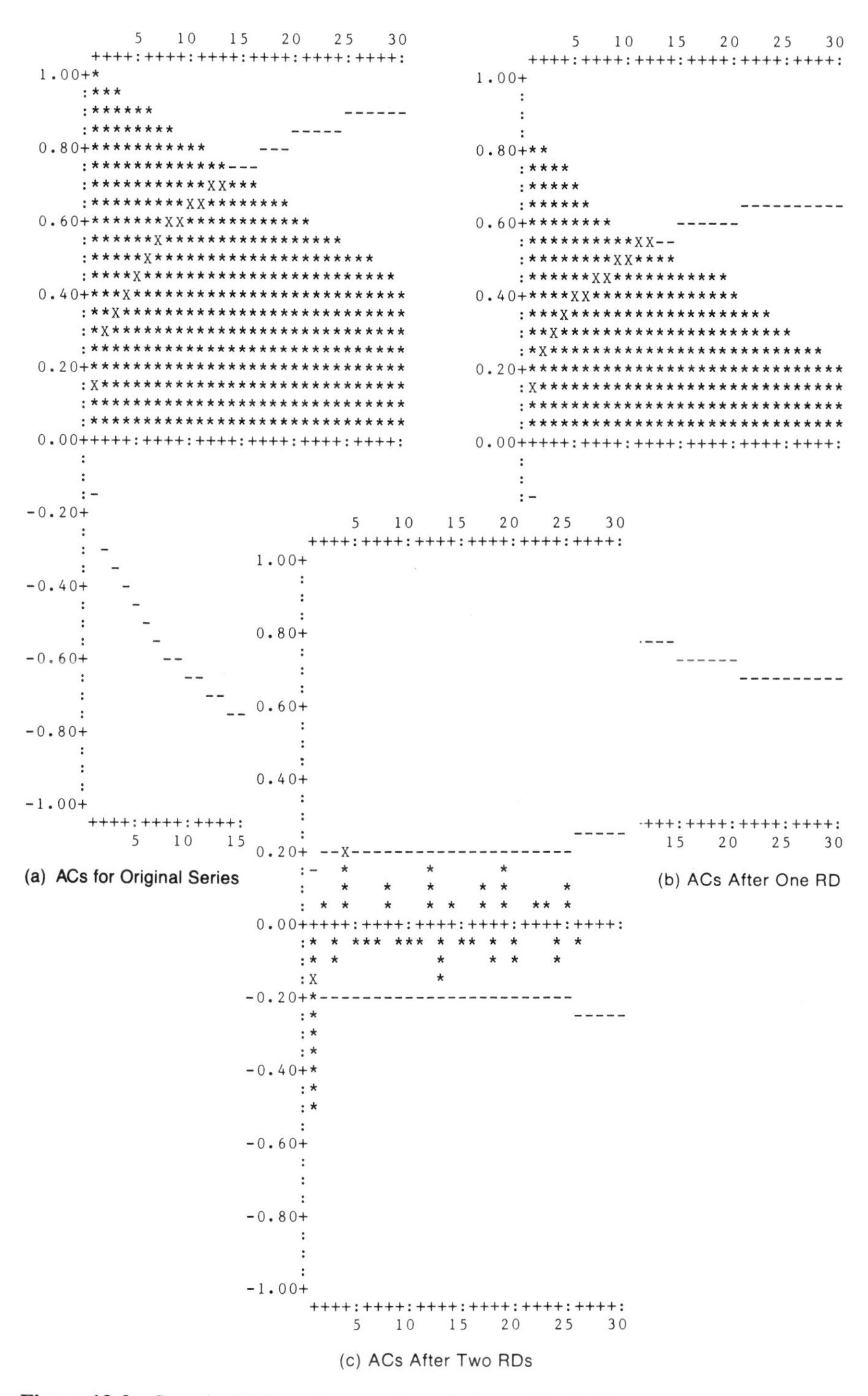

Figure 13.2 Sample AC Correlograms for a Series and Its Differences

Thus in a model with only one AR parameter, the magnitude of the parameter (i.e., disregarding any minus sign) should never be greater than or equal to 1; otherwise, the series would be nonstationary. However, if the AR parameter value is less than, but close to, 1, then the AC pattern of the stationary series could exhibit nonstationary characteristics. In these situations the best approach is to entertain two different models: one with differencing and one without. The two models can then be estimated, tested for validity, and compared.

In general, you should be very careful not to use more differencing than you need in order to achieve stationarity. Overdifferencing only complicates the model while accomplishing nothing. Remember, also, that each time a regular difference is applied, the number of data values available is reduced by one. Usually, only one or at most two regular differences will be needed to achieve stationarity.

13.2 IDENTIFYING THE TREND PARAMETER

As indicated in Chapter 12, you may also need to include a trend parameter in a model for a nonstationary series. This is usually the case when the dominant or only visible pattern (excluding random behavior) in the original series is an overall rising or falling trend. The need for a trend parameter is indicated when the mean value of the differenced series is significantly large.

Some Box-Jenkins programs will tell you whether the mean of the differenced series is significantly large—or will at least provide the raw statistics to allow you to make that determination. The mean of the differenced series is considered to be significantly large if it is greater in magnitude than twice the standard deviation of the differenced series divided by the square root of the number of data values in the differenced series. For example, suppose the original (nonstationary) series contains 94 data values, and two regular differences are applied to achieve stationarity (the differenced series then contains 92 values). If the mean and standard deviation of the differenced series are -2.5 and 9.0, respectively, then $2 \times 9.0 \div \sqrt{92} = 1.88$, which is less than 2.5. Hence the mean value of -2.5 is significantly large (at approximately a 95% level of confidence).

13.3 THE MODEL-BUILDING PROCESS FOR NONSTATIONARY MODELS

The model-building process for stationary series was described in Chapter 11 (see Figure 11.1). This process can now be updated to accommodate building models for nonstationary series, as shown in Figure 13.3.

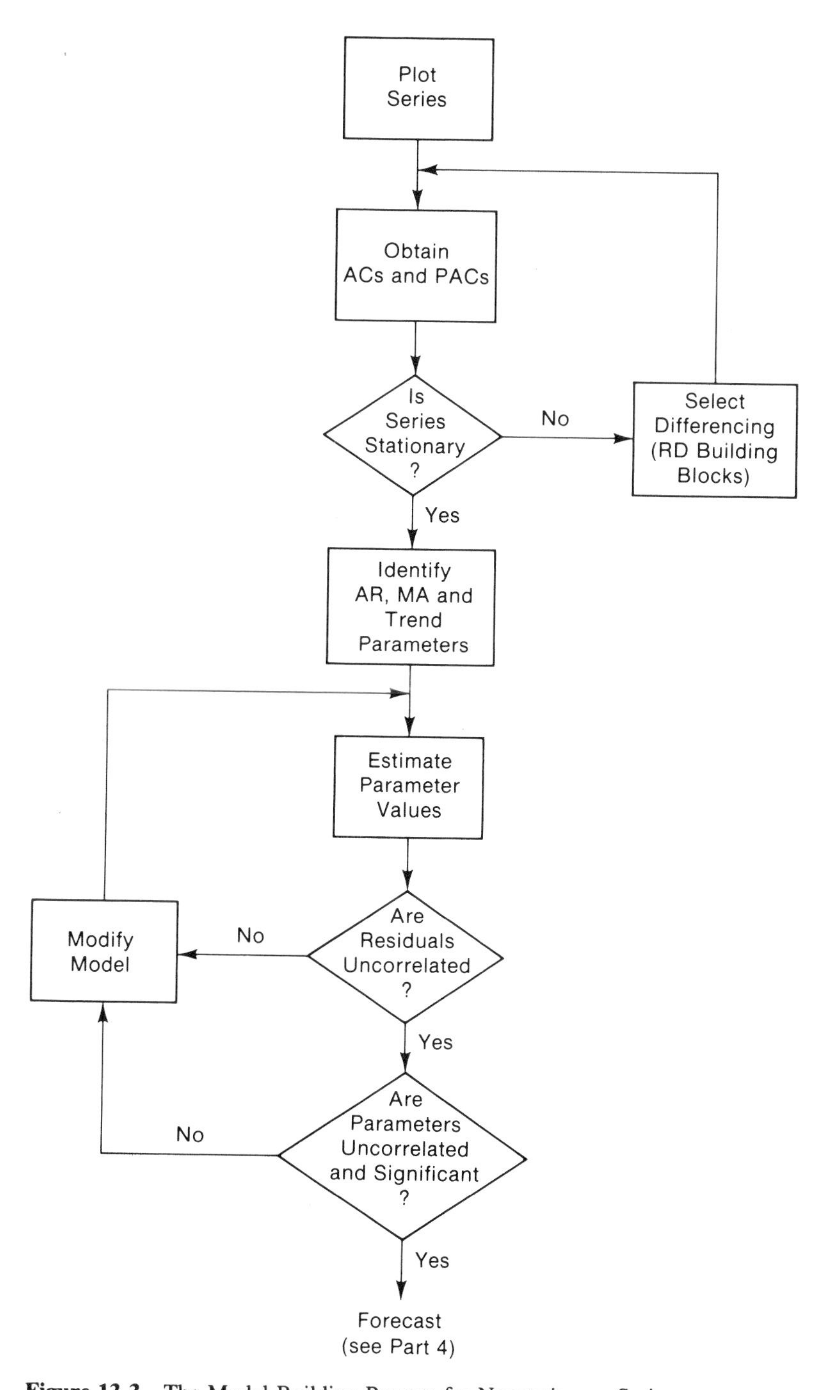

Figure 13.3 The Model-Building Process for Nonstationary Series

In this diagram three new steps have been added to the model-building process:

- Plot the original series.
- Apply regular differencing until stationarity has been achieved.
- Determine whether a trend parameter is needed if regular differencing has been applied.

The last two added steps, of course, correspond to the identification of the RD and trend building blocks that may be required to model a nonstationary series.

The first added step (obtaining a plot of the original series), although not directly related to building a Box-Jenkins model, is nevertheless an important step in building any time series forecasting model. A great deal of useful information can be obtained by visually analyzing the historical behavior of the time series. Recognizing the basic visual patterns in a series beforehand will make you more sensitive to the type of building blocks that may be required in a model. Nonstationary behavior, for example, is usually apparent in the plot of a series. The visual analysis of a series will become even more important in Part 3 when models for more complex series involving seasonal and other trend patterns are discussed. Also, the plot of a series can be useful in identifying extreme values and/or inconsistent data. The modification of such anomalies in the data, of course, is necessary before attempting to build a model.

13.4 EXAMPLES OF MODEL IDENTIFICATION, ESTIMATION, AND VALIDATION FOR NONSTATIONARY SERIES

The estimation and validation of a model for a nonstationary series is accomplished in the same way as it is for stationary series. All you have to do is indicate the number of RDs to be applied and the MA, AR, and trend parameters required to model the differenced (stationary) series. Your Box-Jenkins program will then estimate the model as before, and you may use the same residual and parameter diagnostics to validate the model.

The following four examples illustrate the process of identifying, estimating, and validating models for nonstationary (nonseasonal) series.

Example 13.1

A graph of the nonstationary series for this example is shown in Figure 13.4, along with its AC correlogram.

The AC correlogram for this series shows large autocorrelation spikes at all lags. Although the autocorrelations decrease in magnitude as the lag increases, they do so very slowly compared with autoregressive AC patterns. Thus at least one regular

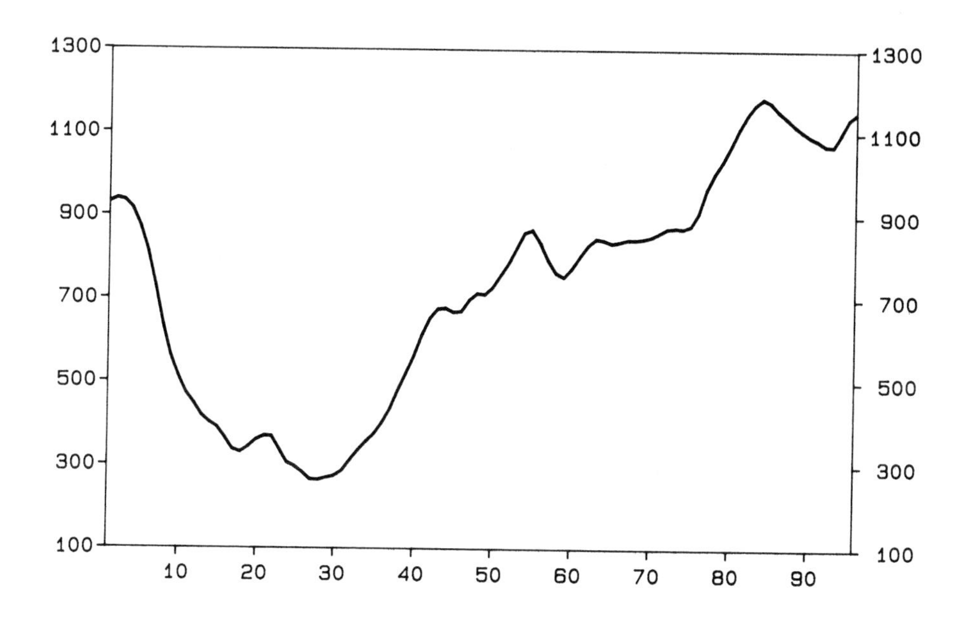


```
AUTOCORRELATIONS
----------------
 RD=0/SD=0/
 MEAN= 7.2273E+02   SDEV= 2.8357E+02
 (MEAN IS SIGNIFICANTLY NONZERO)
```

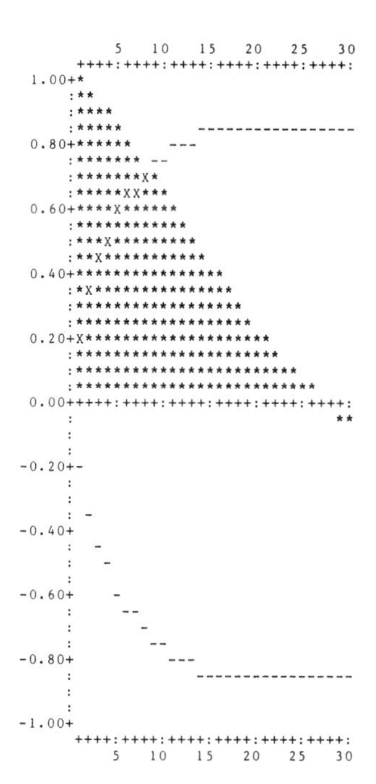

Figure 13.4 Graph and AC Correlogram of the Nonstationary Series for Example 13.1

difference should be applied. Note that in Figure 13.4(b) "RD = 0" is printed beneath the title "AUTOCORRELATIONS," which indicates that the correlogram corresponds to the original series; no differencing has been applied ("SD = 0" will be explained later). Note also that the program prints out the mean (MEAN = 722.73) and the standard deviation (SDEV = 283.57) of the series and indicates whether the mean of the series is significantly nonzero.

The AC correlogram for the differenced series, after applying one regular difference (RD = 1), is shown in Figure 13.5. In this correlogram the heavy autocorrelation pattern of the previous correlogram is no longer present. Instead, an AC pattern associated with one AR parameter has emerged. This result indicates that the differenced series is stationary and that the model for the series contains at least one AR parameter.

As we learned earlier, however, in the situation where AR parameters are called for, the PAC correlogram should also be examined to check the possibility of additional AR parameters or of an MA parameter. Thus the PAC correlogram for the differenced series is needed, and it is shown in Figure 13.6. The PAC pattern indicates the need for one MA parameter in addition to the AR parameter. Note also that the mean of the differenced series was not significantly nonzero (no message was printed indicating otherwise), so a trend parameter is not required.

The model for the original series is thus an ARIMA model with one RD, one AR, and one MA. The estimation results for this model are shown in Figure 13.7. All diagnostics indicate a valid model.

The final estimated model for the series is therefore represented as follows:

$$Z_t = .788Z_{t-1} + .727E_{t-1} + E_t$$

or

$$X_t - X_{t-1} = .788(X_{t-1} - X_{t-2}) + .727E_{t-1} + E_t$$

i.e.,

$$X_t = 1.788X_{t-1} - .788X_{t-2} + .727E_{t-1} + E_t$$

Example 13.2

A graph of the nonstationary series for this example is shown in Figure 13.8, along with its AC correlogram.

The AC correlogram for this series shows large, slowly decreasing autocorrelation spikes at all lags. Thus at least one regular difference should be used.

The AC correlogram for the differenced series, after applying one regular difference (RD = 1), is shown in Figure 13.9. In this correlogram most of the autocorrelations are now insignificant, except at lag 1. This result indicates that the differenced series is stationary and that the model for the series contains one MA parameter.

```
AUTOCORRELATIONS
----------------
 RD=1/SD=0/
 MEAN= 2.3239E+00   SDEV= 2.8914E+01


              5    10    15    20    25    30
         ++++:++++:++++:++++:++++:++++:
   1.00+
       :
       :*
       :*
   0.80+*
       :*
       :*
       :**
   0.60+**
       :**
       :**
       :***    -----------------------
   0.40+**X----
       :*X**
       :****
       :*****
   0.20+X******
       :********
       :*************   ***
       :******************
   0.00+++++:++++:++++:++++:++++:++++:
       :                     **********
       :                       ********
       :                        *******
  -0.20+-                       ****
       :
       :
       : -
  -0.40+  -----
       :       -----------------------
       :
       :
  -0.60+
       :
       :
       :
  -0.80+
       :
       :
       :
  -1.00+
         ++++:++++:++++:++++:++++:++++:
             5    10    15    20    25    30
```

Figure 13.5 AC Correlogram After One Regular Difference (Example 13.1)

In this example, however, the mean of the differenced series is significantly nonzero. Hence the model must also contain a trend parameter (this is evident also by looking at the graph of the original series, since the dominant pattern in the series is a simple overall straight-line trend). The model for the original series is thus an IMA model with one RD, one MA, and a trend parameter. The estimation results shown in Figure 13.10 indicate that this model is correct.

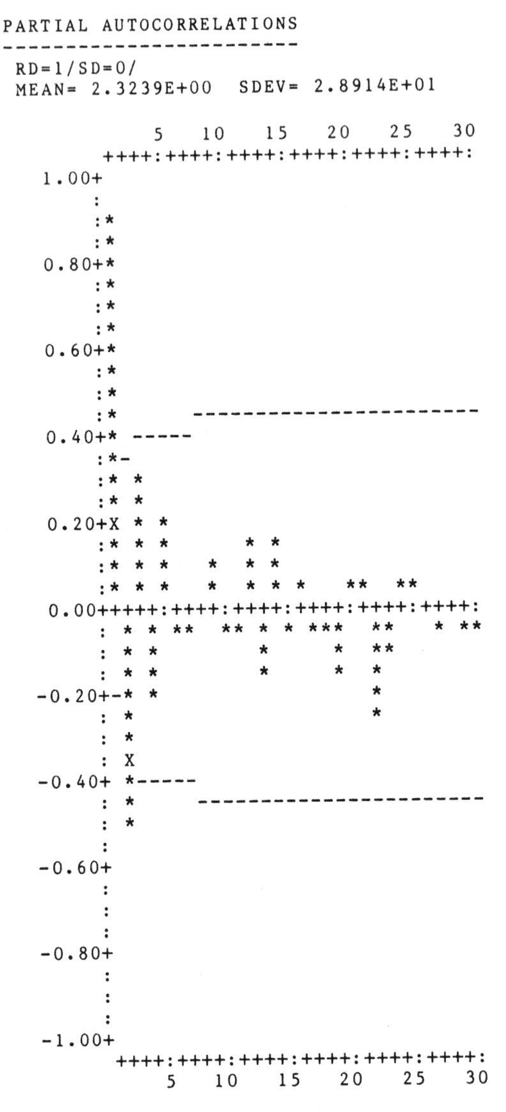

Figure 13.6 PAC Correlogram After One Regular Difference (Example 13.1)

The final estimated model for this series is therefore represented as follows:

$$Z_t = 19.960 - .856E_{t-1} + E_t$$

or

$$X_t = 19.960 + X_{t-1} - .856E_{t-1} + E_t$$

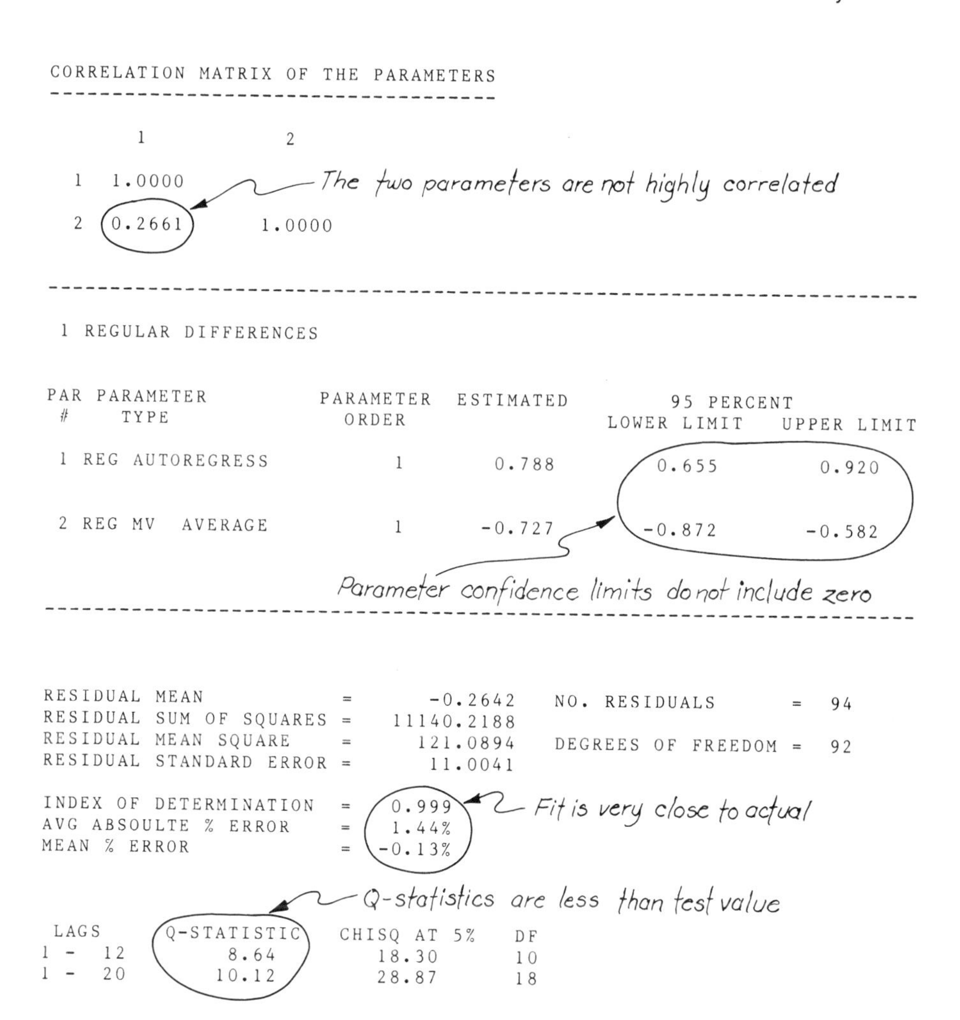

CORRELATION MATRIX OF THE PARAMETERS

	1	2
1	1.0000	
2	0.2661	1.0000

1 REGULAR DIFFERENCES

PAR #	PARAMETER TYPE	PARAMETER ORDER	ESTIMATED	95 PERCENT LOWER LIMIT	95 PERCENT UPPER LIMIT
1	REG AUTOREGRESS	1	0.788	0.655	0.920
2	REG MV AVERAGE	1	-0.727	-0.872	-0.582

RESIDUAL MEAN = -0.2642 NO. RESIDUALS = 94
RESIDUAL SUM OF SQUARES = 11140.2188
RESIDUAL MEAN SQUARE = 121.0894 DEGREES OF FREEDOM = 92
RESIDUAL STANDARD ERROR = 11.0041

INDEX OF DETERMINATION = 0.999
AVG ABSOULTE % ERROR = 1.44%
MEAN % ERROR = -0.13%

LAGS	Q-STATISTIC	CHISQ AT 5%	DF
1 - 12	8.64	18.30	10
1 - 20	10.12	28.87	18

Figure 13.7 Estimation Results for a Model with One RD, One AR of Order 1, and One MA of Order 1 (Example 13.1)

Example 13.3

This series represents the quarterly sales revenue, in thousands of dollars, for a company's established line of machine tools [8]. The series runs for 20 years (80 periods). A graph of the series along with its AC correlogram is shown in Figure 13.11.

The AC correlogram for this series, as in the previous two examples, clearly indicates the need for one regular difference. After applying one regular difference, we obtain the AC correlogram for the differenced series shown in Figure 13.12.

The correlogram in Figure 13.12 demonstrates an interesting pattern not encountered before: The autocorrelations at odd lags (lags 1, 3, 5, . . .) are relatively

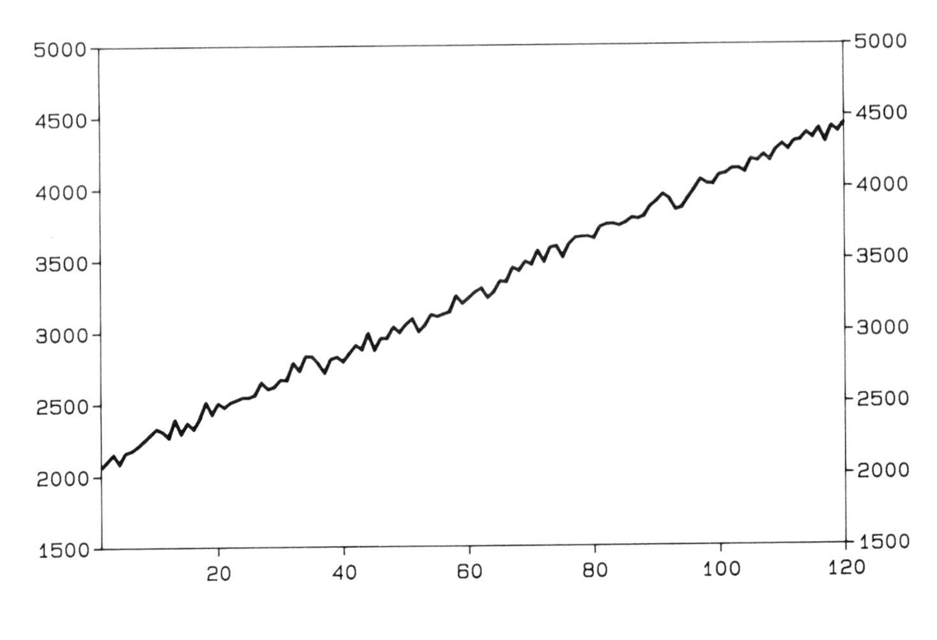

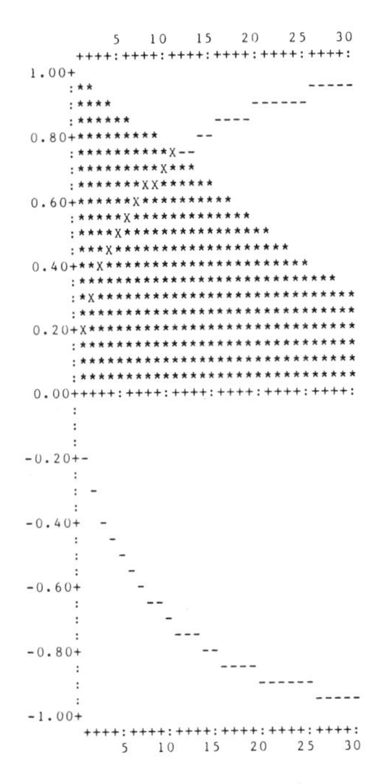

Figure 13.8 Graph and AC Correlogram of the Nonstationary Series for Example 13.2

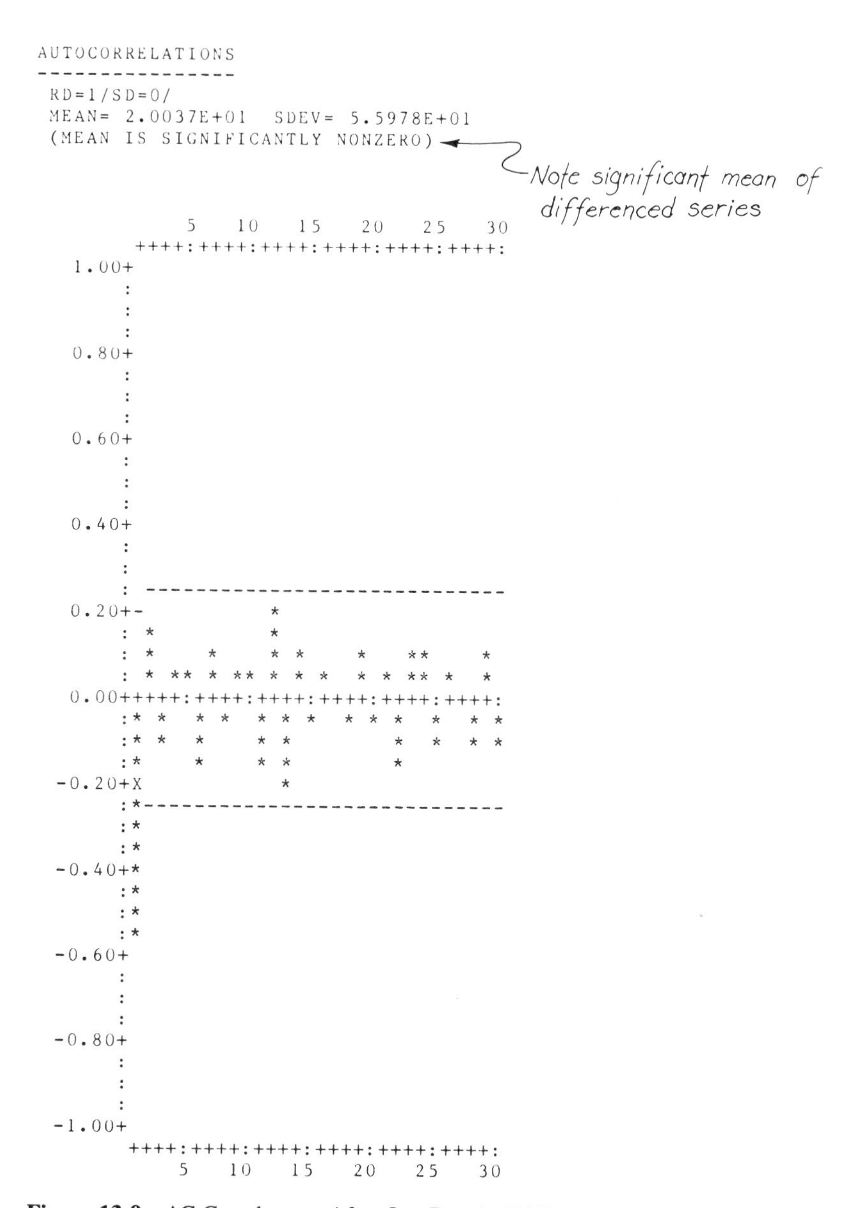

Figure 13.9 AC Correlogram After One Regular Difference (Example 13.2)

nonexistent, while those at even lags (lags 2, 4, 6, . . .) are relatively large with alternating signs, and they decrease in magnitude at longer lags. Although the decrease in magnitude is not that rapid, this pattern does suggest that the differenced series is stationary, and a possible model for the series contains one AR parameter of *order 2*.

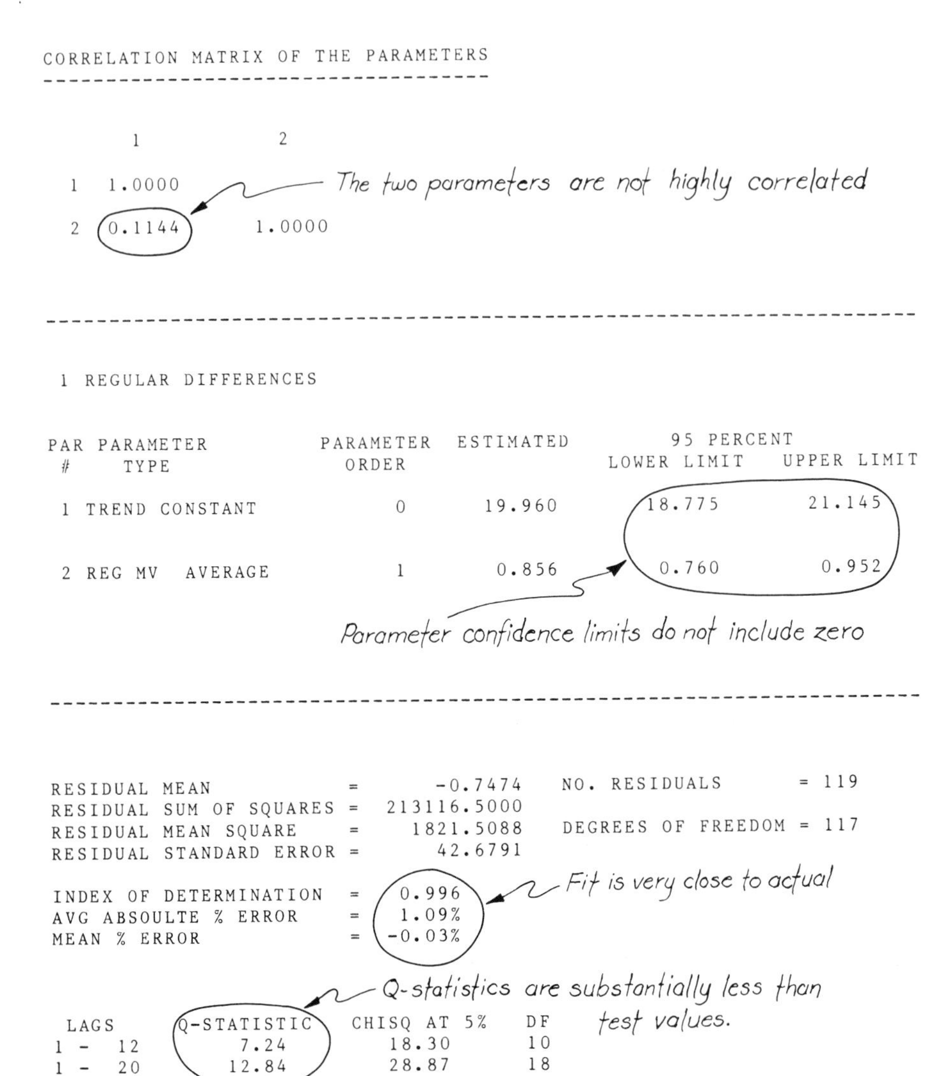

CORRELATION MATRIX OF THE PARAMETERS

	1	2
1	1.0000	
2	0.1144	1.0000

1 REGULAR DIFFERENCES

PAR #	PARAMETER TYPE	PARAMETER ORDER	ESTIMATED	95 PERCENT LOWER LIMIT	95 PERCENT UPPER LIMIT
1	TREND CONSTANT	0	19.960	18.775	21.145
2	REG MV AVERAGE	1	0.856	0.760	0.952

RESIDUAL MEAN	=	-0.7474	NO. RESIDUALS	= 119
RESIDUAL SUM OF SQUARES	=	213116.5000		
RESIDUAL MEAN SQUARE	=	1821.5088	DEGREES OF FREEDOM	= 117
RESIDUAL STANDARD ERROR	=	42.6791		
INDEX OF DETERMINATION	=	0.996		
AVG ABSOULTE % ERROR	=	1.09%		
MEAN % ERROR	=	-0.03%		

LAGS	Q-STATISTIC	CHISQ AT 5%	DF
1 - 12	7.24	18.30	10
1 - 20	12.84	28.87	18

Figure 13.10 Estimation Results for a Model with One RD, One MA of Order 1, and a Trend Parameter (Example 13.2)

This supposition is verified by looking at the PAC correlogram shown in Figure 13.13, where the large spike at lag 2 indicates the presence of the AR parameter of order 2. The model containing one RD and one AR of order 2 is therefore a reasonable possibility for our series.

The estimation results for this model are shown in Figure 13.14. All diagnostics indicate that the model is adequate.

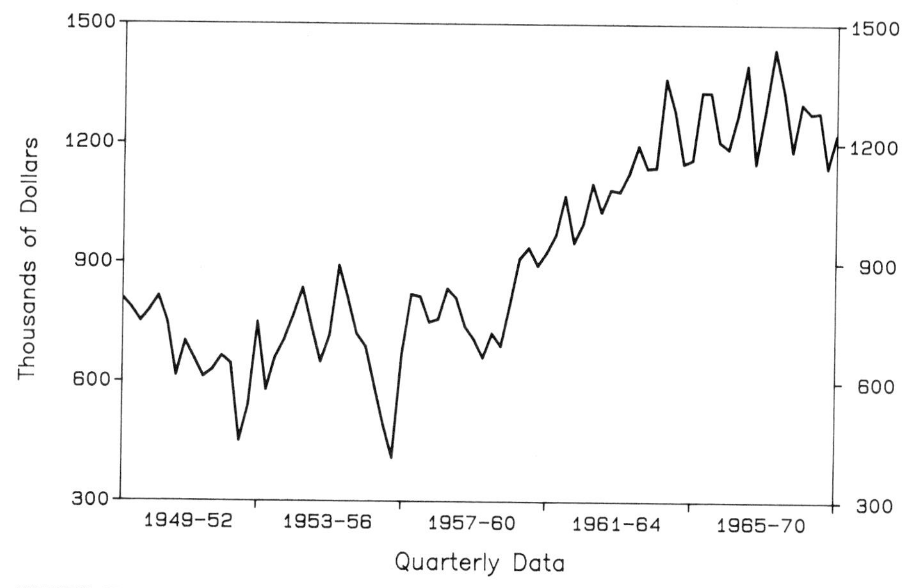

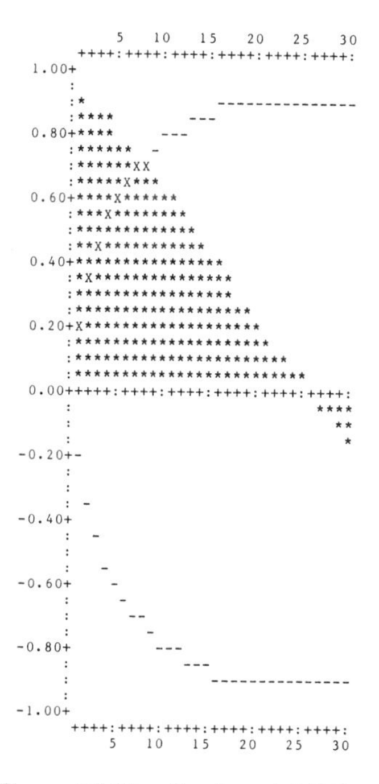

Figure 13.11 Graph and AC Correlogram of the Nonstationary Series for Example 13.3 (Quarterly Sales Revenue for a Line of Machine Tools)

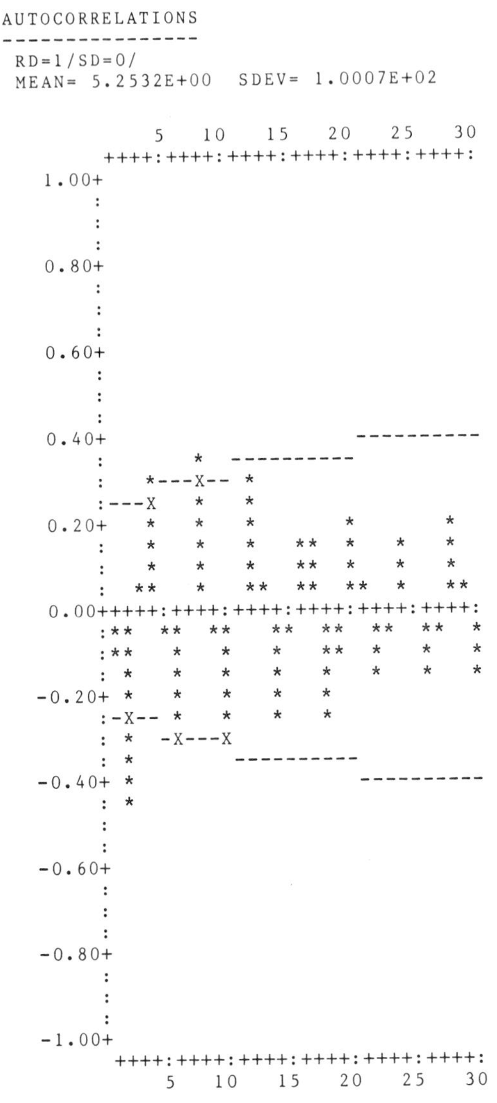

Figure 13.12 AC Correlogram After One Regular Difference (Example 13.3)

The estimated model for this series is therefore represented as follows:

$$Z_t = -.449Z_{t-2} + E_t, \qquad \text{where} \qquad Z_t = X_t - X_{t-1}$$

or

$$X_t = X_{t-1} - .449X_{t-2} + .449X_{t-3} + E_t$$

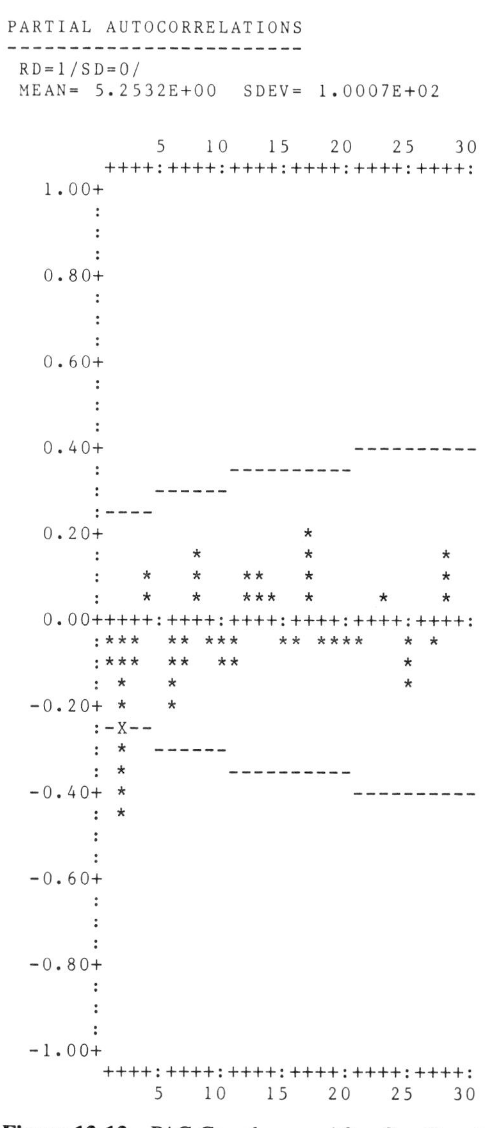

Figure 13.13 PAC Correlogram After One Regular Difference (Example 13.3)

Example 13.4

This series represents quarterly expenditures, in billions of current dollars, for new plant and equipment in the United States durable goods industry (seasonally adjusted). The series runs from the first quarter of 1963 to the fourth quarter of 1982 (80 periods). A graph of the series along with its AC correlogram is shown in Figure 13.15.

```
1 REGULAR DIFFERENCES

PAR PARAMETER          PARAMETER  ESTIMATED          95 PERCENT
 #     TYPE              ORDER                  LOWER LIMIT   UPPER LIMIT

1 REG AUTOREGRESS          2        -0.449        -0.658        -0.240
------------------------------------------------------------------------

RESIDUAL MEAN            =      -8.8821   NO. RESIDUALS        =  77
RESIDUAL SUM OF SQUARES  =  629102.4375
RESIDUAL MEAN SQUARE     =    8277.6641   DEGREES OF FREEDOM   =  76
RESIDUAL STANDARD ERROR  =      90.9817

INDEX OF DETERMINATION   =    0.882
AVG ABSOULTE % ERROR     =    9.00%
MEAN % ERROR             =   -0.03%

LAGS     Q-STATISTIC   CHISQ AT 5%   DF
1 - 12      10.57         19.67      11
1 - 20      16.02         30.14      19
```

Parameter confidence limits exclude zero

Q-statistics are less than their test values

Figure 13.14 Estimation Results for the Model with One RD and One AR of Order 2 (Example 13.3)

The AC correlogram for this series, as in the previous three examples, clearly indicates the need for one regular difference. After applying one regular difference, then, we obtain the AC correlogram shown in Figure 13.16 for the differenced series.

The correlogram in Figure 13.16 presents a couple of possibilities: Either the differenced series is stationary and the model for the series contains one or more AR parameters, or the differenced series is still nonstationary (this result is possible because of the existence of large clumps of autocorrelations at long lags).

Let's assume first that stationarity has been achieved. In this case the PAC correlogram shown in Figure 13.17 can help us determine how many AR parameters to include in the model. In this correlogram the large spike at lag 1 indicates one AR parameter of order 1, and the spike at lag 2, although not very large, might also indicate the need for another AR parameter of order 2. We can therefore entertain a model with one RD and two AR parameters of orders 1 and 2.

The estimation results for this model are shown in Figure 13.18. In these results the parameter diagnostics indicate that the model is definitely overspecified since the two AR parameters are highly correlated and the parameter confidence limits for the AR parameter of order 2 include zero. These diagnostics mean that the AR parameter of order 2 is not significant and should be eliminated. Thus a revised model containing one RD and only one AR parameter of order 1 should be estimated.

The estimation results for this revised model are shown in Figure 13.19. These results show that the one-RD and one-AR-parameter model is adequate. Note, however, that the upper confidence limit for the AR parameter is very close to one. From

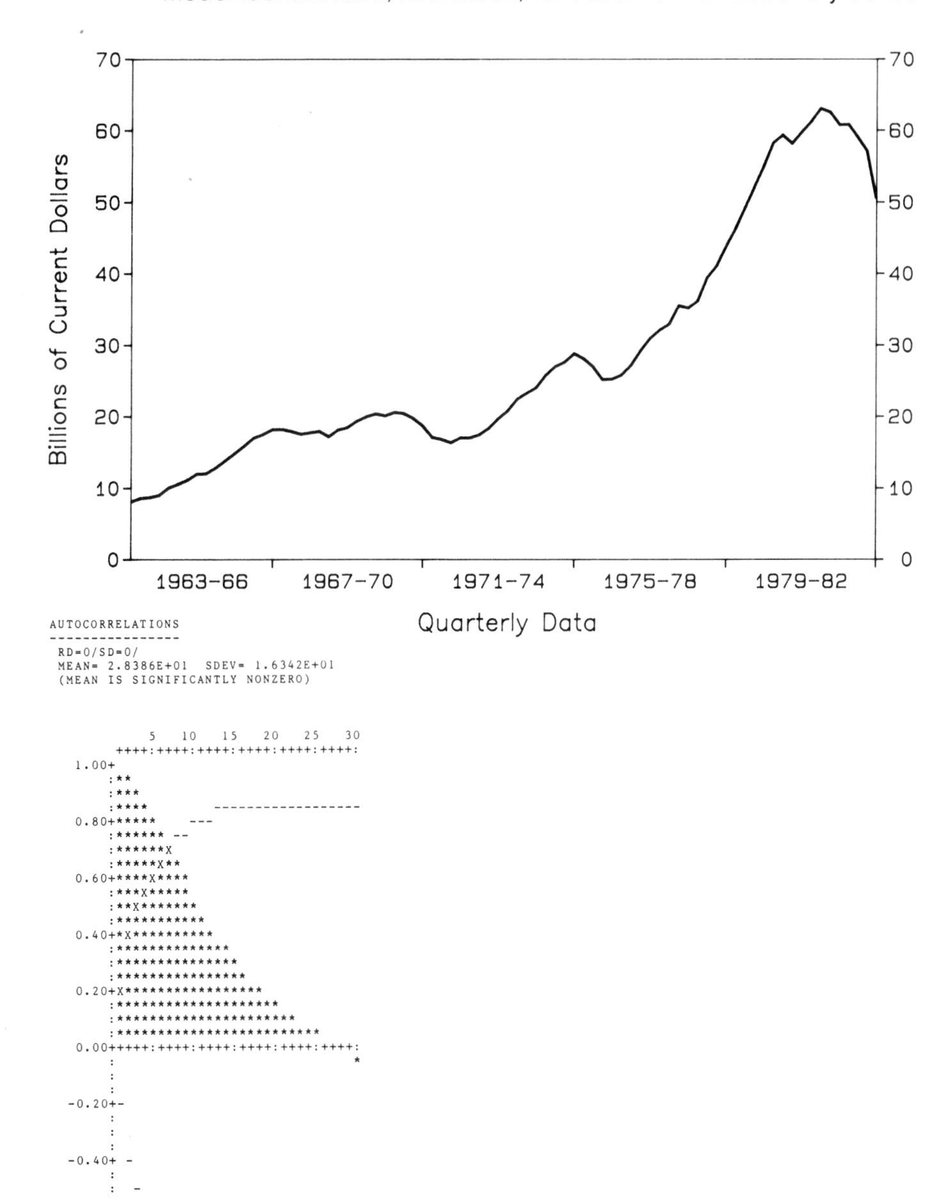

Figure 13.15 Graph and AC Correlogram of the Nonstationary Series for Example 13.4 (Quarterly Expenditures for New Plant and Equipment in the Durable Goods Industry)

Source: EEI Capsule Data Base, Evans Econometrics, Inc.

```
AUTOCORRELATIONS
----------------
 RD=1/SD=0/
 MEAN= 5.3658E-01   SDEV= 1.4348E+00
 (MEAN IS SIGNIFICANTLY NONZERO)

             5    10    15    20    25    30
         ++++:++++:++++:++++:++++:++++:
   1.00+
       :
       :
       :
   0.80+
       :
       :
       :
   0.60+
       :
       :*
       :*
   0.40+**
       :**   --------------------------
       :*X--
       :X***
   0.20+****
       :*****
       :*****                        **
       :*****                       ****
   0.00+++++:++++:++++:++++:++++:++++:
       :        * *********  *****
       :           ********  * **
       :           ******
  -0.20+            ***
       :-
       : ---
       :     --------------------------
  -0.40+
       :
       :
       :
  -0.60+
       :
       :
       :
  -0.80+
       :
       :
       :
  -1.00+
         ++++:++++:++++:++++:++++:++++:
             5    10    15    20    25    30
```

Figure 13.16 AC Correlogram After One Regular Difference (Example 13.4)

the discussion in Section 13.1 this result could mean that the autoregressive relationship represented by the AR parameter is very close to being a nonstationary relationship. Thus it is reasonable to explore the other possibility that the series, after one regular difference, is still nonstationary, and to apply one more regular difference.

The AC correlogram obtained after applying two regular differences is shown in Figure 13.20. This correlogram clearly indicates that the series is stationary. The

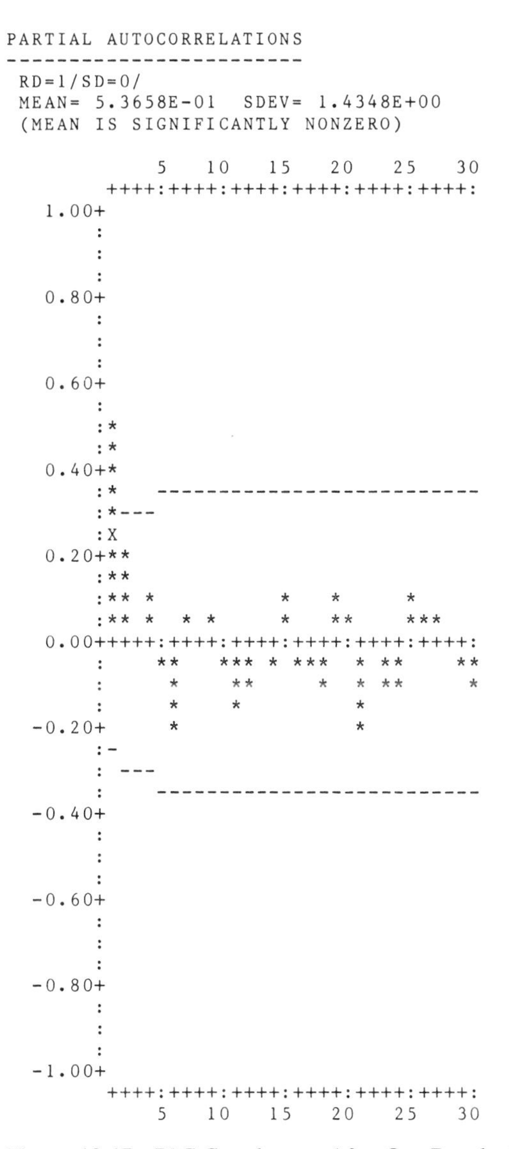

Figure 13.17 PAC Correlogram After One Regular Difference (Example 13.4)

spike at lag 1 indicates the possible need for one MA parameter (the spikes at lags 5 and 11 are ignored since they have no practical significance and are statistically insignificant anyway). However, the PAC correlogram of the twice-differenced series shown in Figure 13.21 does not readily support a one-MA model (if anything, it indicates one AR parameter).

Nevertheless, let's estimate the model containing two RDs and one MA of order 1. The estimation results for this model are shown in Figure 13.22. These results

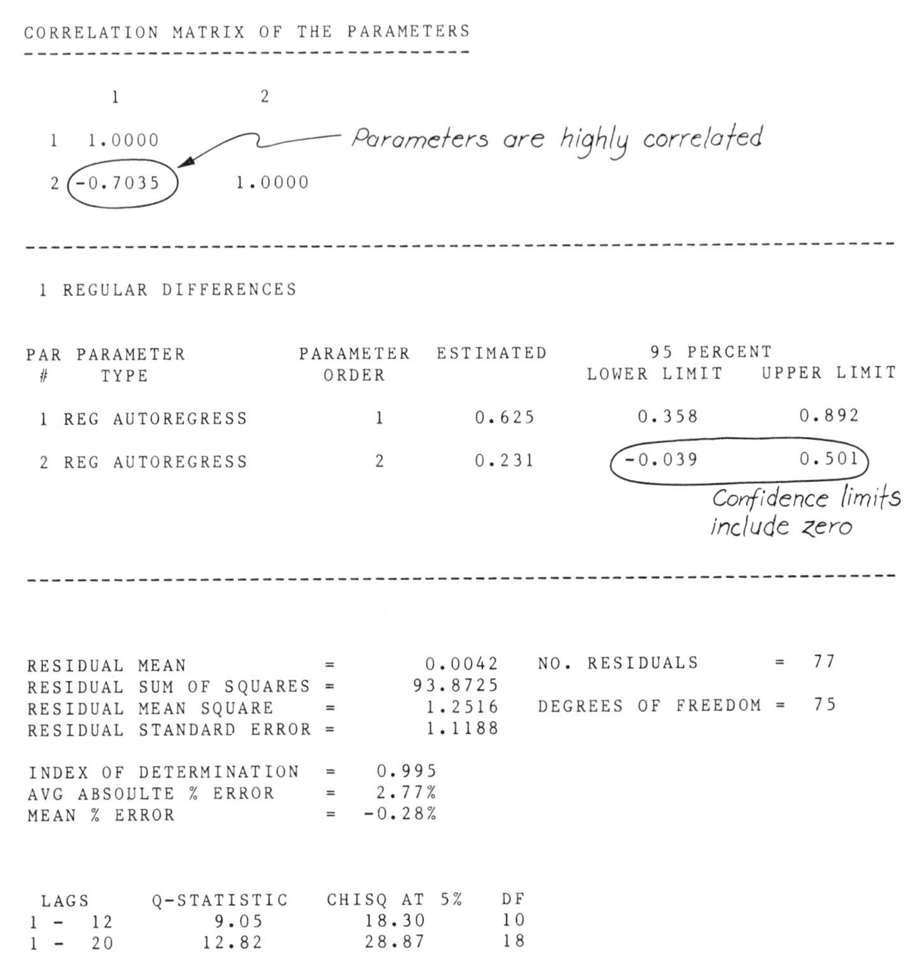

CORRELATION MATRIX OF THE PARAMETERS

	1	2
1	1.0000	
2	-0.7035	1.0000

1 REGULAR DIFFERENCES

PAR #	PARAMETER TYPE	PARAMETER ORDER	ESTIMATED	95 PERCENT LOWER LIMIT	95 PERCENT UPPER LIMIT
1	REG AUTOREGRESS	1	0.625	0.358	0.892
2	REG AUTOREGRESS	2	0.231	-0.039	0.501

RESIDUAL MEAN = 0.0042 NO. RESIDUALS = 77
RESIDUAL SUM OF SQUARES = 93.8725
RESIDUAL MEAN SQUARE = 1.2516 DEGREES OF FREEDOM = 75
RESIDUAL STANDARD ERROR = 1.1188

INDEX OF DETERMINATION = 0.995
AVG ABSOULTE % ERROR = 2.77%
MEAN % ERROR = -0.28%

LAGS	Q-STATISTIC	CHISQ AT 5%	DF
1 - 12	9.05	18.30	10
1 - 20	12.82	28.87	18

Figure 13.18 Estimation Results for the Model with One RD and Two ARs of Orders 1 and 2 (Example 13.4)

indicate that the model is adequate, although the parameter confidence limits for the MA parameter almost include zero.

Comparing these results (Figure 13.22) with those for the one-RD and one-AR model (Figure 13.19), we see that the fit statistics for both models are approximately the same and the Q-statistics for the two-RDs and one-MA model are only slightly better. However, since the residual diagnostics for both models don't reveal a clear winner, and since the MA parameter in the second model is almost statistically insignificant, and since it is desirable to apply as little differencing as possible, the one-RD and one-AR model would be the preferred model for the series. This model is represented as follows:

```
 1 REGULAR DIFFERENCES

PAR PARAMETER           PARAMETER  ESTIMATED         95 PERCENT
 #     TYPE               ORDER                 LOWER LIMIT   UPPER LIMIT

 1 REG AUTOREGRESS          1         0.784         0.593         0.974

----------------------------------------------------------------------

RESIDUAL MEAN              =       -0.0446    NO. RESIDUALS      =  78
RESIDUAL SUM OF SQUARES =          97.5534
RESIDUAL MEAN SQUARE       =        1.2669    DEGREES OF FREEDOM =  77
RESIDUAL STANDARD ERROR =           1.1256

INDEX OF DETERMINATION   =     0.995
AVG ABSOULTE % ERROR     =     2.89%
MEAN % ERROR             =    -0.39%

 LAGS       Q-STATISTIC   CHISQ AT 5%    DF
1 -  12        11.51         19.67        11
1 -  20        15.73         30.14        19
```

Figure 13.19 Estimation Results for the Model with One RD and One AR of Order 1 (Example 13.4)

$$Z_t = .784Z_{t-1} + E_t, \qquad \text{where} \qquad Z_t = X_t - X_{t-1}$$

or

$$X_t = 1.784X_{t-1} - .784X_{t-2} + E_t$$

13.5 REVIEW OF KEY CONCEPTS

The *need for regular differencing* can be identified by looking at the ACs of the original series (and subsequently differenced series). Some important points to remember when determining the number of regular differences to use are as follows:

- Typically, a regular difference is needed if the ACs remain relatively large for many lags.
- The ACs may decrease as the lag increases, but they will do so slowly, usually at a constant rate (versus the rapid rate of decrease associated with AR parameters).
- Visually, a correlogram associated with a nonstationary series will have an overall solid appearance or have large alternating clumps of autocorrelation spikes.

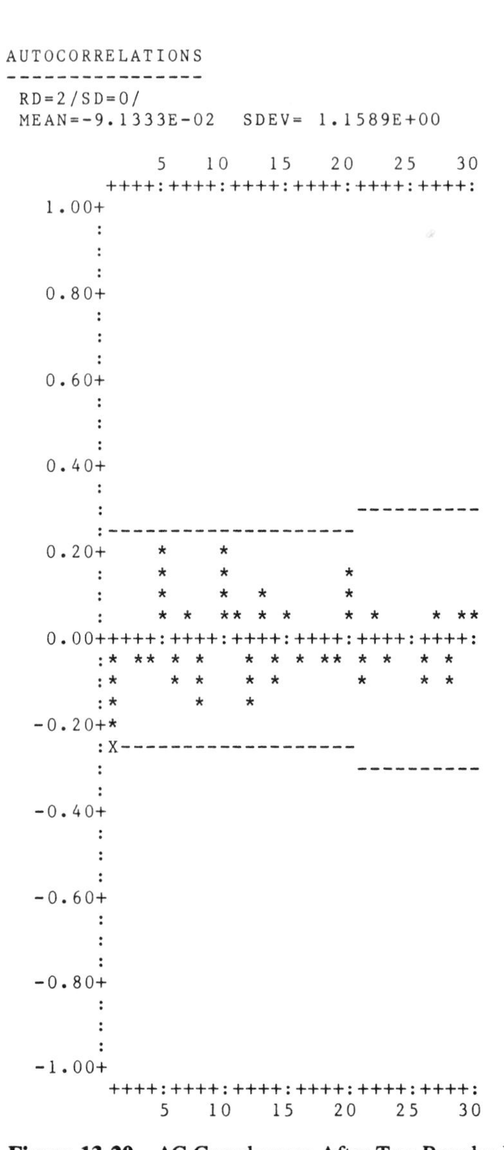

Figure 13.20 AC Correlogram After Two Regular Differences (Example 13.4)

- You will know when you have applied enough regular differences when the AC correlogram of the last differenced series has relatively few significant spikes and when the typical AC patterns associated with the basic AR and MA parameters can be identified.
- You will seldom, if ever, need more than two RDs.
- The need for a regular difference versus an AR parameter can be difficult to determine sometimes. When in doubt, entertain alternative models with and without the regular difference.

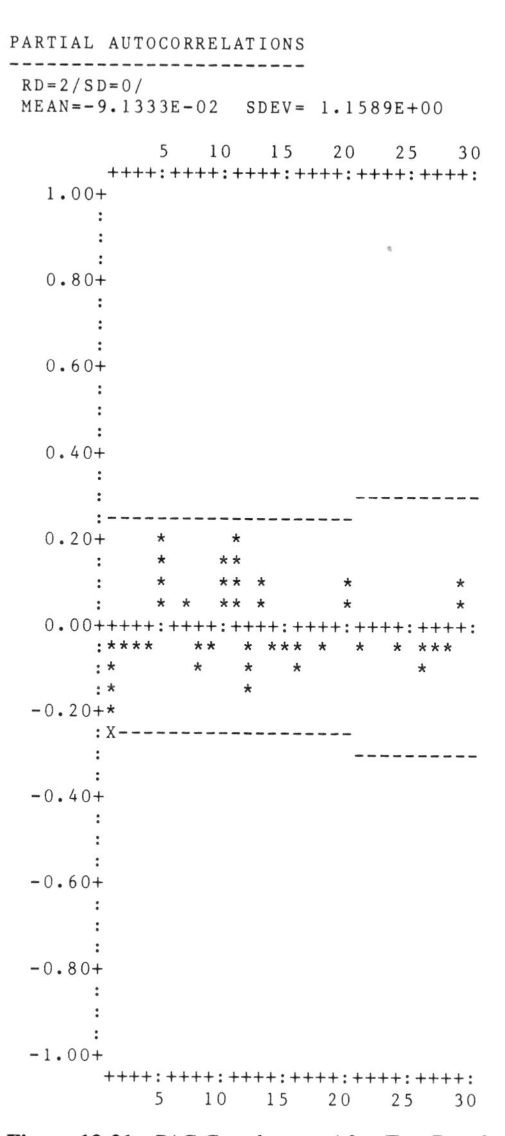

Figure 13.21 PAC Correlogram After Two Regular Differences (Example 13.4)

The *need for a TREND parameter* is evidenced when the mean of the differenced (stationary) series is significantly large. The mean is significantly nonzero if it is larger in magnitude than twice the standard deviation of the series divided by the square root of the number of differenced series values.

The *estimation and validation process for nonstationary models* is the same as the one for stationary models.

```
 2 REGULAR DIFFERENCES

PAR PARAMETER            PARAMETER  ESTIMATED          95 PERCENT
 #     TYPE                ORDER                 LOWER LIMIT   UPPER LIMIT

 1 REG MV   AVERAGE          1         0.339         0.084          0.594


----------------------------------------------------------------------


RESIDUAL MEAN              =       0.1056    NO. RESIDUALS         =  78
RESIDUAL SUM OF SQUARES =         95.8026
RESIDUAL MEAN SQUARE       =       1.2442    DEGREES OF FREEDOM =  77
RESIDUAL STANDARD ERROR =          1.1154

INDEX OF DETERMINATION   =    0.995
AVG ABSOULTE % ERROR     =    2.80%
MEAN % ERROR             =    0.09%


 LAGS      Q-STATISTIC    CHISQ AT 5%    DF
1 -  12        8.93          19.67        11
1 -  20       12.41          30.14        19
```

Figure 13.22 Estimation Results for the Model with Two RDs and One MA of Order 1 (Example 13.4)

CHAPTER 14

Models for Seasonal Series

Seasonal behavior is probably the most prevalent behavior found in economic and business time series. The behavior of many economic activities is tied in some way to the seasons of the year or to the repetition of some other basic time unit. To complete our investigation of Box-Jenkins forecasting models, therefore, we must look at models for seasonal series.

This chapter describes three new building blocks for constructing models for seasonal series. These new building blocks, together with the four building blocks you are already familiar with (MA, AR, RD, and Trend), give you the complete set of Box-Jenkins model-building blocks.

In this chapter you will learn:

- How seasonal series behave.
- How models for stationary seasonal series are constructed.
- How models for nonstationary seasonal series are constructed.

The process of identifying models for seasonal series is covered in Chapter 15.

14.1 SEASONAL SERIES

Seasonal behavior in a time series is simply the tendency of the series to repeat a certain pattern of behavior at regular time intervals called *seasons*. The most common seasonal interval for weekly, monthly, or quarterly economic time series is one year. The number of time series periods within a season is called the *periods per season*. For example, if the season is one year, monthly data has 12 periods per season, weekly data has 52 periods per season, and quarterly data has 4 periods per season. Other "seasons," however, are possible. For example, daily data could have a season of one week with 5, 6, or 7 periods per season; monthly data could have a season of one quarter with 3 periods per season.

Seasonal behavior is a common phenomenon in the economic environment. Many business and economic activities are highly influenced by external events or conditions that occur around the same time every year. Changes in weather temperature and conditions from summer to winter are probably the biggest cause of seasonal behavior. Energy consumption, for example, is always higher in the winter than in the summer, while housing construction has the opposite pattern. Special events, such as holidays or income tax filing, will also produce higher levels of activity during the same time every year for products or services that are uniquely associated with those events. Most time series you will deal with will probably exhibit some kind of seasonal behavior. Figures 14.1 through 14.4 are graphs of time series that illustrate a variety of economic activities exhibiting seasonal behavior.

Although seasonal behavior was described as a tendency to repeat a certain basic pattern from season to season, the basic pattern itself usually undergoes constant change. This behavior is clearly demonstrated in the time series in the figures. For example, the difference between peak levels and trough levels in these series grows larger or smaller from season to season, and in some cases the entire seasonal pattern is constantly changing level where it is superimposed on an overall trend. In addition, the seasonal pattern in a series may not be the only pattern in the series. That is, patterns or relationships of the type we have discussed so far for nonseasonal series may also exist. Models for seasonal series, then, must be able to account for all these patterns, seasonal as well as nonseasonal.

In developing seasonal models, we will first describe models that only account for stationary seasonal behavior. Later we will combine these purely seasonal models with the basic models described in Part 2 and the nonstationary models described in Chapters 12 and 13 to obtain a generalized seasonal model.

14.2 PURELY SEASONAL MODELS

Seasonal behavior in a series implies that some relationship exists among series values that are separated by the number of periods per season. For example, if a monthly series is seasonal, then all January values will tend to be related in some way, all February values will tend to be related, etc. Note that two successive January values, or two successive February values, etc., are separated by 12 months, which is the number of periods per season (year). What kind of models, then, can be used to describe these relationships?

In Part 2 you learned that most *period-to-period relationships* among the values in a stationary series are autoregressive and/or moving-average relationships. From this fact the basic models containing AR and MA parameters were developed. It seems natural, therefore, that the same kind of relationships would exist among

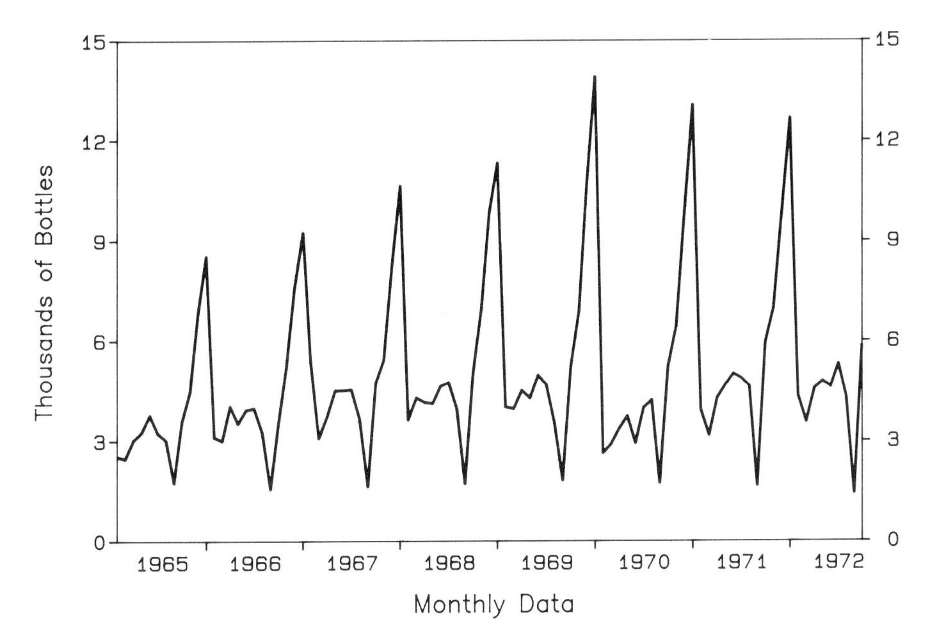

Figure 14.1 Champagne Sales

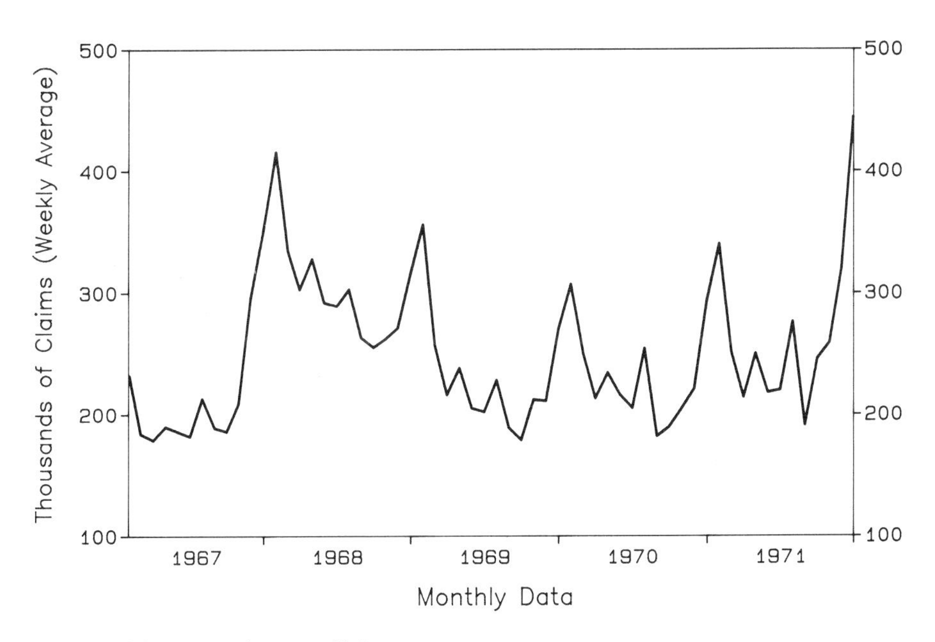

Figure 14.2 Unemployment Claims

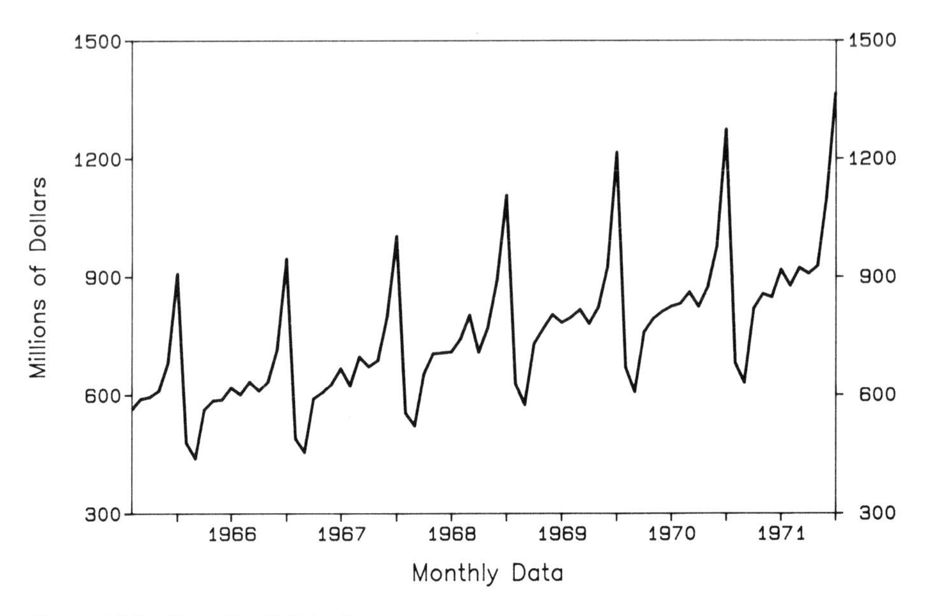

Figure 14.3 Sears Retail Sales Revenue

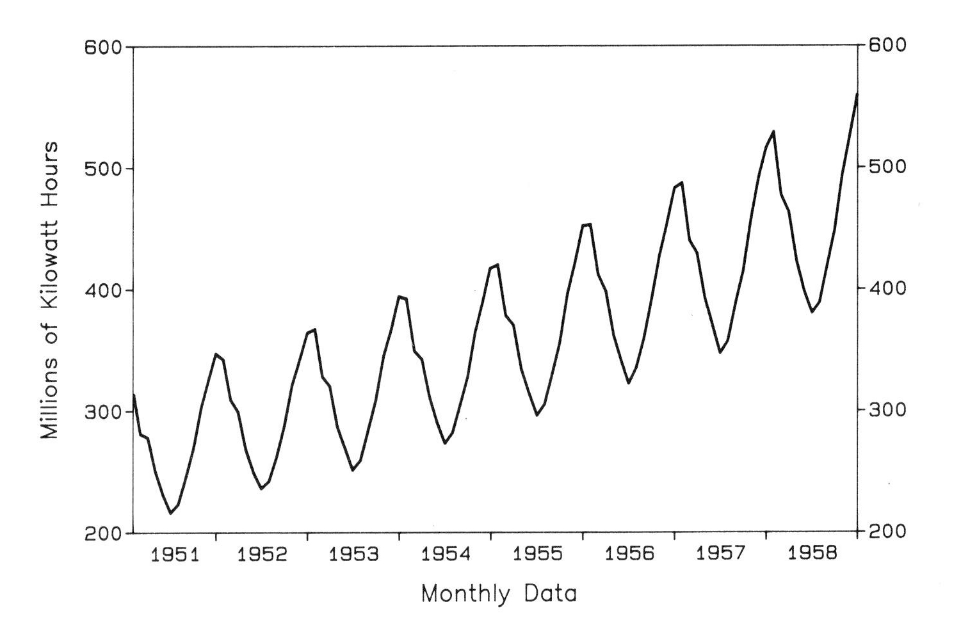

Figure 14.4 Electrical Power Use

values in a seasonal series separated by a multiple of the number of periods per season. For example, if the number of periods per season is 12, then you might try to identify autoregressive and moving-average relationships among the series values $X_1, X_{13}, X_{25}, X_{37}, \ldots$, and among the series values $X_2, X_{14}, X_{26}, X_{38}, \ldots$. If this is the case, then you will be able to use AR and MA parameters to construct models for seasonal series. Such parameters are called *seasonal parameters,* as opposed to the AR and MA parameters discussed in Part 2 for the basic models, which we will now call *regular parameters*. As you will see in a moment, the only difference between seasonal parameters and regular parameters is that the *order of a seasonal parameter* is some multiple of the number of periods per season.

To understand this concept better, let's look at some AR and MA models for *stationary* seasonal series that are *purely seasonal* (i.e., that contain only seasonal AR and MA parameters). Later in this chapter we will discuss how the regular parameters in the basic models of Part 2 can be combined with seasonal parameters to model both seasonal and nonseasonal behavior.

14.2.1 Purely Seasonal Autoregressive Models (SAR Models)

Seasonal autoregressive models are built with parameters called *seasonal autoregressive parameters* (SAR parameters). The SAR parameters represent autoregressive relationships that exist between time series values separated by multiples of the number of periods per season. For example, a model with one SAR parameter is written as follows:

$$X_t = A_1^* X_{t-s} + E_t$$

where s is the number of periods per season. The parameter A_1^* is called an SAR parameter of order s. The asterisk in the parameter symbol A_1^* is used to distinguish it from the regular parameter A_1 in the basic autoregressive model $X_t = A_1 X_{t-1} + E_t$. Note that the order of A_1^* is s and the order of A_1 is one. The above model simply says that any given value X_t is directly proportional to X_{t-s}, i.e., to the time series value s periods ago. For example, if $s = 12$ months, $X_{13} = A_1^* X_1 + E_{13}$ and $X_{14} = A_1^* X_2 + E_{14}$, etc.

The seasonal autoregressive concept can be extended to include more than one SAR parameter. For example, a seasonal autoregressive model with two SAR parameters is written as follows:

$$X_t = A_1^* X_{t-s} + A_2^* X_{t-2s} + E_t$$

where the SAR parameters A_1^* and A_2^* are of order s and $2s$, respectively. This model says that what happens in the current period is directly related to what happened both one season ago and two seasons ago. For example, if $s = 4$, then $X_9 = A_1^* X_5 + A_2^* X_1 + E_9$, etc.

A *general seasonal autoregressive model* with P SAR parameters is written as follows:

$$X_t = A_1^* X_{t-s} + \ldots + A_P^* X_{t-Ps} + E_t$$

where A_1^* is of order s, A_2^* is of order $2s$, . . ., and A_P^* is of order Ps. The order of the model is Ps.

14.2.2 Purely Seasonal Moving-Average Models (SMA Models)

Seasonal moving-average models are built with parameters called *seasonal moving-average parameters* (SMA parameters). SMA parameters represent moving-average relationships that exist among time series values separated by a multiple of the number of periods per season. For example, a model with one SMA parameter is written as follows:

$$X_t = -B_1^* E_{t-s} + E_t$$

The parameter B_1^* is called an SMA parameter of order s. The above model says that X_t is directly related to the random error E_{t-s} that occurred one season ago, i.e., s periods ago. For example, if $s = 4$ quarters, then $X_5 = -B_1^* E_1 + E_5$ and $X_6 = -B_1^* E_2 + E_6$, etc.

As usual, we can generalize this model to include Q SMA parameters, as follows:

$$X_t = -(B_1^* E_{t-s} + \ldots + B_Q^* E_{t-Qs}) + E_t$$

where B_1^* is of order s, B_2^* is of order $2s$, . . ., and B_Q^* is of order Qs. The order of the model is Qs.

14.2.3 Mixed Seasonal AR and MA Models (Seasonal ARMA Models)

Both SAR and SMA parameters can be used in the same model in exactly the same way that regular AR and MA parameters can. A mixed SAR and SMA model would thus be written as follows:

$$X_t = A_1^* X_{t-s} + \ldots + A_P^* X_{t-Ps} - (B_1^* E_{t-s} + \ldots + B_Q^* E_{t-Qs}) + E_t$$

The order of this seasonal ARMA model is expressed in terms of both Ps and Qs.

14.3 COMBINED REGULAR AND SEASONAL MODELS

Purely seasonal models, unfortunately, are not sufficient for modeling most seasonal economic and business time series. These models can only describe relationships among series values separated by a multiple of the number of periods per season; the relationships among series values from one period to the next are ignored in purely seasonal models. For most seasonal series, of course, both seasonal interactions and period-to-period interactions take place simultaneously. Therefore we need to be able to *combine* the purely seasonal models discussed earlier in this chapter with the basic models described in Part 2.

As noted previously, for convenience in distinguishing between the seasonal AR and MA parameters and the AR and MA parameters in the basic models, the basic model parameters are called regular parameters. In addition, the following acronyms will be used in the remainder of this book to refer to each type of parameter:

RAR	Regular autoregressive
RMA	Regular moving average
SAR	Seasonal autoregressive
SMA	Seasonal moving average

Unless otherwise noted, if we are talking about n RAR parameters, their orders are assumed to be 1, 2, . . ., n; likewise for n RMA parameters. If we are talking about n SAR parameters, their orders are assumed to be s $2s$, . . ., ns, where s is the number of periods per season; likewise for n SMA parameters.

Now, there are two ways to go about combining seasonal and regular parameters to construct a seasonal Box-Jenkins model. One approach is to simply *add* seasonal and regular terms together, as illustrated below in a model containing one RAR and one SAR parameter.

$$X_t = A^1\mathbf{X}_{t-1} + A_1^*X_{t-12} + E_t$$

Such a combined model is called an *additive seasonal model*.

It turns out, however, that most seasonal models for economic and business time series are not additive but multiplicative. In a *multiplicative seasonal model*, additional terms appear in the model that represent the *interaction* of seasonal behavior with period-to-period behavior. For example, a multiplicative seasonal model containing one RAR and one SAR is written as follows:

$$X_t = A_1X_{t-1} + A_1^*X_{t-12} - A_1A_1^*X_{t-13}\ E_t$$

This model says that what happened in a given period is directly related not only to what happened last period and 12 periods ago but *also* to what happened 13 periods ago. This relationship makes sense when you consider the fact that what happened in period 12 is directly related to what happened in period 13 because of the RAR parameter A_1. Note that the third term, $-A_1A_1^*X_{t-13}$, does not introduce a new parameter but simply uses the negative product of the parameter values A_1 and A_1^* as its "parameter"—the model still contains only two parameters, but the form of the model is different.

The number of additional terms that appear in a multiplicative seasonal AR model depends on how many SAR and RAR parameters are used. There will be one additional term for every unique SAR/RAR pair. In practice, however, it is usually not important to be able to write down the precise model form, so you don't need to know what all the additional terms are. (See Appendix D for details on the mathematical representation of multiplicative seasonal models.)

Of course, regular and seasonal moving-average parameters may be combined in a multiplicative model just as autoregressive parameters can be. The following model is a multiplicative seasonal model containing one RMA and one SMA parameter:

$$X_t = -B_1E_{t-1} - B_1^*E_{t-12} + B_1B_1^*E_{t-13} + E_t$$

Furthermore, all four types of parameters (RMA, RAR, SMA, SAR) may be used in one seasonal model: the moving-average terms are simply added to the autoregressive terms. For example, the following multiplicative model contains one RMA, one RAR, one SMA, and one SAR parameter:

$$\begin{aligned} X_t &= (A_1X_{t-1} + A_1^*X_{t-12} - A_1A_1^*X_{t-13}) \\ &\quad - (B_1E_{t-1} + B_1^*E_{t-12} - B_1B_1^*E_{t-13}) + E_t \end{aligned}$$

In practice, a seasonal model will rarely require more than a total of three or four parameters. More often than not, a seasonal model will contain only one regular and one seasonal parameter. The identification of seasonal models is discussed in Chapter 15.

14.4 MODELS FOR NONSTATIONARY SEASONAL SERIES (SEASONAL ARIMA MODELS)

As discussed in Chapter 12, a nonstationary series must first be transformed into a stationary series before any AR or MA parameters can be identified. This transformation can usually be accomplished by applying one or more regular differences to the series. A seasonal series may also be nonstationary, but the form of its nonstationarity can be more complex. Thus we will need to explore the nature of seasonal nonstationarity and find out how we can transform nonstationary seasonal series into stationary series.

14.4.1 Nonstationary Seasonal Series

A seasonal series can be nonstationary in the same sense as a nonseasonal series; i.e., its seasonal pattern may be superimposed on an overall trend or on a pattern of behavior that exhibits random changes in established levels and/or slopes. Thus the application of one or more regular differences (differences of order 1) may be used to transform some seasonal series into stationary ones.

Seasonal series, however, can also exhibit nonstationary behavior in another way: namely, *the seasonal pattern itself may be nonstationary.* For example, if the number of periods per season is 4, then the series of values $X_1, X_5, X_9, \ldots$, and $X_2, X_6, X_{10}, \ldots$, etc. may also exhibit an overall trend or changes in established levels and/or slopes. In this case the application of regular differences will not be sufficient to achieve stationarity; i.e., the seasonal pattern will still be nonstationary. Figure 14.5 illustrates a series that has a nonstationary seasonal pattern. Note how the amplitude, or differences between successive peak and trough levels, increases as time moves on; this property indicates a trend, and hence nonstationarity, in the seasonal pattern.

14.4.2 Seasonal Differencing (The SD Building Block)

As you may have already guessed, nonstationary behavior in the seasonal pattern may be transformed into stationary behavior by differencing the series with differences whose order is equal to the number of periods per season. Such differences

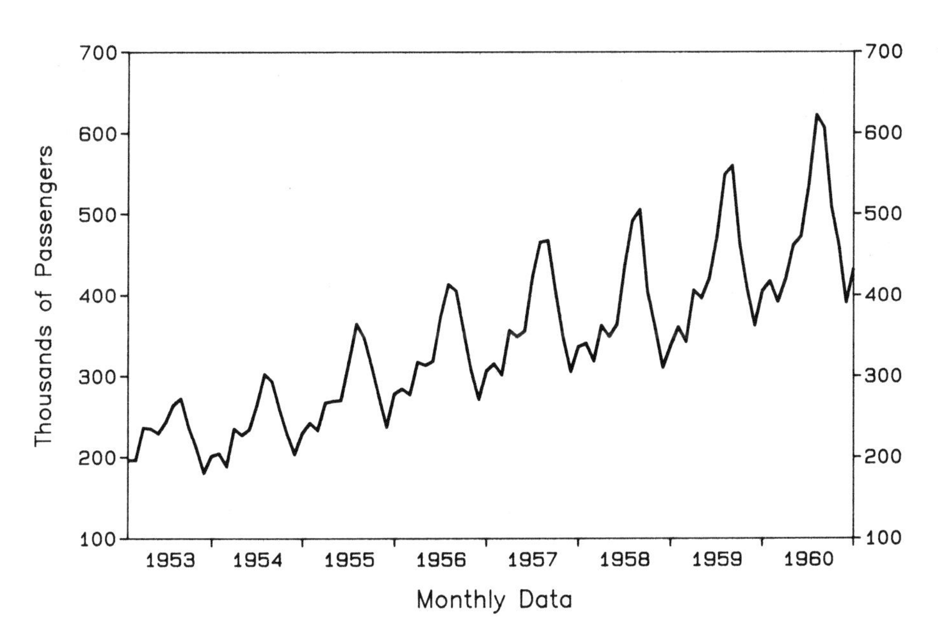

Figure 14.5 Airline Passengers: A Seasonal Series with Nonstationary Behavior in the Seasonal Pattern

are called *seasonal differences*. Thus one seasonal difference of order 12 would be computed as follows, where Z_t represents the differenced series:

$$Z_t = X_t - X_{t-12}, \qquad t = 13, 14, \ldots, N$$

For example, $Z_{13} = X_{13} - X_1$ and $Z_{14} = X_{14} - X_2$, etc.

A seasonal series will usually require both regular differencing and seasonal differencing. Table 14.1 shows an example of differencing a series with both one regular difference and one seasonal difference of order 4 (it doesn't matter which difference is applied first—the resulting differenced series is the same in either case). Note that no differences are computed for the first five periods in the final differenced series.

We now have two types of model-building blocks that are used to convert a nonstationary series into a stationary one: regular differences and seasonal differences. The following acronyms will be used from now on to refer to these building blocks:

RD Regular difference

SD Seasonal difference

The order of any RD is always one, and the order of any SD is always s, where s is the number of periods per season.

In Chapter 15, you will learn how to identify the need for seasonal differencing and how to select the correct number of seasonal differences. Most series will usually not require more than one RD or one SD to achieve stationarity. Only infrequently

Table 14.1 Differencing a Series with One Regular Difference and One Seasonal Difference of Order 4

Original Series	Series After One Seasonal Difference	Series After One Seasonal and One Regular Difference
110	—	—
115	—	—
90	—	—
130	—	—
140	30	—
150	35	5
120	30	−5
115	−15	−45
130	−10	5
135	−15	−5

are two RDs and/or two SDs required. More than two differences of either type are almost never required. In general, you should use as few differences as possible to achieve stationarity. Unnecessary differencing only complicates matters and decreases the amount of data you have to work with.

14.5 REVIEW OF KEY CONCEPTS

Seasonal behavior is an extremely common phenomenon in business and economic time series. Seasonal behavior simply means that a certain basic pattern tends to be repeated at regular intervals called *seasons*. The most common seasonal interval is one year. The number of time series periods within a season is called the *periods per season*. The basic seasonal pattern, however, usually undergoes constant change as time moves on. Models constructed for seasonal time series must be able to account for the basic season-to-season relationship as well as changes in seasonal behavior.

Purely seasonal models are Box-Jenkins models that model *only* the seasonal relationships in a series. These models are constructed with two new model-building blocks:

- *Seasonal autoregressive parameters* (SAR parameters) that are used to represent autoregressive relationships among series values separated by the number of periods per season.
- *Seasonal moving-average parameters* (SMA parameters) that are used to represent moving-average relationships among series values separated by the number of periods per season.

Purely seasonal models look almost exactly like the basic models described in Part 2, except that the order of each parameter in a purely seasonal model is some multiple of the number of periods per season. The purely seasonal models are written as follows, where s is the number of periods per season:

1. *Purely seasonal AR models.*

   $$X_t = A_1^* X_{t-s} + \dots + A_P^* X_{t-Ps} + E_t$$

   where X_t is directly related to one or more past series values that occurred a multiple of s periods ago.

2. *Purely seasonal MA models.*

   $$X_t = -B_1^* E_{t-s} - \dots + B_Q^* E_{t-Qs} + E_t$$

   where X_t is directly related to one or more past random errors that occurred a multiple of s periods ago.

3. *Purely seasonal ARMA models.*

   $$X_t = (A_1^*X_{t-s} + \ldots + A_P^*X_{t-Ps}) - (B_1^*E_{t-s} \ldots + B_Q^*E_{t-Qa}) + E_t$$

   where X_t is directly related to both past series values and past random errors that occurred a multiple of s periods ago.

The A_k^* are called seasonal autoregressive parameters and the B_k^* are seasonal moving-average parameters. The order of A_k^* or B_k^* *is ks*.

Combined regular and seasonal models are models that contain both seasonal parameters and the basic AR and MA parameters described in Part 2 for the basic models. These basic AR and MA parameters are generally called *regular parameters* (RAR and RMA parameters) to distinguish them from the seasonal parameters (SAR and SMA parameters). Most models for seasonal series will be combined with regular and seasonal models since purely seasonal models can only model season-to-season relationships and ignore any period-to-period relationships. Period-to-period relationships, of course, are addressed by the regular parameters. A combined regular and seasonal model may be constructed as either an additive seasonal model or a multiplicative seasonal model. These models are described as follows:

1. *Additive seasonal model.* The regular and seasonal terms are simply added together. For example, an additive model with one RAR and one SAR parameter is written as follows:

   $$X_t = A_1X_{t-1} + A_1^*X_{t-s} + E_t$$

2. *Multiplicative seasonal model.* The regular and seasonal terms are added together, and additional terms representing the interaction of the seasonal and period-to-period relationships are also included. For example, a multiplicative model with one RMA and one SMA parameter is written as follows:

   $$X_t = -B_1E_{t-s} - B_1^*E_{t-s} + B_1B_1^*E_{t-s-1} + E_t$$

   And a multiplicative model with one RAR and one SAR parameter is written as follows:

   $$X_t = A_1X_{t-1} + A_1^*X_{t-s} - A_1A_1^*X_{t-s-1} + E_t$$

   Most seasonal models for business and economic series are multiplicative.

An *integrated seasonal model* is a model for a nonstationary seasonal series. A seasonal series may be nonstationary in two ways:

1. On a period-to-period basis. This type of nonstationarity requires the application of regular differences (the RD building block) to achieve stationarity, as described in Chapter 12.
2. On a season-to-season basis. This type of nonstationarity requires the application of a new building block called seasonal differencing. A *seasonal dif-*

ference (SD building block) simply computes the differences between every two successive series values separated by the number of periods per season.

A nonstationary seasonal series may require both RDs and SDs to achieve stationarity, or it may just require one or the other. Rarely, if ever, will you need to use more than one or two of either type. If any differencing is used, the resulting integrated model is called a seasonal ARI, seasonal IMA, or seasonal ARIMA model, depending on the type of parameters included. Remember that if any differencing is used and the mean of the differenced (stationary) series is significantly nonzero, you will have to include a trend parameter also.

With the addition of the SAR, SMA, and SD building blocks, we now have the complete set of Box-Jenkins building blocks. With these building blocks you will be able to construct models for virtually any type of time series, whether stationary, nonstationary, or seasonal. The complete repertoire of Box-Jenkins model-building blocks is listed below:

RAR	Regular autoregressive parameter
RMA	Regular moving-average parameter
SAR	Seasonal autoregressive parameter
SMA	Seasonal moving-average parameter
RD	Regular difference
SD	Seasonal difference
Trend	Trend parameter

In the next chapter we will learn how to identify the need for seasonal parameters and combined regular and seasonal parameters. We will also learn how to determine the requirements for seasonal differencing.

CHAPTER 15

Identifying Seasonal Models

The identification of seasonal parameters, like the identification of regular parameters, is accomplished by examining the autocorrelation patterns (and partial-autocorrelation patterns) of an appropriately differenced (stationary) series.

In this chapter you will learn:

- What the theoretical AC and PAC patterns are for purely seasonal AR, MA, and ARMA models.
- What the theoretical AC and PAC patterns are for some combined regular and seasonal models.
- How to select the correct amount of regular and seasonal differencing for nonstationary seasonal series.
- How to read the sample AC and PAC for identifying seasonal models.

15.1 INTRODUCTION TO SEASONAL AC AND PAC PATTERNS

The autocorrelation patterns associated with purely seasonal models are analogous to those for nonseasonal models—the only difference is that the nonzero autocorrelations that form the pattern occur at lags that are multiples of the number of periods per season. In Chapter 8 we saw the autocorrelation patterns of some models that are actually purely seasonal models. For example, the AC pattern in Figure 8.3 is associated with a model that contains just one regular MA parameter of order 3; this model is the same as a purely seasonal MA model containing one SMA parameter of order 3 (see also the theoretical AC patterns in Examples 8.8 and 8.9).

Autocorrelation patterns for combined regular and seasonal models, whether additive or multiplicative, are a bit more complicated. For one thing, the individual

AC patterns associated with the regular and seasonal components of the model are present simultaneously. Second, additional nonzero autocorrelations appear as part of the total AC pattern for a combined model.

In this chapter you will learn to identify both purely seasonal models and combined regular and seasonal models. You can then test your identification skills with the examples given in Section 15.7.

15.2 THEORETICAL AC PATTERNS FOR PURELY SEASONAL MA MODELS

The autocorrelation patterns for models containing only SMA parameters are similar to those for regular MA models, except that the nonzero autocorrelations occur only at lags that are multiples of the number of periods per season. More precisely, if s is the number of periods per season, and the model contains exactly Q seasonal SMA parameters, then the autocorrelations associated with the model will be nonzero at lags s, $2s$, $3s$, . . . , sQ and zero elsewhere.

For example, in the seasonal MA model

$$X_t = -B_1^* E_{t-12} + E_t$$

it is clear that X_t is directly related to the random error that occurred 12 periods ago and hence to $X_{t-12} = -B_1^* E_{t-24} + E_{t-12}$; i.e., the autocorrelation at lag 12 is nonzero. On the other hand, $X_{t-1}, X_{t-2}, \ldots, X_{t-11}$ do not contain E_{t-12}, and neither does any other X_{t-i} for i greater than 12. Therefore the autocorrelation at any lag, except 12, is zero.

The correlograms in Figures 15.1 and 15.2 show the autocorrelation patterns for two simple seasonal MA models where $s = 12$. The first model contains one SMA parameter, and hence it shows one isolated spike at lag 12. The second model contains two SMA parameters, and hence it shows two isolated spikes at lags 12 and 24.

15.3 THEORETICAL AC PATTERNS FOR PURELY SEASONAL AR MODELS

The autocorrelation patterns for models containing only SAR parameters are similar to those for regular AR models, except that the only nonzero autocorrelations occur at lags s, $2s$, $3s$, . . . , where s is the number of periods per season. These autocorrelations will then exhibit the typical autoregressive autocorrelation patterns. For example, the correlogram in Figure 15.3 shows the autocorrelation pattern for a

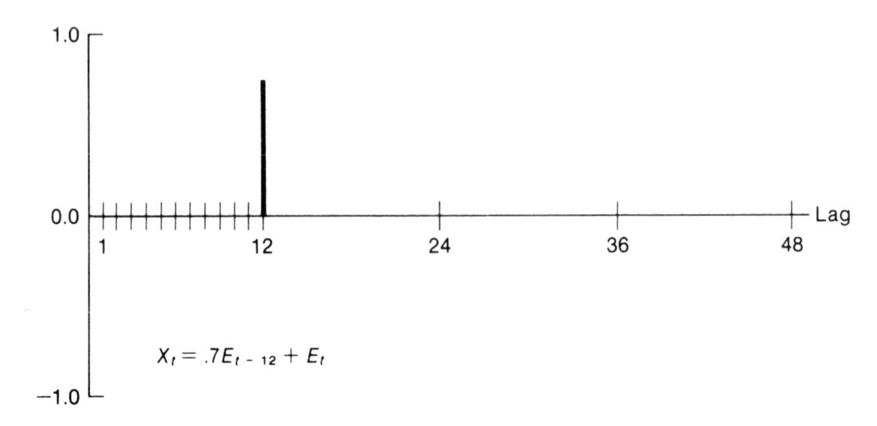

Figure 15.1 AC Pattern for One SMA Parameter of Order 12

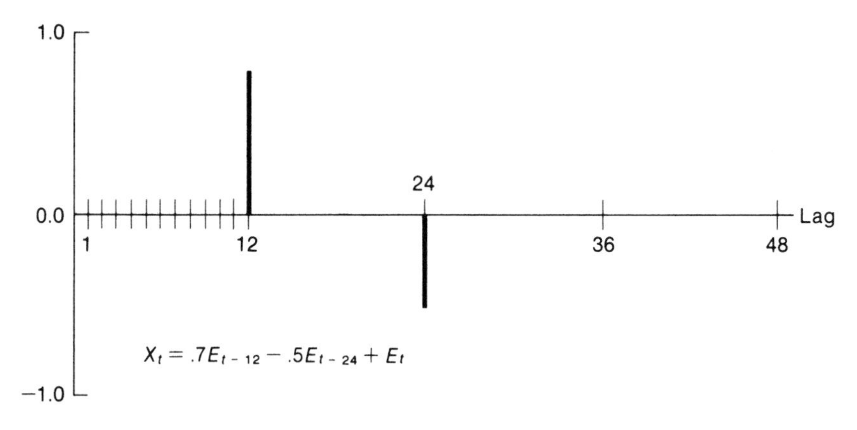

Figure 15.2 AC Pattern for Two SMA Parameters of Orders 12 and 24

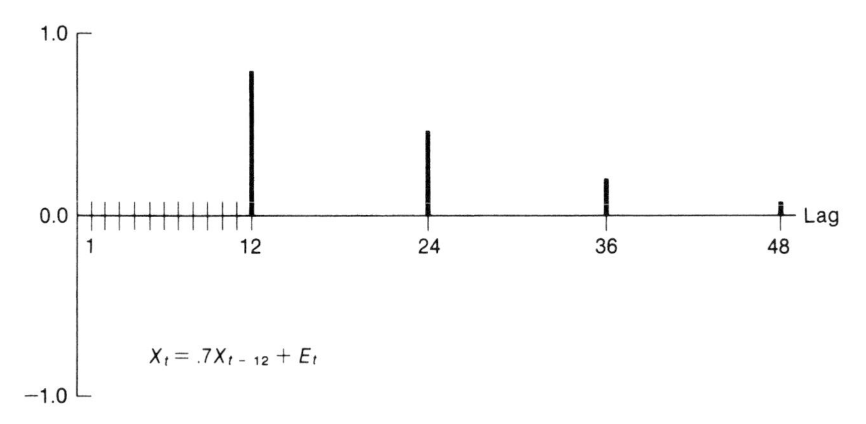

Figure 15.3 AC Pattern for One SAR Parameter of Order 12

seasonal model with one SAR parameter, where $s = 12$. Note that autocorrelation spikes occur only at lags 12, 24, 36, etc. The rapidly decreasing pattern of these spikes, however, is the same as the pattern generated by a regular AR model (see Figure 8.5 part (a))—the spikes are just 12 lags apart in the seasonal case.

15.4 THEORETICAL PAC PATTERNS FOR PURELY SEASONAL AR AND MA MODELS

Partial-autocorrelation patterns for purely seasonal models are also analogous to the partial-autocorrelation patterns for regular models. Thus the PAC patterns for purely seasonal MA models are the same as the AC patterns for purely seasonal AR models, and vice versa. Hence, just like their regular counterparts, PAC patterns for purely seasonal models are useful for identifying SAR parameters. The AC and PAC patterns shown in Figure 15.4 correspond to a model with one SMA parameter, and those in Figure 15.5 correspond to a model with one SAR parameter (compare these patterns with the AC and PAC patterns for regular models in Figure 8.7).

15.5 THEORETICAL AC AND PAC PATTERNS FOR PURELY SEASONAL ARMA MODELS

The AC and PAC patterns of mixed seasonal relationships, involving both SMA and SAR parameters, are combinations of the individual patterns already discussed. For example, Figure 15.6 shows the AC and PAC correlograms for a model containing one SMA and one SAR parameter.

15.6 THEORETICAL AC PATTERNS FOR COMBINED REGULAR AND SEASONAL MODELS

Most Box-Jenkins models for seasonal time series usually contain both regular and seasonal parameters; i.e., they are combined models. The AC and PAC patterns for combined models are slightly more complex than those for the regular models or purely seasonal models previously discussed. However, an easy way to get a feeling for the types of patterns associated with combined models is to look at a few examples.

The following examples represent some of the more common combined models you will encounter. Keep in mind, however, that the pattern shown in the example

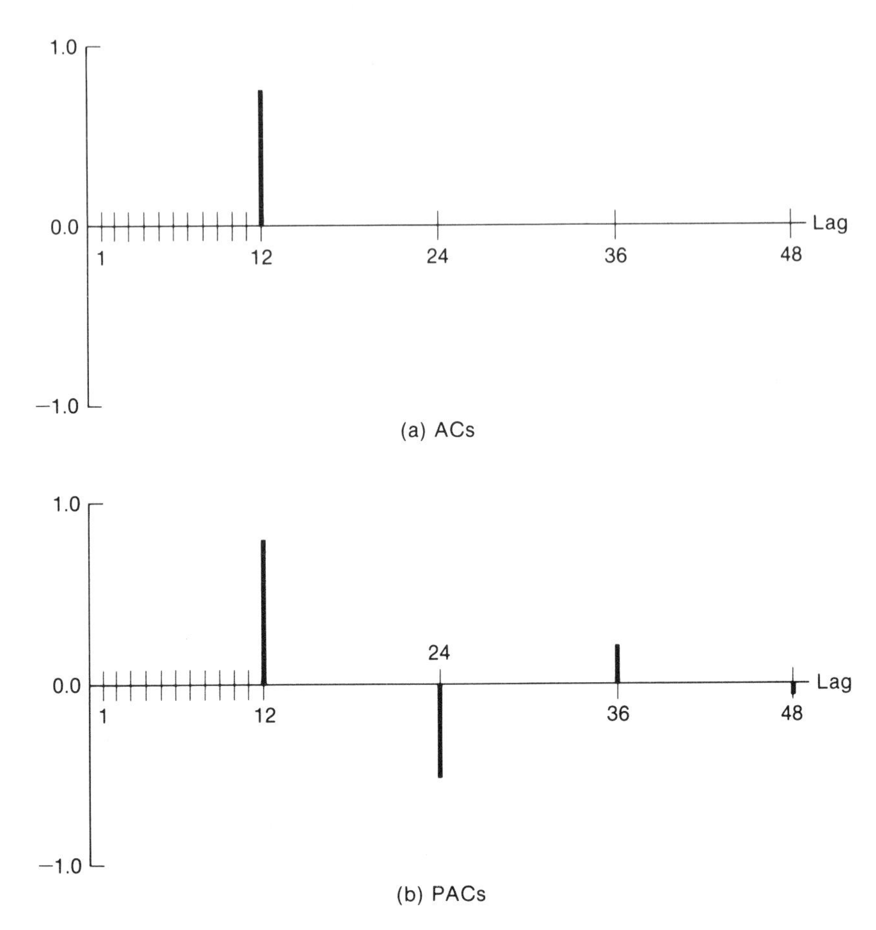

Figure 15.4 Comparison of AC and PAC Correlograms for a Model with One SMA Parameter

for a particular model is only one of several similar patterns possible for such a model. Several patterns are possible because of the many different combinations of positive and negative parameter values. All examples assume that the number of periods per season is 12. All examples, with the exception of the first, are assumed to be multiplicative.

Example 15.1: One RMA and One SMA (Additive)

A typical AC correlogram for this model is shown in Figure 15.7. The autocorrelations are nonzero only at lags 1, 11, and 12. The relatively large spikes at lags 1 and 12 are due to the one RMA and one SMA parameter, respectively. The nonzero autocorrelation at lag 11 is something new. It is due to the combined effect of the RMA and SMA.

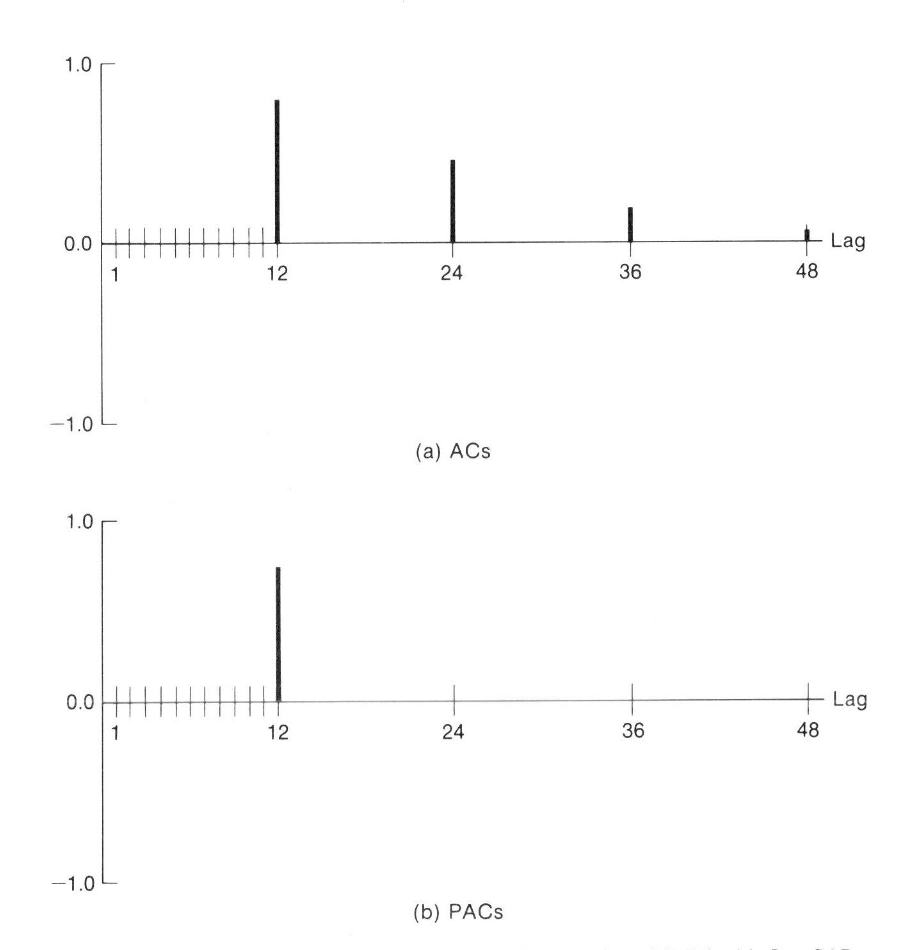

Figure 15.5 Comparison of AC and PAC Correlograms for a Model with One SAR Parameter

To see why this is so, consider this model written in equation form:

$$X_t = - B_1 E_{t-1} - B_1^* E_{t-12} + E_t$$

The series value 11 periods ago is then represented as

$$X_{t-11} = -B_1 E_{t-12} - B_1^* E_{t-23} + E_t$$

which contains a term with E_{t-12}. However, X_t also contains E_{t-12} and must thus be directly related to X_{t-11}. This result means, of course, that the autocorrelation at lag 11 is nonzero.

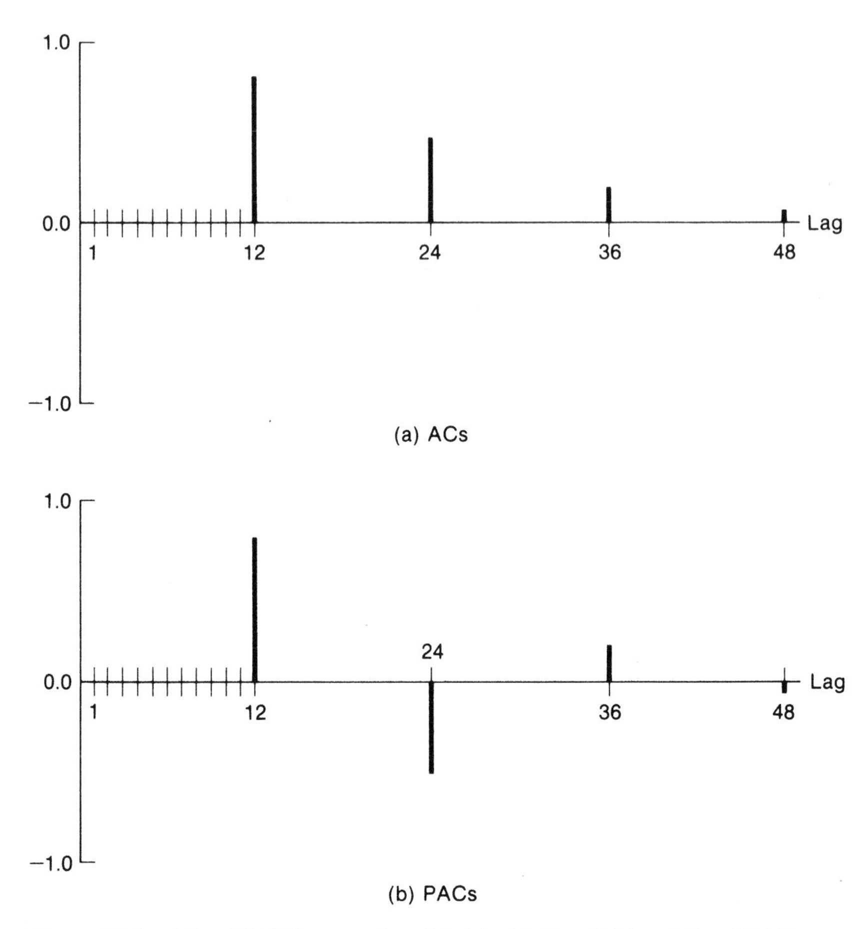

Figure 15.6 AC and PAC Patterns for a Model with One SMA and One SAR Parameter

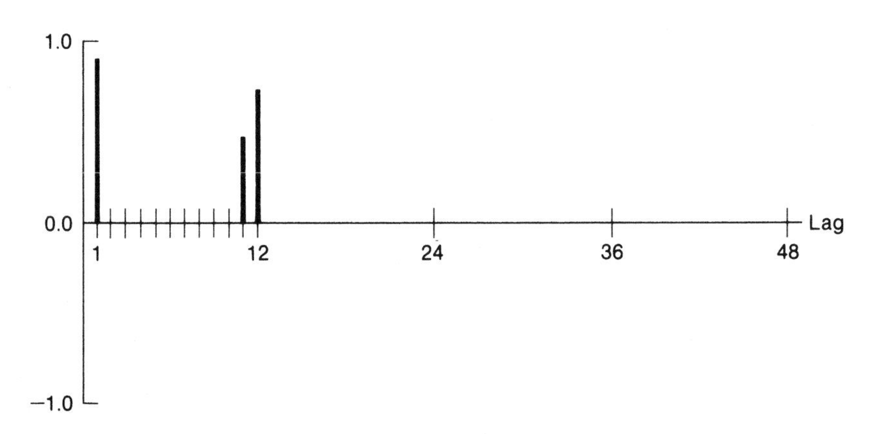

Figure 15.7 AC Correlogram for a Model with One RMA and One SMA (Additive)

Example 15.2: One RMA and One SMA (Multiplicative)

A typical AC correlogram for this model is shown in Figure 15.8. As in Example 15.1, the nonzero autocorrelations at lags 1 and 12 are due to the RMA and SMA parameters, respectively. The autocorrelations at lags 11 and 13 are due to the combined effect of the RMA and SMA. The difference between this example and Example 15.1 is the appearance of a new nonzero autocorrelation at lag 13. This AC is due to the fact that the model is multiplicative. To understand why, recall that a multiplicative model gives rise to additional terms in the model equation. The one-RMA and one-SMA multiplicative model is thus written as follows:

$$X_t = -B_1 E_{t-1} - B_1^* E_{t-12} + B_1 B_1^* E_{t-13} + E_t$$

The additional term involving E_{t-13} is the cause of the nonzero autocorrelation at lag 13.

Note also the apparent equal size of the autocorrelations at lags 11 and 13. This equal size is, in fact, the case. If we use the notation AC(n) to denote the autocorrelation value at lag n, then it can be shown that AC(11) = AC(13) = AC(1) times AC(12).

Example 15.3: Two RMAs and One SMA

A typical AC correlogram for this model is shown in Figure 15.9. The autocorrelations are nonzero at lags 1, 2, 10, 11, 12, 13, and 14. The large spikes at lags 1, 2, and 12 are due to the two RMAs and one SMA. The combined effects of the two RMA and one SMA parameters produce the smaller nonzero autocorrelations at lags 10, 11, 13, and 14. Those at lags 11 and 13 are due to the interaction of the first RMA (order 1) and the SMA, and those at lags 10 and 14 are due to the interaction of the second RMA (order 2) and the SMA. Note that AC(10) = AC(14) and AC(11) = AC(13).

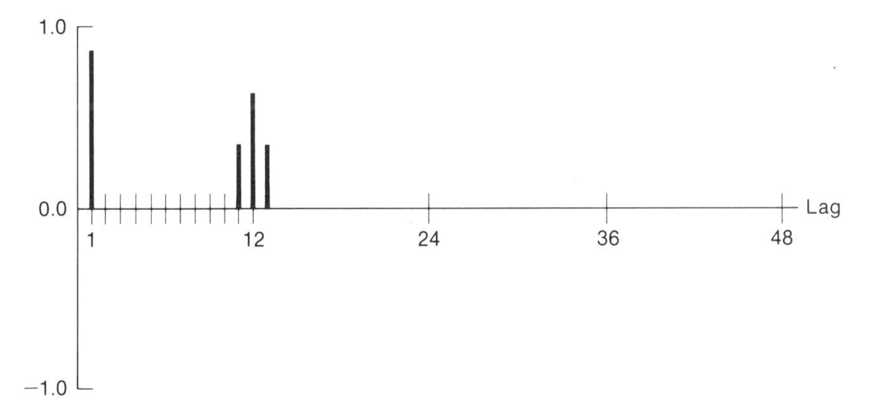

Figure 15.8 AC Correlogram for a Model with One RMA and One SMA (Multiplicative)

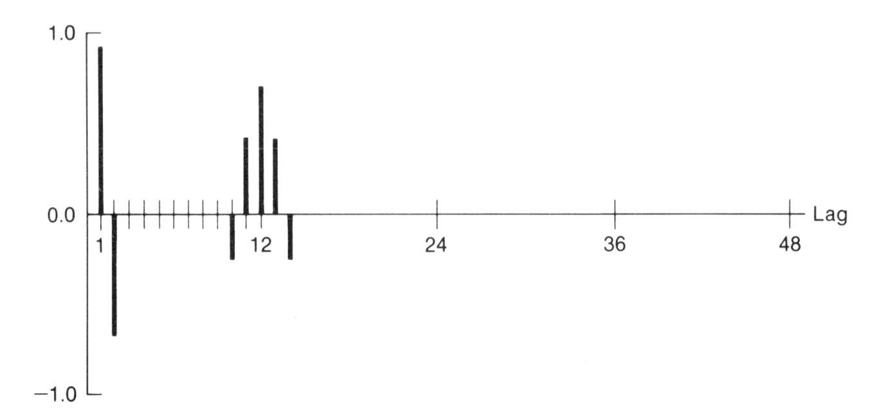

Figure 15.9 AC Correlogram for a Model with Two RMAs and One SMA

Example 15.4: One RMA and Two SMAs

A typical AC correlogram for this model is shown in Figure 15.10. The autocorrelations are nonzero at lags 1, 11, 12, 13, 23, 24, and 25. The large spikes at lags 1, 12, and 24 are due to the RMA and two SMAs. The combined effects of the one regular and two seasonal MA parameters produce the smaller autocorrelations at lags 11, 13, 23 and 25. Those at lags 11 and 13 are due to the interaction of the RMA and the first SMA (order 12), and those at lags 23 and 25 are due to the interaction of the RMA and second SMA (order 24). Note that AC(11) = AC(13) and AC(23) = AC(24).

Example 15.5: Two RMAs and Two SMAs

A typical AC correlogram for this model is shown in Figure 15.11. The autocorrelations are nonzero at lags 1, 2, 10 through 14, and 22 through 26. The large spikes at lags 1, 2, 12, and 24 are due to the individual RMA and SMA. The combined effects of the two regular and two seasonal MA parameters produce the smaller nonzero autocorrelations at lags 10, 11, 13, 14, 22, 23, 25, and 26. The first RMA and first SMA produce the ACs at lags 11 and 13; the first RMA and second SMA at lags 23 and 25; the second RMA and first SMA at lags 10 and 14; and the second RMA and second SMA at lags 22 and 26. As usual, AC(11) = AC(13), AC(10) = AC(14), AC(23) = AC(25), and AC(22) = AC(26).

Example 15.6: One RMA, One SMA, and One SAR

A typical AC correlogram for this model is shown in Figure 15.12. The autocorrelations are nonzero at lags 1, 11, 12, 13, 23, 24, 25, 35, 36, 37, etc. More precisely, they are nonzero at lag 1 and at lags $12k - 1$, $12k$, and $12k + 1$, for $k = 1, 2, 3, \ldots$. The large spikes at lags 1 and 12 are due to the one RMA and one SMA parameter, respectively. An exponentially decreasing pattern occurs at

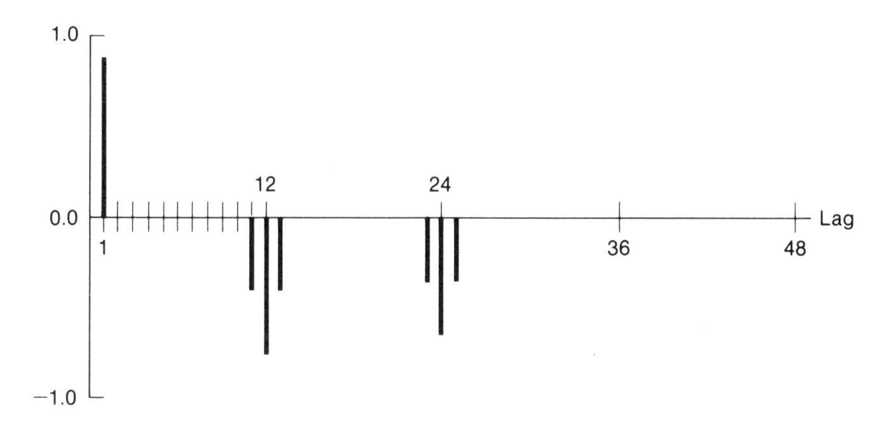

Figure 15.10 AC Correlogram for a Model with One RMA and Two SMAs

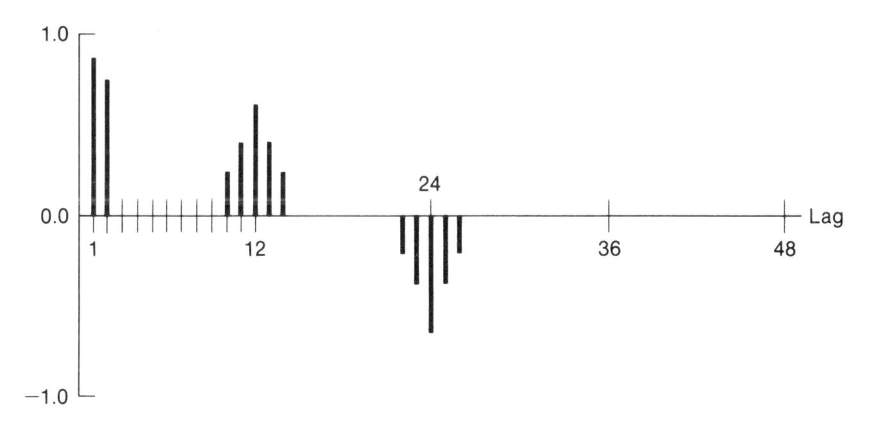

Figure 15.11 AC Correlogram for a Model with Two RMAs and Two SMAs

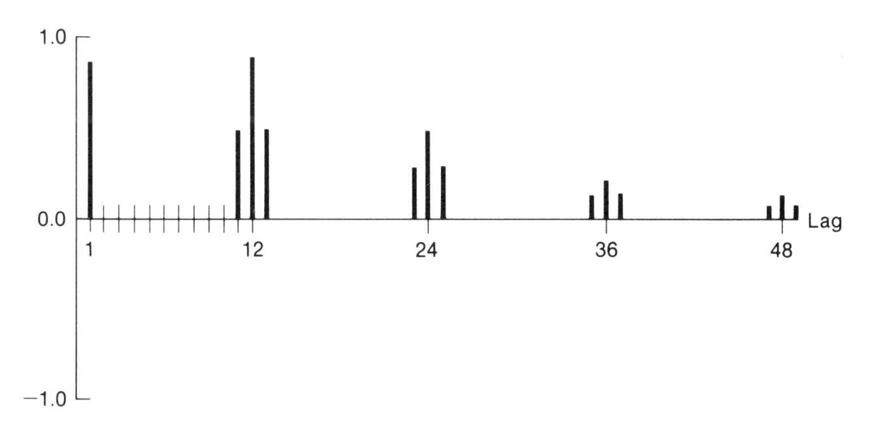

Figure 15.12 AC Correlogram for a Model with One RMA, One SMA, and One SAR

lags 12, 24, 36, . . . , which reflects the presence of the SAR parameter. The combined effects of the RMA and the SAR parameters also produce nonzero autocorrelations at lags $12k - 1$ and $12k + 1$, for $k = 1, 2, \ldots$. These effects produce an exponentially decreasing pattern at lags 11, 23, 35, . . . , and a similar one at lags 13, 25, 37, ...

15.7 SAMPLE ACs FOR SEASONAL MODELS—SOME EXAMPLES TO TEST YOURSELF

The sample ACs of the stationary series are, of course, the only means you have for identifying the seasonal AC patterns. As you can appreciate, when seasonal relationships are involved, it can be a little more difficult to make the proper interpretation of the AC pattern, especially if the number of periods per season is small (e.g., four quarters per year). As always, though, you can entertain a number of different possible models and then use the estimation phase to select the appropriate one. The following observations, however, can make it easier for you to identify the correct model:

1. You should always look at a minimum of $3s$ autocorrelations (preferably more), where s is the number of periods per season. All these autocorrelations are needed in order to get a good reading of the seasonal AC patterns.
2. You should generally assume a multiplicative seasonal model, because the sample autocorrelations generated by the combined effects of regular and seasonal relationships are usually not helpful in determining whether the model is additive or multiplicative. Those autocorrelations that should theoretically be of equal size, if the model were multiplicative, usually are not, and often one of them appears to be essentially zero (which would indicate an additive model). Concentrate on the first few lags to determine AC patterns for regular parameters, and concentrate on lags that are multiples of the number of periods per season for determining seasonal parameters. Since additive models are rare, the multiplicative assumption is usually a safe one.*
3. You should ignore large autocorrelations at lags that are inappropriate for most economic and business time series. Thus a large autocorrelation at lag 7 for a monthly time series would generally have no practical significance in real life; i.e., seasonality over a 7-month "season" is improbable.

*Note that most Box-Jenkins programs assume a multiplicative model when you specify a parameter as a seasonal parameter. To specify an additive seasonal model, you would have to specify the seasonal parameter(s) as a regular parameter(s) of order equal to some multiple of the number of periods per season.

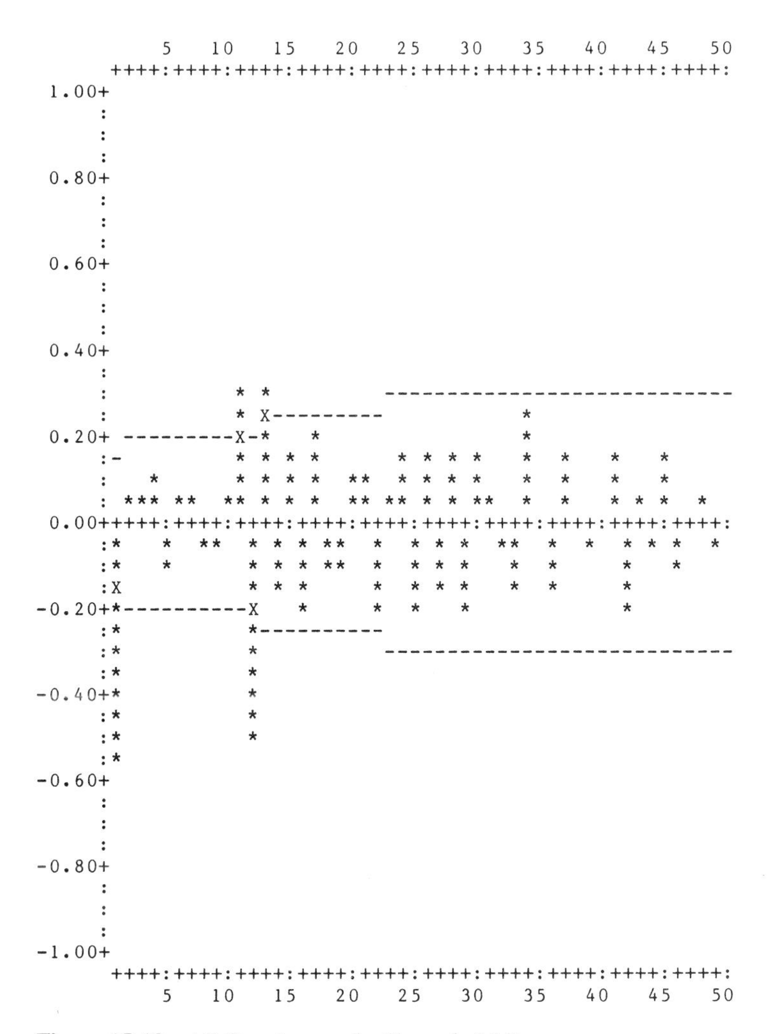

Figure 15.13 AC Correlogram for Example 15.7

With these observations in mind, try to identify the seasonal model suggested by the correlogram in the following five examples. The answers are given in Appendix G.

Example 15.7

Identify the seasonal model for the correlogram in Figure 15.13.

Example 15.8

Identify the seasonal model for the correlogram in Figure 15.14.

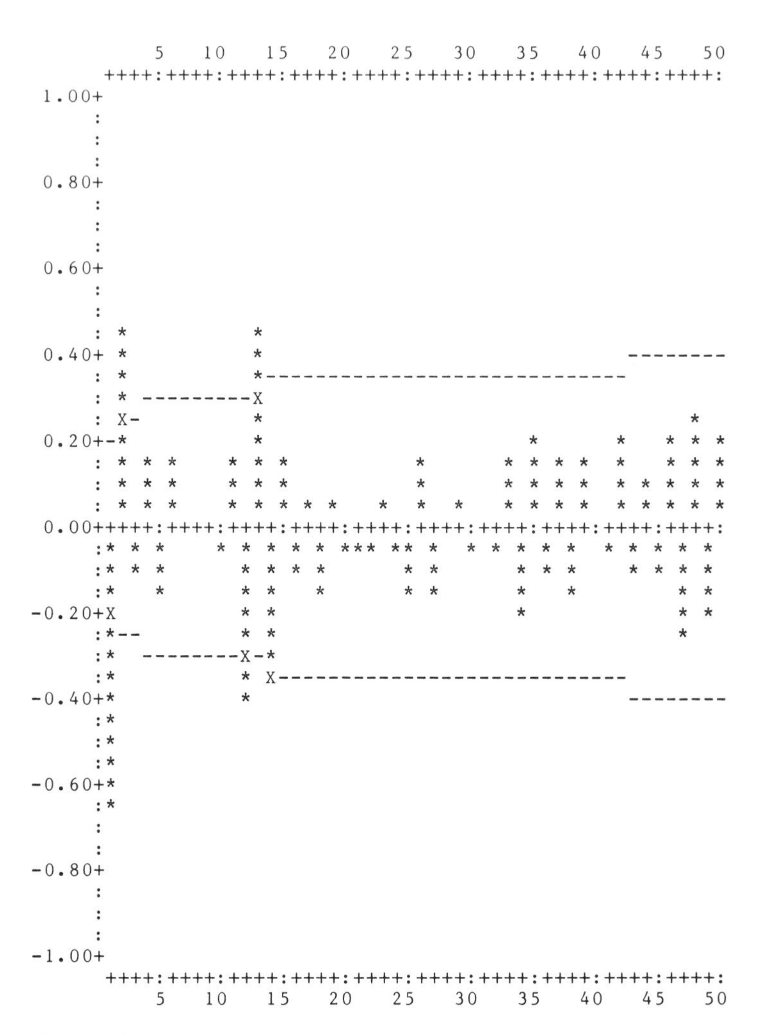

Figure 15.14 AC Correlogram for Example 15.8

Example 15.9

Identify the seasonal model for the correlogram in Figure 15.15.

Example 15.10

Identify the seasonal model for the correlogram in Figure 15.16.

Example 15.11

Identify the seasonal model for the correlogram in Figure 15.17.

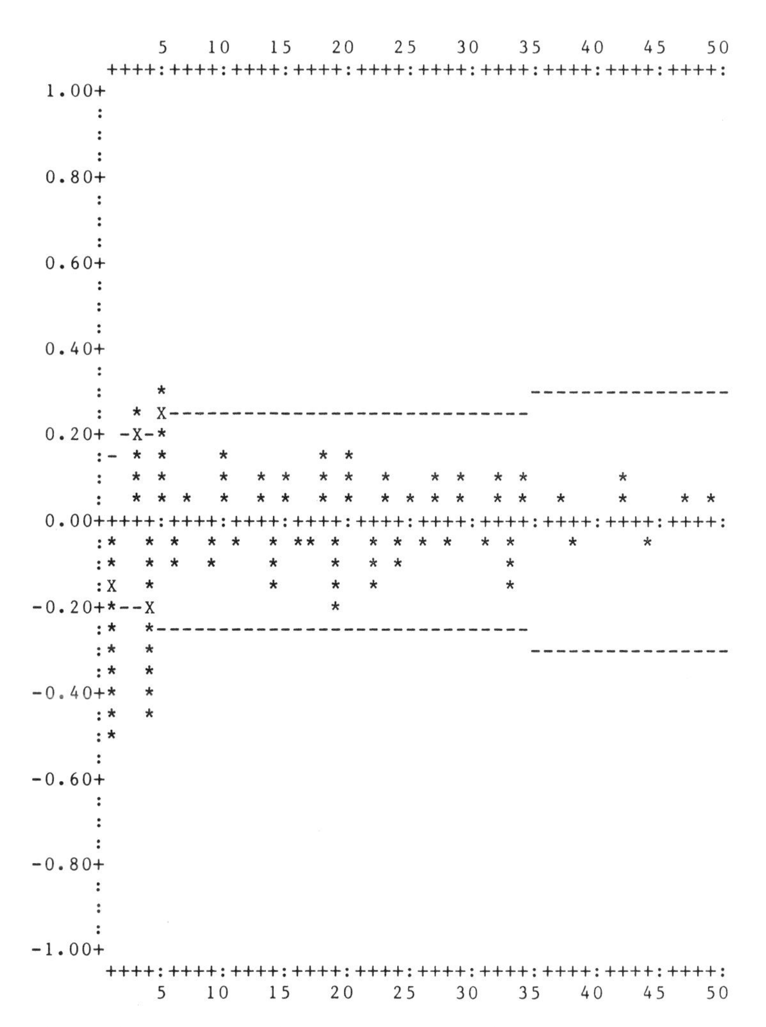

Figure 15.15 AC Correlogram for Example 15.9

15.8 SELECTING THE CORRECT AMOUNT OF SEASONAL DIFFERENCING

If the seasonal pattern in a series is nonstationary, then you will need to use seasonal differencing to achieve stationarity. The need for seasonal differencing can be determined by again looking at the AC patterns of the series. Just as you look for seasonal AR and MA autocorrelation patterns by examining only the autocorrelations at lags that are multiples of the number of periods per season, you also examine the same

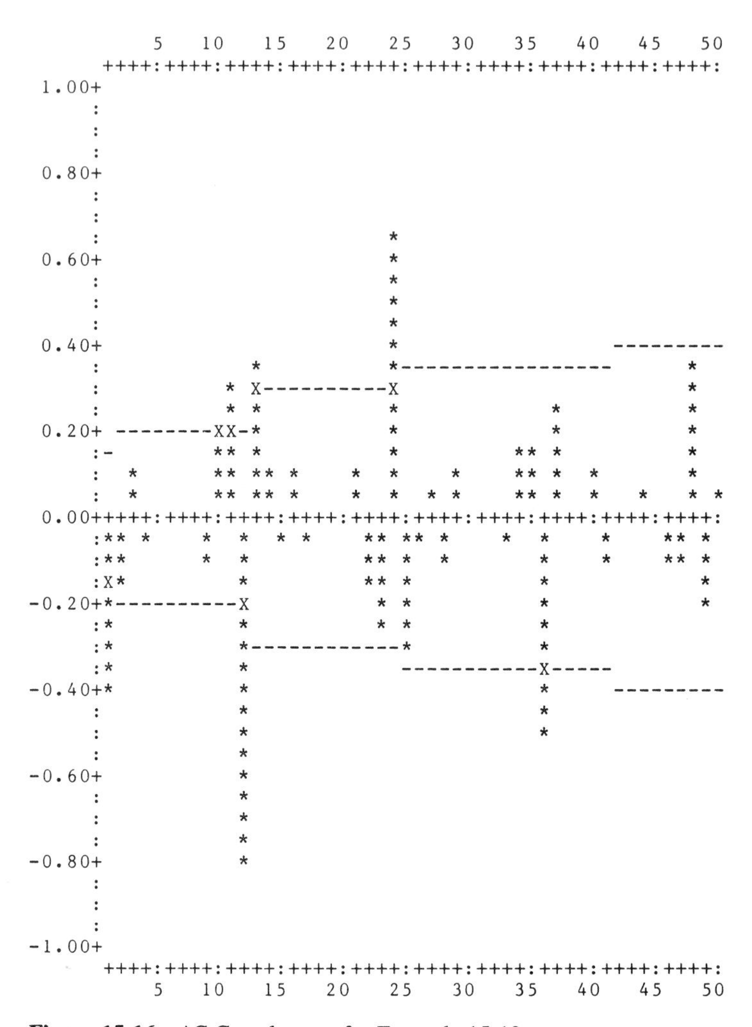

Figure 15.16 AC Correlogram for Example 15.10

autocorrelations to determine whether you need seasonal differencing. Seasonal differencing is required if these autocorrelations tend to remain large at longer lags (although they will generally decrease *slowly* in magnitude). Figure 15.18 shows a sample AC pattern indicating the need for seasonal differencing.

Usually, a series that requires seasonal differencing will also require regular differencing. In this case the need for seasonal differencing often is not apparent until the regular differencing has been applied. The sequence of correlograms in Figure 15.19 illustrates this situation where a stationary series is obtained through successive applications of both regular and seasonal differencing.

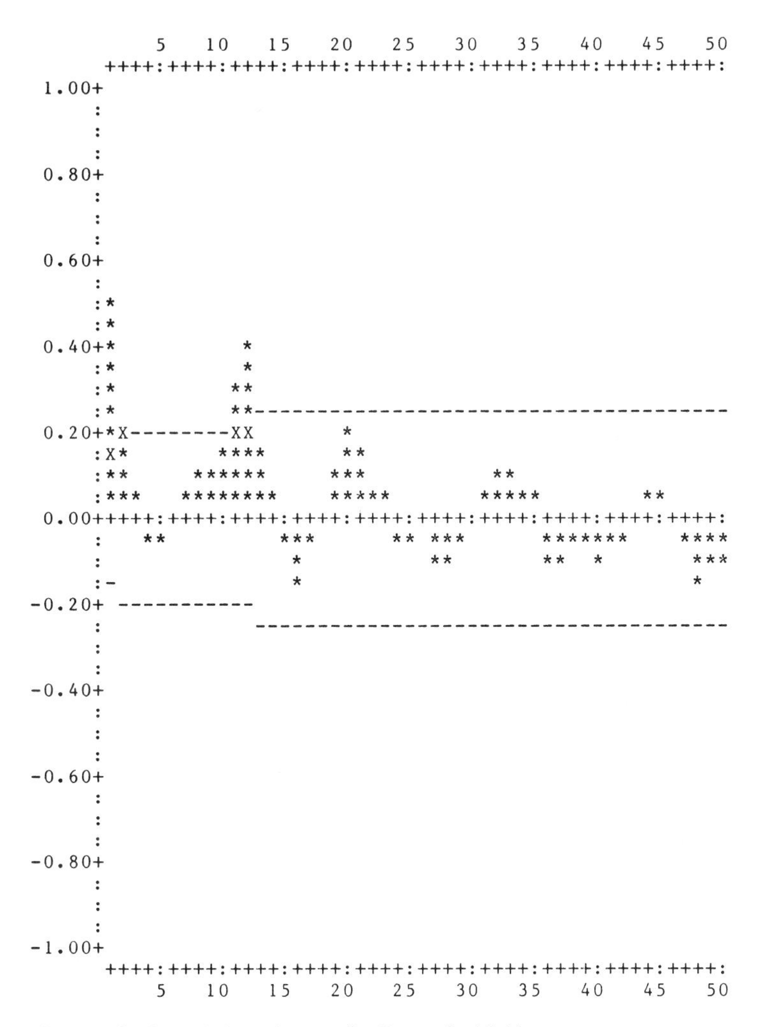

Figure 15.17 AC Correlogram for Example 15.11

The correlogram in part (a) shows the ACs of an undifferenced series. Note that there is no hint of seasonal patterns. The heavy, slowly decreasing autocorrelation pattern, however, indicates the need for at least one regular difference.

The correlogram in part (b) shows the ACs of the same series after applying one regular difference. Much of the heavy pattern has now been eliminated, which suggests that no further regular differencing is needed. A seasonal AC pattern, though, has emerged at lags 12, 24, 36, etc. These autocorrelations remain quite large and decrease slowly as the lag increases. This result suggests that at least one seasonal difference of order 12 is required.

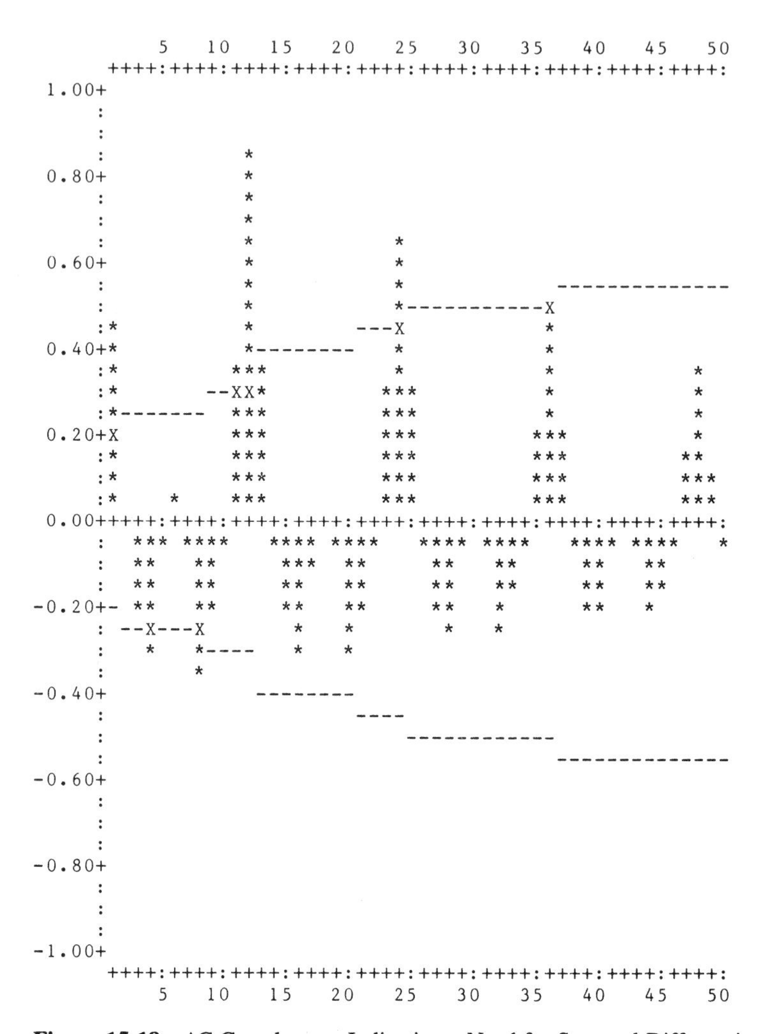

Figure 15.18 AC Correlogram Indicating a Need for Seasonal Differencing

The correlogram of part (c) shows the correlogram of the same series after both one RD and one SD have been applied. Here most autocorrelations are now not significantly large, which indicates that stationarity has been achieved. The remaining autocorrelation pattern demonstrates the need for one MA parameter of order 12, as indicated by the significant spike at lag 12.

On the other hand, the seasonal behavior in a nonstationary series may be so strong that the AC pattern of the original undifferenced series will provide an indication that seasonal differencing may be required, despite the need for regular differencing. Figure 15.20 shows an example of such an AC correlogram. Note the

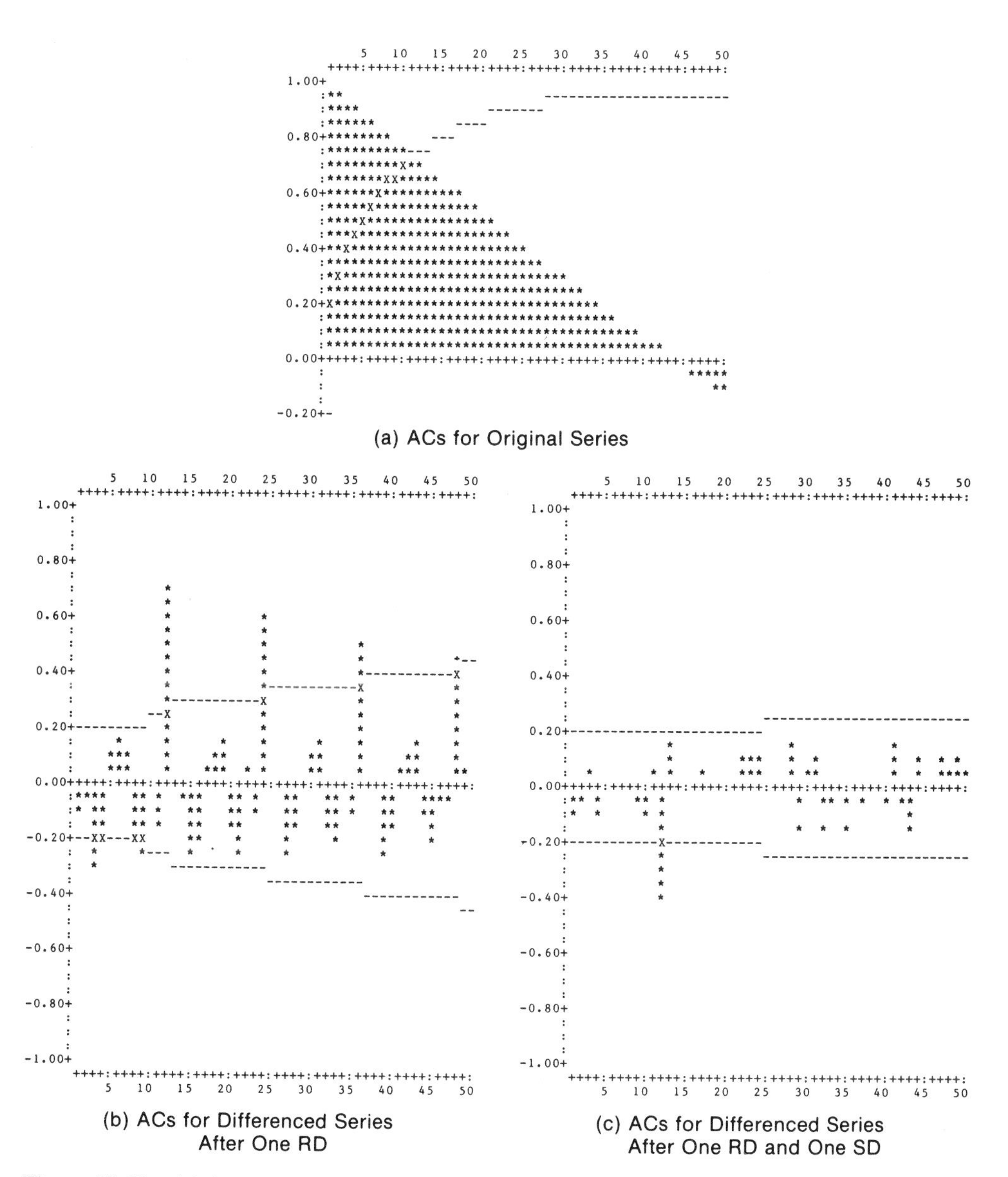

(a) ACs for Original Series

(b) ACs for Differenced Series After One RD

(c) ACs for Differenced Series After One RD and One SD

Figure 15.19 AC Correlogram for a Nonstationary Seasonal Series and Its Differences

stair step effect, where each step is 12 lags long, indicating the presence of seasonal behavior with periods per season of 12.

In this situation it doesn't matter which difference is applied first: The resulting AC correlogram of the differenced series after applying one type of difference should still indicate the need for the other type. Figure 15.21 shows the AC correlogram

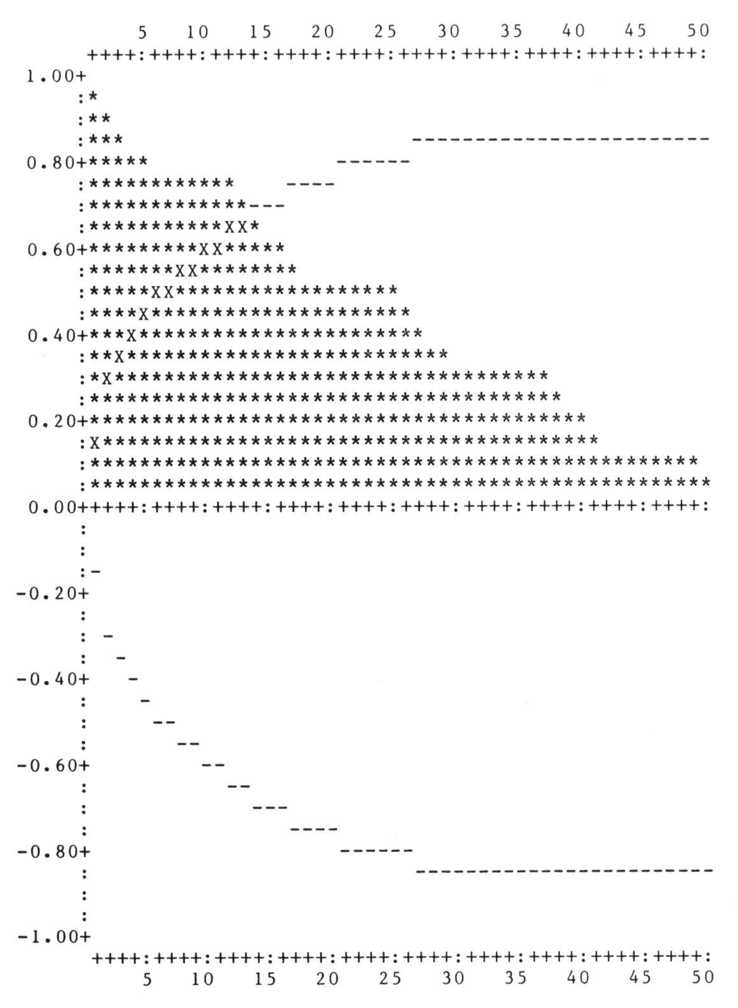

Figure 15.20 AC Correlogram for a Nonstationary Series Indicating a Need for Regular and Seasonal Differencing

obtained after first applying one seasonal difference to the series associated with the AC pattern in Figure 15.20 (note that the large alternating "clumps" of autocorrelation indicate the need for regular differencing). And Figure 15.22 shows the AC correlogram obtained after first applying one regular difference (note that the large spikes at lags 12, 24, etc. indicate the need for a further seasonal difference).

The AC correlogram for the same series after applying one RD and one SD (in either order) is shown in Figure 15.23. In Chapter 16 you will see some additional examples illustrating the selection of differences in the construction of models for some real-life series.

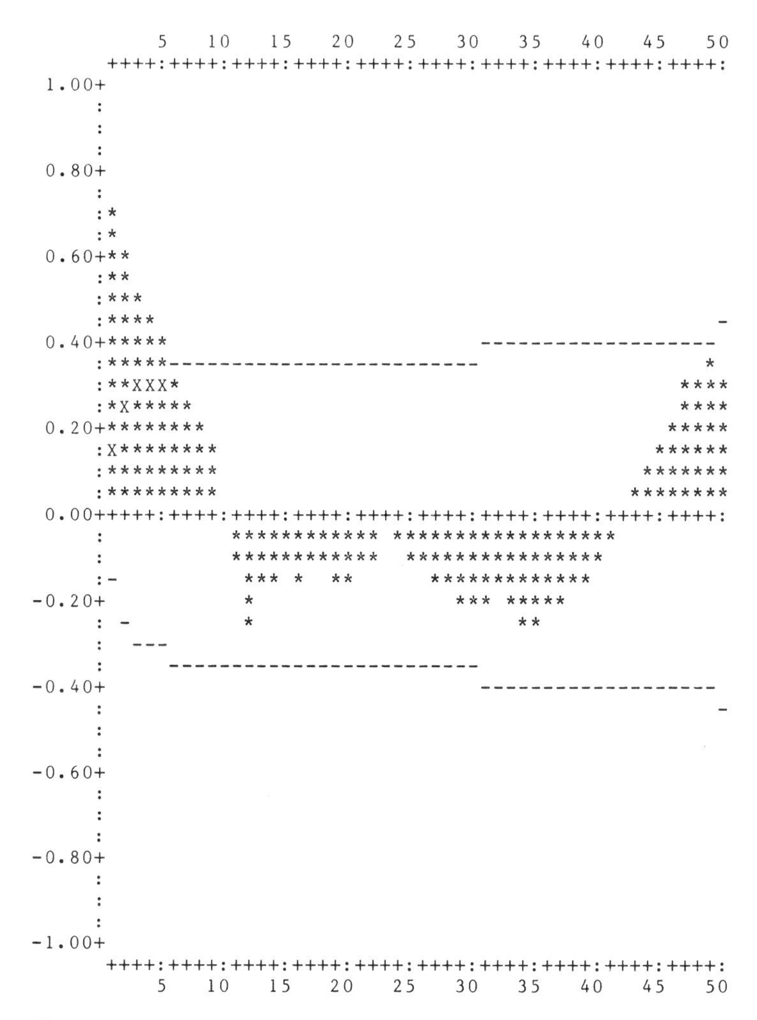

Figure 15.21 AC Correlogram After Applying One SD First

15.9 REVIEW OF KEY CONCEPTS

You can identify *the need for seasonal AR and MA parameters* by examining only the autocorrelations and partial-autocorrelations of a stationary series at lags that are multiples of the number of periods per season. Thus SAR and SMA parameters are required if the ACs and PACs at lags s, $2s$, $3s$, etc., where s is the number of periods per season, are nonzero and exhibit the typical patterns associated with AR and MA parameters. A summary of the theoretical AC and PAC patterns associated

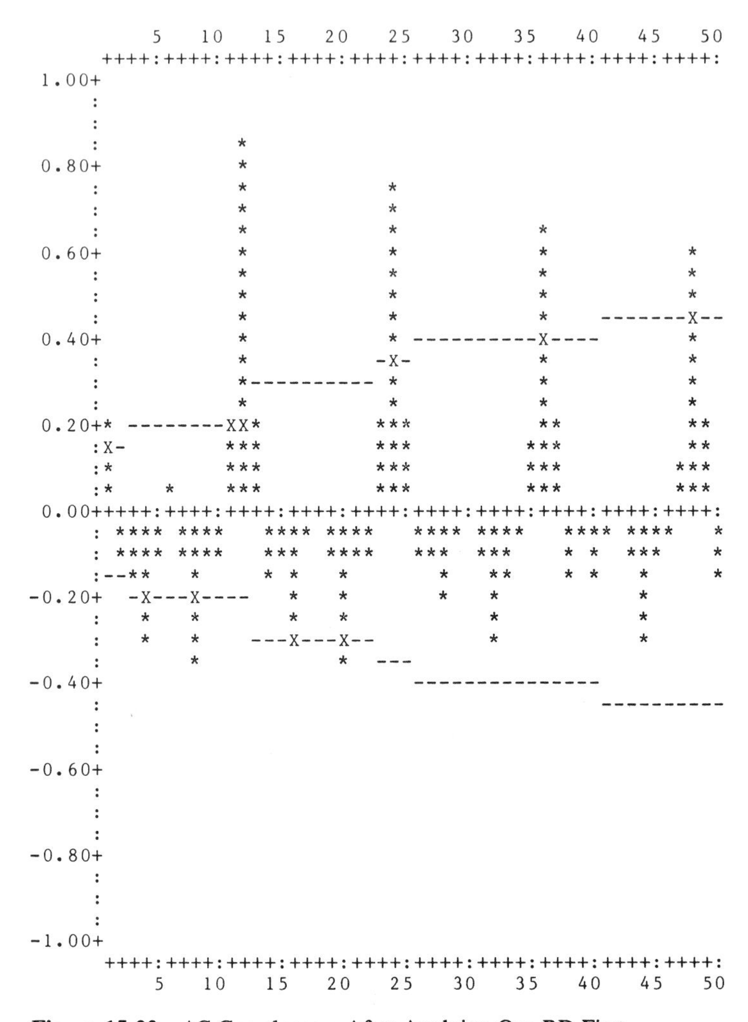

Figure 15.22 AC Correlogram After Applying One RD First

with both regular and *purely seasonal AR, MA, and ARMA models* is shown in Table 15.1.

The *need for both regular and seasonal parameters* is almost always the case in a seasonal model; i.e., you will generally end up with a *combined regular and seasonal model* for a seasonal series. The AC and PAC patterns for combined models have the following properties:

- The AC (or PAC) patterns associated with both the regular and seasonal parameters individually appear simultaneously. For example, if a series requires a model with one RMA and one SMA of order 12, then the AC pattern for the series will show large autocorrelations at lags 1 and 12.

Table 15.1 Summary of AC and PAC Patterns for Regular and Purely Seasonal Models

Type of Model	Number and Orders of Parameters	AC Pattern	PAC Pattern
Regular MA	q parameters of orders 1, 2, . . . , q	Spikes at lags 1 through q, zeros elsewhere	Spikes decreasing exponentially beginning at lag 1; if q is greater than 1, damped sine waves may be superimposed on the pattern
Regular AR	p parameters of orders 1, 2, . . . , p	Spikes decreasing exponentially beginning at lag 1; if p is greater than 1, damped sine waves may be superimposed on the pattern	Spikes at lags 1 through p, zeros elsewhere
Regular ARMA	q MA and p AR parameters of orders 1, 2, . . . , q and 1, 2, . . . , p, respectively	Irregular pattern of spikes at lags 1 through q; remaining autocorrelations as in regular AR	Same as AC pattern, with irregular pattern of spikes at lags 1 through p
Purely seasonal MA	Q parameters of orders s, $2s$, . . . , Qs	Spikes at lags s, $2s$, . . . , Qs, zeros elsewhere	Spikes decreasing exponentially at lags that are multiples of s, beginning at lag s
Purely seasonal AR	P parameters of orders s, $2s$, . . . , Ps	Spikes decreasing exponentially at lags that are multiples of s, beginning at lag s	Spikes at lags s, $2s$, . . . , Ps, zeros elsewhere
Purely seasonal ARMA	Q MA and P AR parameters of orders s, $2s$, . . . , Ps and s, $2s$, . . . , Qs, respectively	Irregular pattern of spikes at lags 1 through Q; remaining autocorrelations as in seasonal AR	Same as AC pattern, with irregular pattern of spikes at lags 1 through P

Note: When regular and seasonal parameters are combined in a model, additional nonzero autocorrelations are produced.

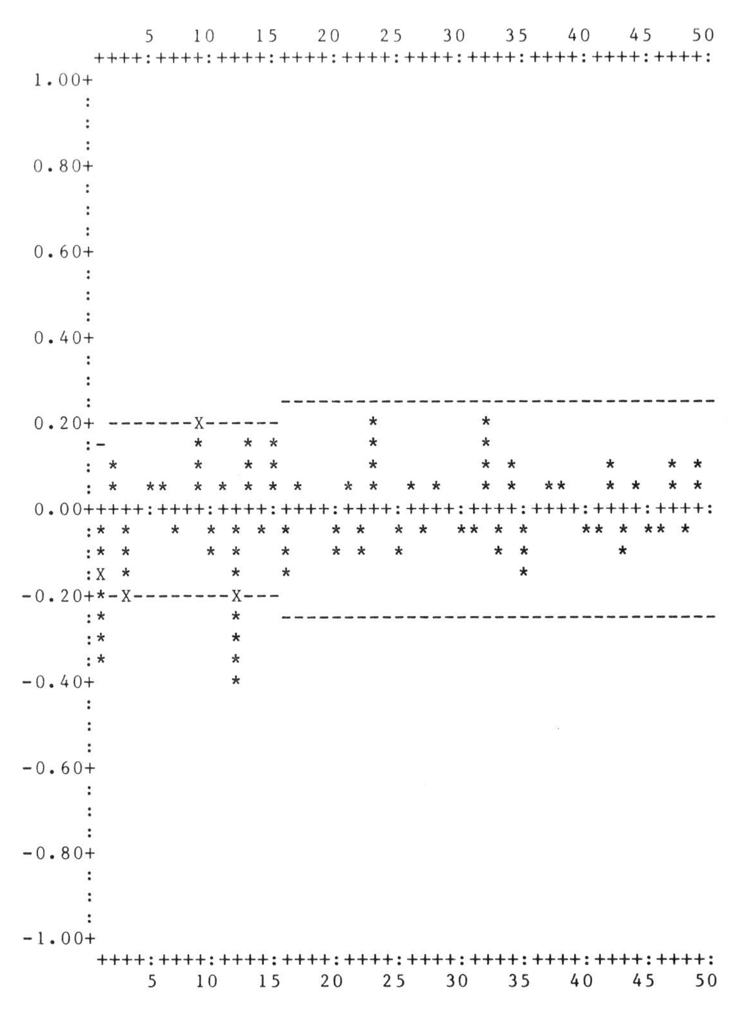

Figure 15.23 AC Correlogram for Differenced Series After Applying One RD and One SD

- Additional nonzero, but smaller, autocorrelations appear on either side of the nonzero autocorrelations associated with the seasonal parameters. The presence of these autocorrelations is due to the interaction of the regular and seasonal relationships in the series.
- If the additional autocorrelations occur on both sides of the seasonal autocorrelation, the model is multiplicative; if they occur only on the left side, the model is additive.
- Usually it is difficult to recognize the correct additional autocorrelation patterns in the sample AC of a series. Fortunately, almost all seasonal models are multiplicative, and you can generally assume this is the case.

Most Box-Jenkins programs assume your model is multiplicative when you specify both regular and seasonal parameters. In fact, to specify an additive model, you must specify all your seasonal parameters as regular parameters with orders equal to some appropriate multiple of the number of periods per season.

The need for seasonal differencing is indicated when the autocorrelations at lags s, $2s$, $3s$, etc. remain relatively large; i.e., they do not decrease rapidly. Usually, both regular and seasonal differencing will be required for most seasonal economic time series. In this case the need for seasonal differencing is sometimes not apparent until regular differencing has been applied. Thus you should generally apply regular differencing first before looking for seasonal nonstationarity.

In summary, when you are identifying models for seasonal series, keep the following points in mind:

- You should examine at least $3s$ to $4s$ autocorrelations, where s is the number of periods per season.
- You can generally assume that a combined regular and seasonal model is multiplicative. Don't get too carried away looking for the equal-sized "side" autocorrelations.
- Concentrate just on lags that are multiples of the number of periods per season for identifying the seasonal parameters.
- You should place less emphasis on significant sample autocorrelations that occur at lags that have no apparent practical significance in real life (e.g., a large autocorrelation at lag 7 in a monthly series).
- You should place greater emphasis on autocorrelations that are almost, or just barely, significant when they occur at lags equal to multiples of the number of periods per season.
- You will seldom need to apply more than one RD and one SD.
- You will seldom require more than one or two regular parameters and one or two seasonal parameters for a seasonal model.

CHAPTER 16

Computing and Validating Some Seasonal Models: Eight Examples

In this chapter you will see examples of building models for some real-life seasonal time series. These models represent some of the more common types of Box-Jenkins models for seasonal economic time series.

As you go through the examples, keep in mind some of the objectives and caveats in building Box-Jenkins models mentioned in earlier chapters:

- Don't overdifference the series.
- Use as few parameters as possible.
- Don't attach undue importance to sample autocorrelations at lags that have no practical meaning for the series (e.g., lag 7 for a monthly series).
- Place more importance on autocorrelations at the first few lags and lags that are multiples of the periods per season.
- Use the autocorrelation confidence limits only as guides, not as a rigid test for the inclusion or exclusion of parameters in a model.
- Remember that combined regular and seasonal models are almost always multiplicative.

Following these guidelines will help you build better models and identify models easier in the long run.

16.1 MODELS FOR REAL-LIFE SEASONAL SERIES

Example 16.1: Printing and Writing Paper Industry Sales (France)

This series represents the sales revenue, in thousands of francs, of the French printing and writing paper industry [27]. The complete series runs from January 1963 to December 1972 (120 periods). A graph of the last six years is shown in Figure 16.1.

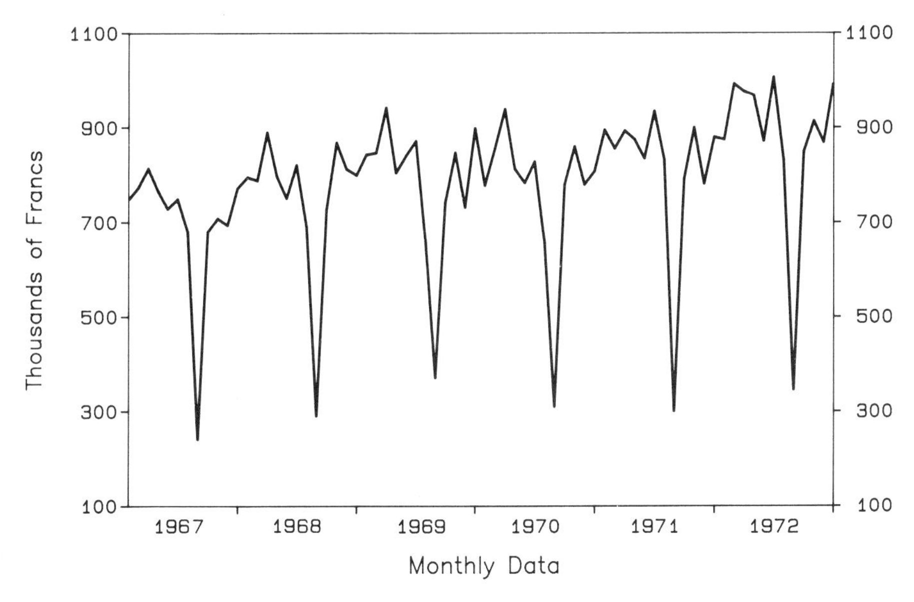

Figure 16.1 Paper Industry Sales Revenue

The model developed for this series turns out to be an extremely common model for seasonal economic time series. You will see this type of model, with slight modifications, occur again and again. One of its most common variations is the model developed in Example 16.7 (airline passenger series).

The AC correlogram of Figure 16.2 corresponds to the original undifferenced series. The graph of the series and its AC correlogram both clearly indicate a strong seasonal pattern with 12 periods per season. The large, slowly decreasing autocorrelations at lags 12, 24, 36, etc. suggest the need for one seasonal difference. There is also an indication of the need for one regular difference because of the clumps of autocorrelations from lags 1 through 4 and 8 through 15.

The most natural action to take at this step, however, is to apply one seasonal difference first. The resulting AC correlogram after applying one seasonal difference appears in Figure 16.3. The autocorrelations at lags 24, 36, etc. have now essentially disappeared, indicating that stationarity in the seasonal pattern has been achieved. The large autocorrelation at lag 12 indicates the need for one SMA of order 12.

But is the series, as a whole, really stationary yet? A strict interpretation of the autocorrelation confidence bands would tend to indicate that most of the autocorrelations are not significant, and hence the currently differenced series is stationary. However, experience indicates that when autocorrelations appear in clumps, as from lags 1 to 10 and again from lags 14 to 30, then regular differencing is probably required.

Let's assume, however, that the seasonally differenced series is stationary, and proceed to estimate the model suggested by the correlogram of Figure 16.3, namely,

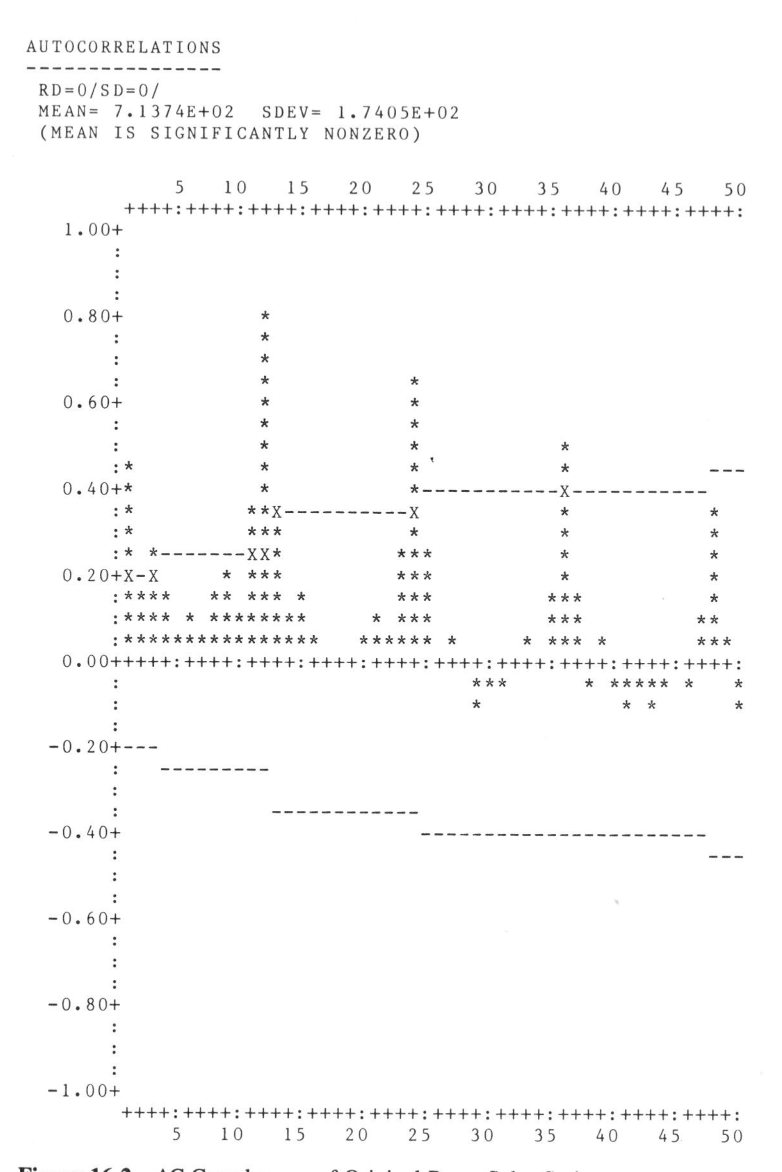

Figure 16.2 AC Correlogram of Original Paper Sales Series

one SD of order 12 and one SMA of order 12. The estimation results for this model are given in Figure 16.4.

The large Q-statistics in Figure 16.4 immediately tell us that something is wrong. The large Q-statistics indicate highly autocorrelated residuals as a whole through lag 12, 24, 36, and 48 which means something has been left out of the model. A look at the residual AC correlogram in Figure 16.5 also indicates a totally inadequate

```
AUTOCORRELATIONS
----------------
 RD=0/SD=1/PPS=12/
 MEAN= 3.5602E+01   SDEV= 5.0813E+01
 (MEAN IS SIGNIFICANTLY NONZERO)

          5    10   15   20   25   30   35   40   45   50
      ++++:++++:++++:++++:++++:++++:++++:++++:++++:++++:
 1.00+
     :
     :
     :
 0.80+
     :
     :
     :
 0.60+
     :
     :
     :
 0.40+
     :
     :                                          --------
     :            -----------------------------X
 0.20+--X---------                           * *
     : ***  *                        *       * *
     :**** ** *                      *  **  ** *   *
     :**********                     ** ****** *  **
 0.00+++++:++++:++++:++++:++++:++++:++++:++++:++++:++++:
     :           * *** ******** ****   *      *  *     *
     :           * *** **** *** ** *
     :           *  ** ****
-0.20+-----------X
     :           *------------------------------
     :           *                              --------
     :           *
-0.40+           *
     :
     :
     :
-0.60+
     :
     :
     :
-0.80+
     :
     :
     :
-1.00+
      ++++:++++:++++:++++:++++:++++:++++:++++:++++:++++:
          5    10   15   20   25   30   35   40   45   50
```

Figure 16.3 AC Correlogram of Paper Sales Series After Applying One SD (Order 12)

model. Note that the residual AC correlogram looks almost like the AC correlogram for the seasonally differenced series. This result is typical when the model is estimated from a nonstationary series; i.e., nonstationary behavior in the series overpowers the underlying autoregressive and moving-average relationships to the extent that any attempt to model these relationships has little or no impact.

```
 0 REGULAR DIFFERENCES
 1 SEASONAL DIFFERENCES OF ORDER 12

PAR PARAMETER            PARAMETER  ESTIMATED           95 PERCENT
 #     TYPE                ORDER                  LOWER LIMIT   UPPER LIMIT

 1 SEAS MV AVERAGE          12        -0.015        -0.226         0.195

-----------------------------------------------------------------------


RESIDUAL MEAN           =      -35.1620    NO. RESIDUALS        = 108
RESIDUAL SUM OF SQUARES =  413039.3125
RESIDUAL MEAN SQUARE    =    3860.1809    DEGREES OF FREEDOM   = 107
RESIDUAL STANDARD ERROR =      62.1304

(RESIDUAL MEAN IS SIGNIFICANTLY NONZERO)

INDEX OF DETERMINATION   =    0.867
AVG ABSOULTE % ERROR     =    6.75%
MEAN % ERROR             =   -4.64%


 LAGS       Q-STATISTIC    CHISQ AT 5%    DF
1 -   12       30.64          19.67       11
1 -   24       49.20          35.17       23
1 -   36       58.75          49.81       35
1 -   48       73.59          64.01       47
```

Figure 16.4 Model Estimation Results for Paper Sales Series (One SD Order 12 and One SMA of Order 12)

We therefore go back to our original suspicion that a regular difference is required. The AC correlogram in Figure 16.6 was obtained after applying one RD and one SD of order 12. The correlogram now clearly indicates that the differenced series is stationary. The large significant spikes at lags 1 and 12 (along with the smaller, but still significant, spikes at lags 11 and 13) identify a multiplicative model with one RMA and one SMA of order 12 (note how the spike at lag 1 suddenly appeared after using the regular difference). The spikes at lags 41, 42, and 43 also catch the eye, but these spikes would seem to have no practical significance, and they are statistically insignificant anyway. They are therefore ignored.

We have thus identified a model with one RD, one SD of order 12, one RMA, and one SMA of order 12. The estimation results for the identified model are shown in Figure 16.7. All diagnostics point to a valid and acceptable model. Note that the Q-statistics are all significantly less than their test value, indicating that the model is not underspecified. Also, both parameters are significant since their confidence intervals do not include zero, and they are not highly correlated to each other. This result indicates the model is not overspecified. The fit statistics indicate a reasonably close fit with over 93% of the total variation in the series accounted for by the model.

The final estimated model for the paper sales series is therefore

$$Z_t = -.812E_{t-1} - .6E_{t-12} + .4872E_{t-13} + \mathrm{E}_t$$

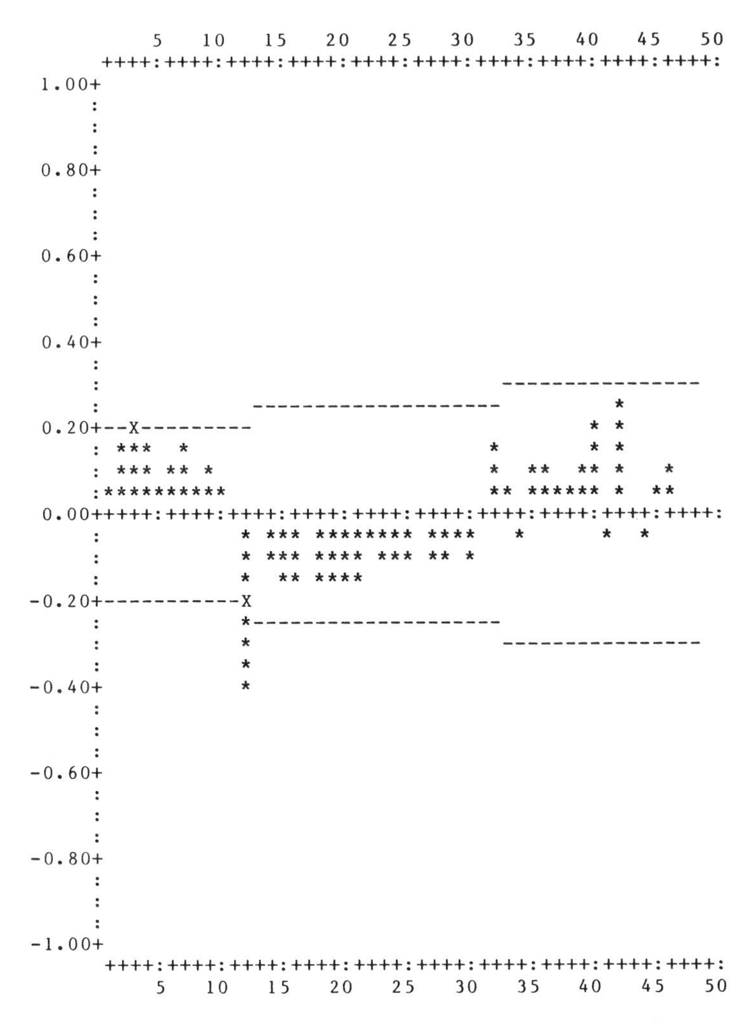

Figure 16.5 Residual AC Correlogram from Model of Figure 16.4

where $Z_t = X_t - X_{t-1} - X_{t-12} + X_{t-13}$ is the differenced series. [*Note:* .4872 = (.812) × (.6).]

Example 16.2: Regional Postal Service Volume

This series represents the volume of mail, in thousands of pieces, for a U.S. Postal Service region. The time periods are 4-week intervals. Thus there are *13 periods* per year. The series runs from approximately May 1979 to August 1982 (42 periods). A graph of the series is shown in Figure 16.8.

The model developed for this series is quite similar to the model for the series in Example 16.1. The only difference is the periods per season.

```
AUTOCORRELATIONS
----------------
 RD=1/SD=1/PPS=12/
 MEAN= 3.8967E-01   SDEV= 6.8870E+01
```

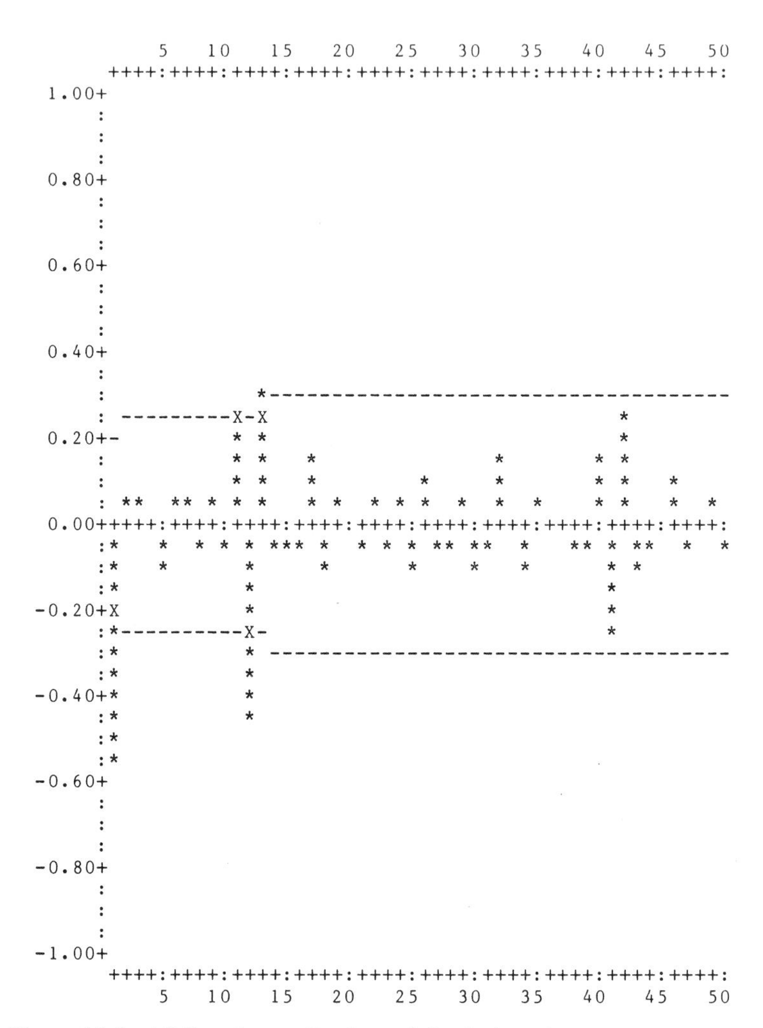

Figure 16.6 AC Correlogram For Paper Sales Series After Applying One RD and One SD of Order 12

The AC correlogram for the original undifferenced series is shown in Figure 16.9. As in Example 16.1, the graph of the series along with its AC correlogram clearly indicates seasonal behavior with periods per season equal to 13. The large, slowly decreasing autocorrelations at lags 13 and 26 suggest the need for one seasonal difference of order 13. Note that no further autocorrelations could be computed for lags $13s$ when $s > 3$, since there are only 42 periods of data. Note also that

```
CORRELATION MATRIX OF THE PARAMETERS
------------------------------------

        1           2

  1  1.0000

  2 -0.0770       1.0000

----------------------------------------------------------------------

 1 REGULAR DIFFERENCES
 1 SEASONAL DIFFERENCES OF ORDER 12

PAR PARAMETER              PARAMETER  ESTIMATED          95 PERCENT
 #     TYPE                  ORDER                  LOWER LIMIT   UPPER LIMIT

 1 REG MV   AVERAGE            1        0.812           0.698         0.925


 2 SEAS MV AVERAGE            12        0.600           0.425         0.776

----------------------------------------------------------------------


RESIDUAL MEAN            =        3.7610    NO. RESIDUALS        = 107
RESIDUAL SUM OF SQUARES =   208353.1875
RESIDUAL MEAN SQUARE     =     1984.3167    DEGREES OF FREEDOM = 105
RESIDUAL STANDARD ERROR =        44.5457

INDEX OF DETERMINATION   =    0.933
AVG ABSOULTE % ERROR     =    5.67%
MEAN % ERROR             =    1.76%


 LAGS        Q-STATISTIC    CHISQ AT 5%    DF
1 -   12          2.91          18.30      10
1 -   24          8.38          33.93      22
1 -   36         12.72          48.61      34
1 -   48         19.76          62.84      46
```

Figure 16.7 Model Estimation Results for Paper Sales Series (One RD, One SD of Order 12, One RMA, and One SMA of Order 12)

since there are so few periods of data, the confidence bands for the autocorrelations are relatively wide. This is often the case when there is little data to work with; i.e., the less data available, the less confidence there is in the statistics computed from the data.

The confidence bands notwithstanding, the need for one seasonal difference is still apparent in the AC correlogram. The need for one regular difference is also suggested by the small clumps of autocorrelations at lags 1 to 3, 4 to 9, 17 to 22, etc., but the evidence is not that strong. At any rate, the resulting AC correlogram after applying one seasonal difference of order 13 is as shown in Figure 16.10.

The AC correlogram of the series after applying one seasonal difference of order 13 now clearly indicates the need for a regular difference. This example shows how

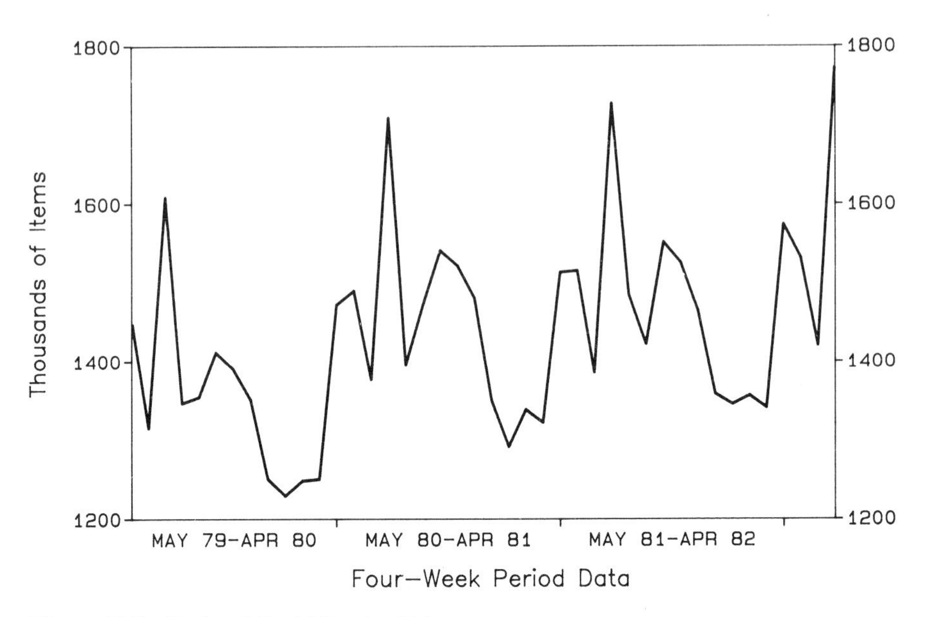

Figure 16.8 Regional Postal Service Volume

nonstationarity in the seasonal pattern of a series can often obscure the overall nonstationary behavior of a series (the reverse is also true). The resulting AC correlogram after applying one seasonal difference and one regular difference appears in Figure 16.11.

The AC correlogram of Figure 16.11 now indicates that stationarity has been achieved. (Note that since a regular difference and a seasonal difference have been applied, the maximum number of autocorrelations that can be computed is 26.) The large significant spike at lag 1 immediately identifies one RMA parameter of order 1. The remaining autocorrelations, however, are within the confidence bands. Under a strict interpretation of significance (at 95% confidence) these autocorrelations would not be significant. But since we know we are working with seasonal data with 13 periods per season, the next largest autocorrelation at lag 13 probably really does have significance. This spike, along with the autocorrelations at lags 12 and 14, would therefore indicate one SMA of order 13. There are other autocorrelations at lags 3, 6, and 9 almost as large as the one at lag 13, but there is no reason for suspecting that these autocorrelations have any practical significance (if the periods per season were 12, there might be). Thus a tentative model for the postal volume series would be one RD, one SD of order 13, one RMA, and one SMA of order 13.

The estimation results for this model are shown in Figure 16.12. The parameter diagnostics indicate that the two parameters are uncorrelated and insignificant. The model is therefore not overspecified. Note, however, the large confidence intervals for the parameters. This result, again, is largely due to the fact that we are working

```
AUTOCORRELATIONS
----------------
 RD=0/SD=0/
 MEAN= 1.4348E+06   SDEV= 1.2746E+05
 (MEAN IS SIGNIFICANTLY NONZERO)

              5    10   15   20   25   30   35   40
         ++++:++++:++++:++++:++++:++++:++++:++++:
   1.00+
        :
        :
        :
   0.80+
        :
        :
        :
   0.60+           *
        :          *                ------
        :          *-----------------
        :          *
   0.40+      -----X
        : -----    *
        :--        *            *
        :  *       *            *
   0.20+***        *            *
        :***     * *            *
        :***     * *          * *
        :***     **** **      * *          *
   0.00+++++:++++:++++:++++:++++:++++:++++:++++:
        :   ******     ******  * * ******
        :   ******     ******      ****
        :   * ****     **** *      *
  -0.20+    * ** *     * *
        :
        :--
        :  -----
  -0.40+       ------
        :            -----------------
        :                             ------
        :
  -0.60+
        :
        :
        :
  -0.80+
        :
        :
        :
  -1.00+
         ++++:++++:++++:++++:++++:++++:++++:++++:
              5    10   15   20   25   30   35   40
```

Figure 16.9 AC Correlogram of Original Postal Volume Series

with only a small amount of data. Thus there is a high degree of uncertainty that the computed values of the parameters are representative of the "true" values. As far as the residual diagnostics are concerned, the Q-statistics are all much less than their test values, so the residual autocorrelations are not significant. Since the mean percent error is practically zero, we can then conclude that the residuals represent random error and that the model is not incorrectly specified or underspecified. The

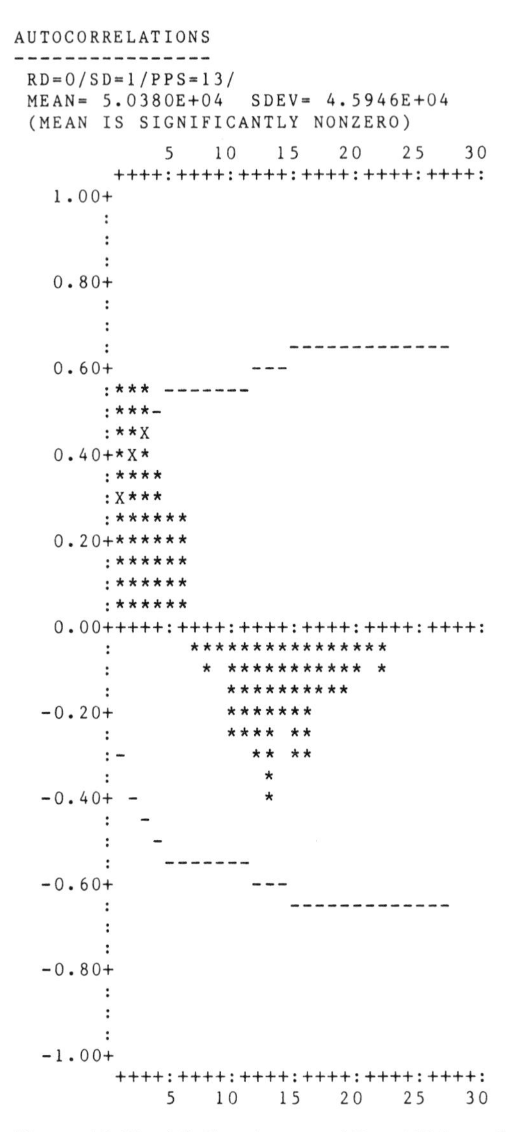

Figure 16.10 AC Correlogram of Postal Volume Series After Applying One SD of Order 13

fit statistics also indicate a close fit, with 93% of the series variation explained by the model.

The final estimated model for the postal volume series is therefore

$$Z_t = -.491E_{t-1} - .674E_{t-13} + .331E_{t-14} + \mathrm{E}_t$$

where $Z_t = X_t - X_{t-1} - X_{t-13} + X_{t-14}$ is the differenced series. (*Note:* $.331 = .491 \times .674$)

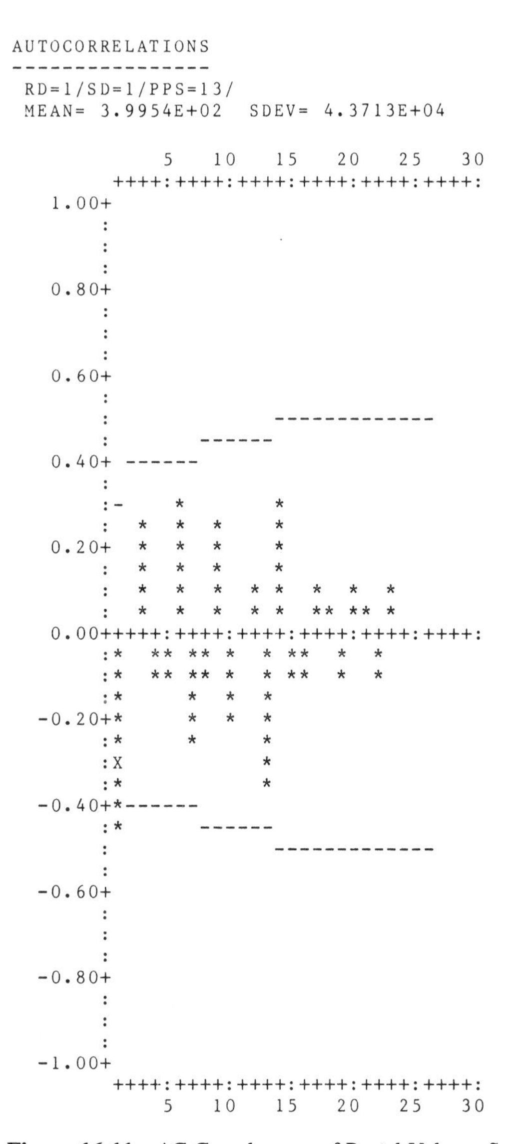

Figure 16.11 AC Correlogram of Postal Volume Series After Applying One RD and One SD of Order 13

Example 16.3: Lawn Chair Sales Revenue

This series represents the monthly sales revenue, in dollars, of a company's brand of lawn chairs [17]. The series runs for four years from January 1971 to December 1974 (48 periods). A graph of the series is shown in Figure 16.13.

The model developed for this series is also a common one for seasonal economic time series, but it is not as prevalent as the one-RMA and one-SMA model constructed for the series in Examples 16.1 and 16.2.

```
CORRELATION MATRIX OF THE PARAMETERS
------------------------------------

        1           2

  1  1.0000

  2  0.1235      1.0000

----------------------------------------------------------------------

 1 REGULAR DIFFERENCES
 1 SEASONAL DIFFERENCES OF ORDER 13

PAR PARAMETER          PARAMETER  ESTIMATED          95 PERCENT
 #     TYPE              ORDER                 LOWER LIMIT   UPPER LIMIT

 1 REG MV   AVERAGE          1          0.491         0.146          0.836


 2 SEAS MV AVERAGE          13          0.674         0.186          1.161


----------------------------------------------------------------------


RESIDUAL MEAN              =      -904.2856   NO. RESIDUALS        =  28
RESIDUAL SUM OF SQUARES = 0.295471E+11
RESIDUAL MEAN SQUARE     = 0.113643E+10   DEGREES OF FREEDOM =  26
RESIDUAL STANDARD ERROR = 0.337109E+05

INDEX OF DETERMINATION   =    0.930
AVG ABSOULTE % ERROR     =    1.85%
MEAN % ERROR             =    0.02%


 LAGS       Q-STATISTIC    CHISQ AT 5%    DF
1 -   12        6.29          18.30       10
1 -   24       10.60          33.93       22
1 -   26       10.87          36.42       24
```

Figure 16.12 Model Estimation Results for Postal Volume Series One RD, One SD of Order 13, and One RMA and One SMA of Order 13

The AC correlogram for the original undifferenced series is shown in Figure 16.14. Since there are only 48 periods of data, autocorrelations at lags 12s for $s >$ 3 cannot be computed. Note also the wide autocorrelation confidence bands. As in Example 16.2, this wide band is primarily due to the lack of data.

The graph of the series and its AC pattern clearly demonstrate the existence of a seasonal pattern. But first we will consider the large alternating clumps of autocorrelations centered at lags 6, 12, 18, 24, etc. Clumps such as these, you'll recall, usually indicate the need for regular differencing, especially since the last negative clump actually gets larger. The alternative interpretation is an autoregressive pattern, but this is unlikely.

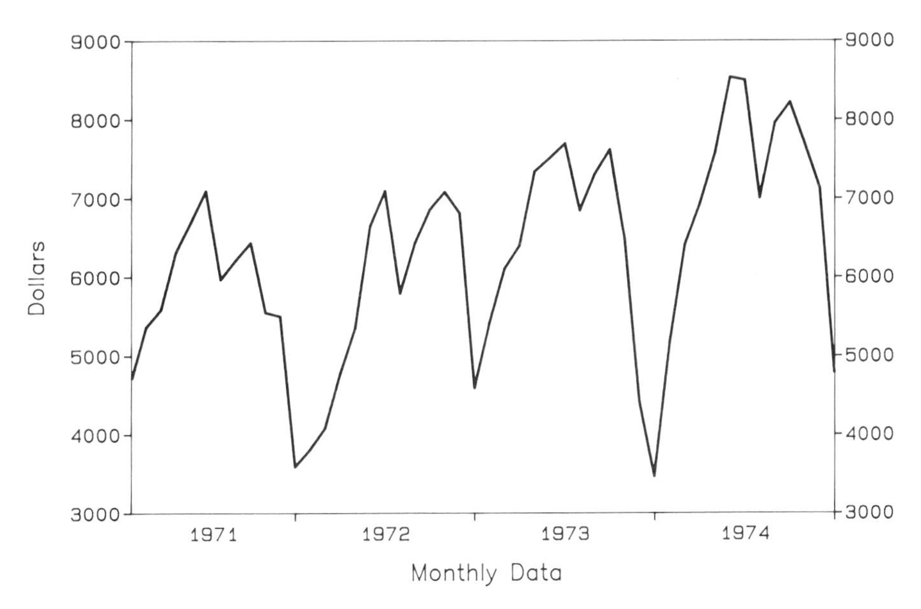

Figure 16.13 Lawn Chair Sales Revenue

After applying one RD, then, we obtain the AC correlogram of Figure 16.15. Most of the autocorrelations have now died out, except at lags 6, 12, 24, and 36. Since the ones at lags 12, 24, and 36 decrease slowly, they indicate the need for one seasonal difference of order 12.

After applying both one RD and one SD of order 12, we obtain the AC correlogram of Figure 16.16. The AC correlogram now indicates that the differenced series is stationary, and it shows two significant autocorrelations at lags 5 and 12, as well as two fairly large (but insignificant) ones at lags 4 and 7. The spike at lag 12 clearly indicates the need for one SMA of order 12. Because this series represents monthly data, it is unlikely that the spikes at lags 4, 5, and 7 have any real significance. Nevertheless, let's assume the spike at lag 5 really is significant, and hence we'll estimate a model with one RD, one SD of order 12, one RMA of order 5, and one SMA of order 12.

The estimation results for this model are shown in Figure 16.17. From these results we see immediately that the RMA of order 5 is *not* significant, since its confidence interval includes zero (although just barely). Therefore the model containing one RD, one SD of order 12, and one SMA of order 12 is probably the correct one.

The estimation results for this revised model are shown in Figure 16.18. The diagnostics indicate an acceptable model: The parameter confidence interval, although fairly large, does not include zero, and the Q-statistics are much less than their test values. The index of determination, however, is only .782, which means that the random error component in the series is almost 22% of the total variation in the

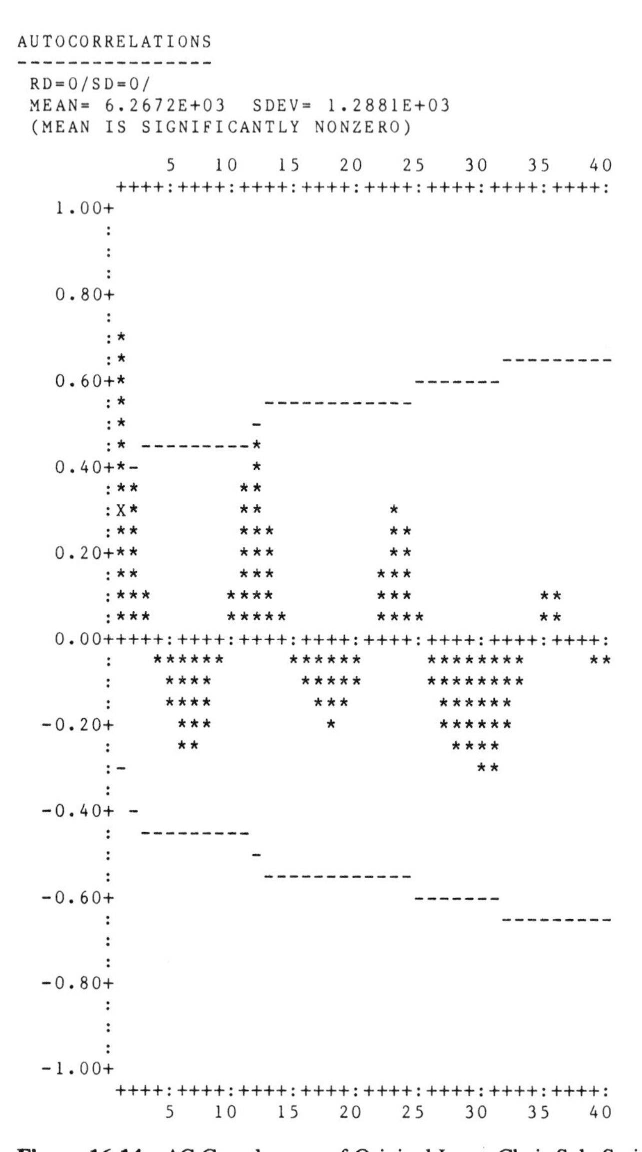

Figure 16.14 AC Correlogram of Original Lawn Chair Sale Series

series. Thus we can expect a fairly high degree of uncertainty in forecasts computed from the model. (See Chapter 17, Figure 17.5, for the forecasting results from this model.)

The final estimated model for the lawn chair sales series is

$$Z_t = -.789E_{t-12} + E_t$$

where $Z_t = X_t - X_{t-1} - X_{t-12} + X_{t-13}$ is the differenced series.

```
AUTOCORRELATIONS
----------------
 RD=1/SD=0/
 MEAN= 2.0638E+00   SDEV= 9.6014E+02

          5    10   15   20   25   30   35   40
      ++++:++++:++++:++++:++++:++++:++++:++++:
 1.00+
     :
     :
     :
 0.80+
     :
     :
     :
 0.60+
     :
     :           *
     :           *           -----------------
 0.40+           *-----------*
     :      -----X           *
     :------     *           *
     :           *           *
 0.20+           *           *           *
     :*         **          **           *
     :*         **          **           *
     :*   *     ***   *     **          ** *
 0.00+++++:++++:++++:++++:++++:++++:++++:++++:
     : *** *****   **  * **   ***  ****
     : *** * ***       * *     *   ***
     :  *  * ***       * *         *
-0.20+     *           *
     :     *
     :-----X
     :      ------
-0.40+            -----------
     :                       -----------------
     :
     :
-0.60+
     :
     :
     :
-0.80+
     :
     :
     :
-1.00+
      ++++:++++:++++:++++:++++:++++:++++:++++:
          5    10   15   20   25   30   35   40
```

Figure 16.15 AC Correlogram of Lawn Chair Sales After Applying One RD

Example 16.4: Champagne Sales

This series represents the monthly unit sales of champagne, in thousands of bottles, for a wine company in France [23]. The series runs from January 1965 to September 1972 (93 periods). The graph of the series is shown in Figure 16.19.

The AC correlogram of the original undifferenced series is shown in Figure 16.20. The graph of the series and its AC correlogram both clearly indicate a strong seasonal pattern with periods per season equal to 12. The larger autocorrelations at

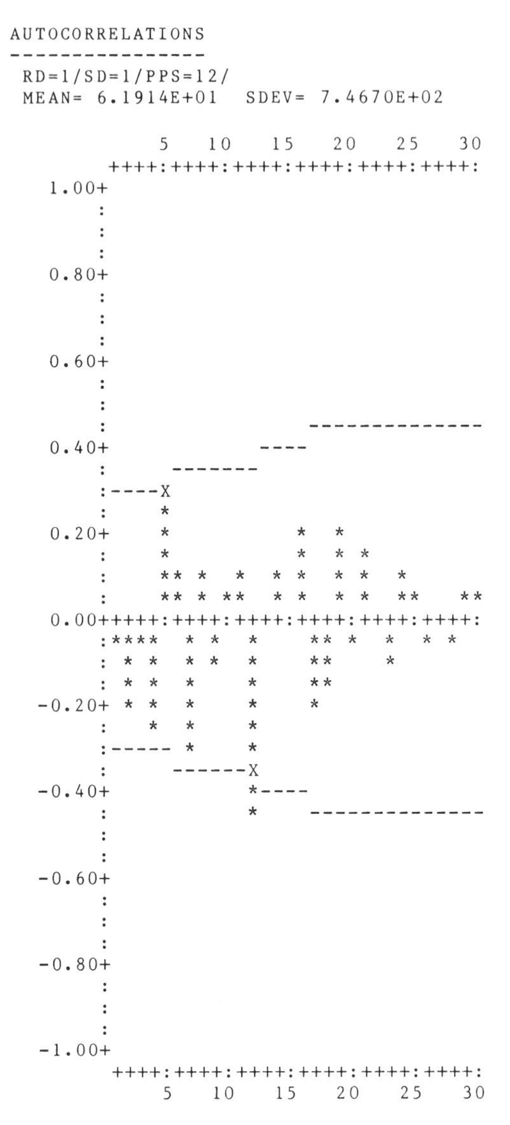

Figure 16.16 AC Correlogram of Lawn Chair Sales After Applying One RD and One SD of Order 12

lags 12, 24, 36, etc. are decreasing slowly, which indicates the need for at least one SD of Order 12. No regular differencing, on the other hand, appears to be needed.

After applying one SD of order 12, we obtain the AC correlogram of Figure 16.21. In this correlogram most autocorrelations now appear insignificant, indicating that stationarity has been achieved and that there apparently really is no need for regular differencing is definitely not required in this case. The only ACs that appear

```
CORRELATION MATRIX OF THE PARAMETERS
------------------------------------

        1           2

  1   1.0000

  2 -0.2190       1.0000

-----------------------------------------------------------------------

 1 REGULAR DIFFERENCES
 1 SEASONAL DIFFERENCES OF ORDER 12

PAR PARAMETER             PARAMETER  ESTIMATED          95 PERCENT
 #     TYPE                 ORDER                   LOWER LIMIT   UPPER LIMIT

 1 REG MV   AVERAGE             5        -0.341         -0.690          0.007


 2 SEAS MV AVERAGE             12         0.828          0.403          1.254

-----------------------------------------------------------------------

RESIDUAL MEAN             =      -70.8911    NO. RESIDUALS        =   35
RESIDUAL SUM OF SQUARES =11382063.0000
RESIDUAL MEAN SQUARE     =    344911.0000    DEGREES OF FREEDOM =   33
RESIDUAL STANDARD ERROR =        587.2913

INDEX OF DETERMINATION    =    0.799
AVG ABSOULTE % ERROR      =    7.54%
MEAN % ERROR              =   -1.07%

CORRELATION OF RESIDUALS WITH DATA =    0.020

 LAGS      Q-STATISTIC    CHISQ AT 5%    DF
1 -   12          4.15         18.30      10
1 -   20          6.82         28.87      18
```

Figure 16.17 Model Estimation Results for Law Chair Sales (One RD, One SD of Order 12, and One RMA of Order 6 and One SMA of Order 12)

significant are those at lag 1 and lag 12. Although the autocorrelations at these lags are barely significant, it is highly probable that they do indicate the need for one RMA and one SMA of order 12, since they occur at lags 1 and 12. Note, also, that the mean of the differenced stationary series is significantly nonzero as indicated above the correlogram in Figure 16.21. This result identifies the need for a trend parameter. A tentative model for this series is therefore one SD of order 12 and one RMA, one SMA of order 12, and a trend parameter.

The estimation results for this model are shown in Figure 16.22. All diagnostics indicate an acceptable model. The final estimated model for the champagne sales series is therefore

$$Z_t = .267 + .281E_{t-1} - .357E_{t-12} - .1E_{t-13} + E_t$$

where $Z_t = X_t - X_{t-12}$

[*Note:* $.1 = (.281)(.357)$.]

```
1 REGULAR DIFFERENCES
1 SEASONAL DIFFERENCES OF ORDER 12

PAR PARAMETER            PARAMETER  ESTIMATED         95 PERCENT
 #     TYPE                ORDER                 LOWER LIMIT   UPPER LIMIT

1 SEAS MV AVERAGE            12         0.789         0.392          1.186

----------------------------------------------------------------------------

RESIDUAL MEAN              =      -89.1834    NO. RESIDUALS        =  35
RESIDUAL SUM OF SQUARES =12353086.0000
RESIDUAL MEAN SQUARE       =    363326.0625   DEGREES OF FREEDOM =  34
RESIDUAL STANDARD ERROR =        602.7654

INDEX OF DETERMINATION   =    0.782
AVG ABSOULTE % ERROR     =    7.67%
MEAN % ERROR             =   -1.27%

 LAGS      Q-STATISTIC   CHISQ AT 5%   DF
1 -  12        6.61         19.67       11
1 -  24       10.75         35.17       23
1 -  30       11.19         42.56       29
```

Figure 16.18 Model Estimation Results for Lawn Chair Sales Series One RD, One SD of Order 12, and One SMA of Order 12

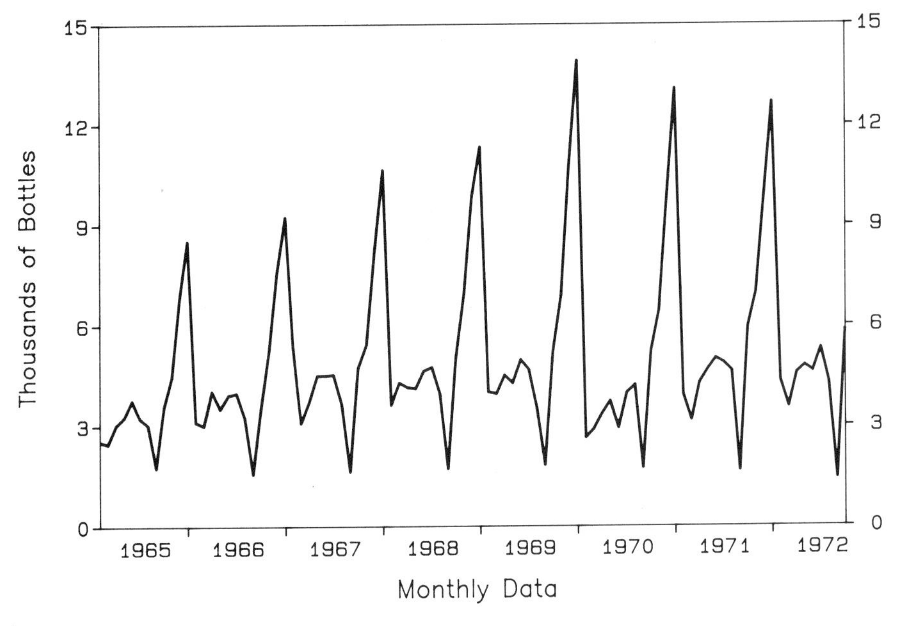

Figure 16.19 Champagne Sales

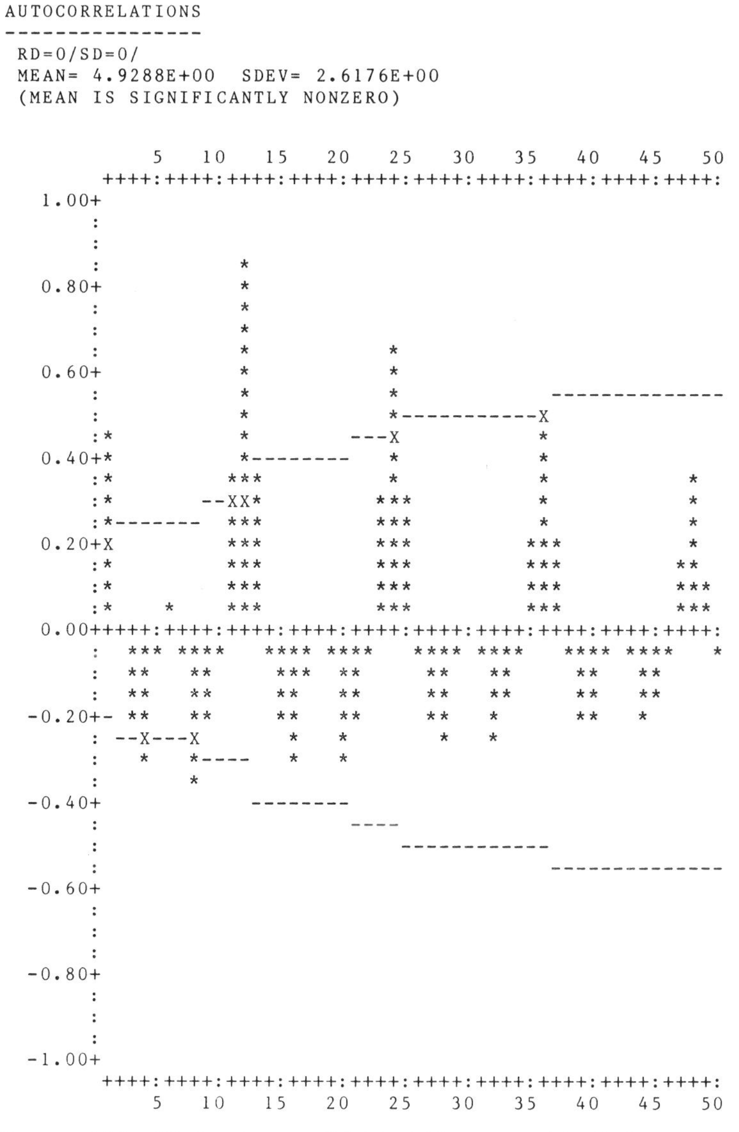

Figure 16.20 AC Correlogram of Original Champagne Sales Series

This is a rather interesting seasonal model since no regular differencing was used and an overall trend parameter was included. This situation is fairly uncommon, since regular differencing is usually required in most models for economic time series. It turns out, for this series, that one seasonal difference and a trend parameter are sufficient to model the nonstationary behavior in the series. The trend parameter is required since the seasonally differenced series has a significantly nonzero mean,

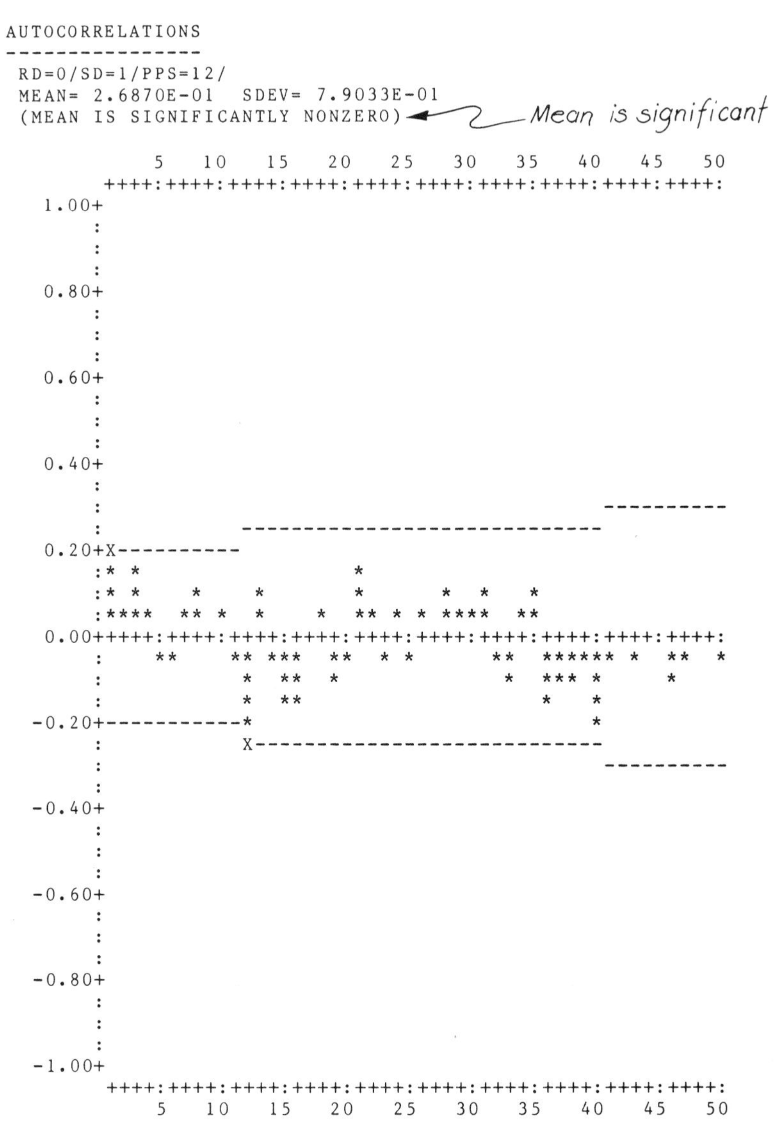

Figure 16.21 AC Correlogram for Champagne Sales Series After Applying One SD of Order 12

and therefore accounts for the overall trend in the seasonal pattern.

Example 16.5: Insurance Claim Frequency

This series represents the frequency of insurance claims for a large insurance company recorded on a quarterly basis. The series runs from the first quarter of 1955

```
CORRELATION MATRIX OF THE PARAMETERS
------------------------------------

        1            2            3

  1  1.0000

  2  0.0010       1.0000

  3  0.0881       0.0087       1.0000

----------------------------------------------------------------------

 0 REGULAR DIFFERENCES
 1 SEASONAL DIFFERENCES OF ORDER 12

PAR PARAMETER            PARAMETER  ESTIMATED          95 PERCENT
 #     TYPE                ORDER                  LOWER LIMIT   UPPER LIMIT

 1 TREND CONSTANT            0         0.267          0.120          0.414


 2 REG MV  AVERAGE           1        -0.281         -0.498         -0.064


 3 SEAS MV AVERAGE          12         0.357          0.140          0.573

----------------------------------------------------------------------

RESIDUAL MEAN            =       -0.0149   NO. RESIDUALS       =  81
RESIDUAL SUM OF SQUARES  =       42.8519
RESIDUAL MEAN SQUARE     =        0.5494   DEGREES OF FREEDOM =  78
RESIDUAL STANDARD ERROR  =        0.7412

INDEX OF DETERMINATION   =   0.925
AVG ABSOULTE % ERROR     =  12.56%
MEAN % ERROR             =   3.58%


 LAGS       Q-STATISTIC    CHISQ AT 5%    DF
1 -  12         2.38          16.91        9
1 -  24        12.32          32.67       21
1 -  36        16.50          47.41       33
1 -  48        20.30          61.66       45
```

Figure 16.22 Model Estimation Results for Champagne Sales Series One SD of Order 12 and One RMA, One SMA of Order 12, and a Trend Parameter

to the fourth quarter of 1976 (88 periods). A graph of the series is shown in Figure 16.23. Although the graph shows no apparent seasonal behavior, you will see later that seasonal patterns do exist.

The AC correlogram of the original undifferenced series is shown in Figure 16.24. This heavy autocorrelation pattern with large, slowly decreasing spikes unquestionably indicates the need for at least one regular difference.

After applying one RD, we obtain the AC correlogram of Figure 16.25. This AC pattern now indicates two possibilities: (1) the differenced series is stationary and the model contains one RAR parameter, or (2) the differenced series is not

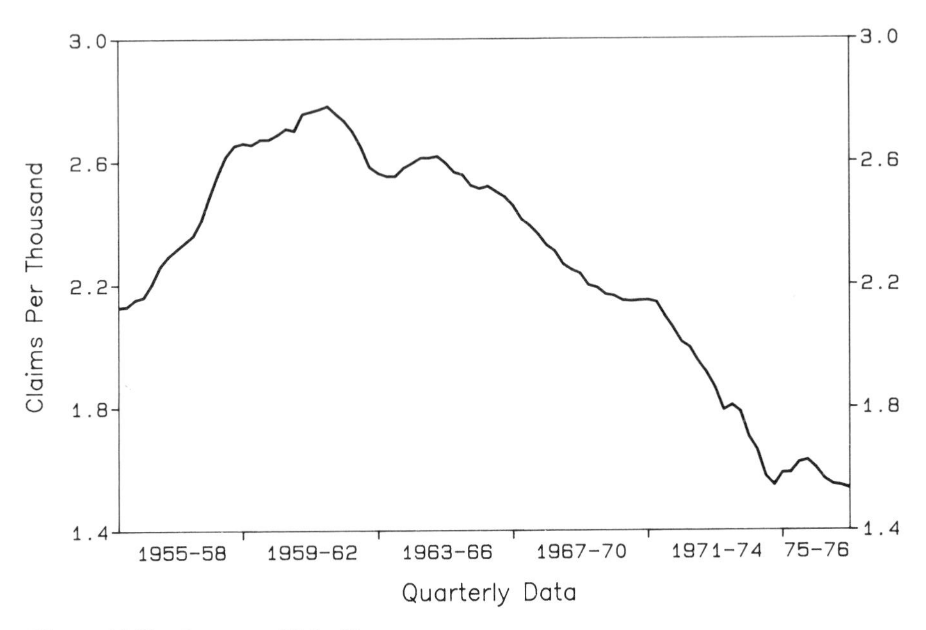

Figure 16.23 Insurance Claim Frequency

stationary and requires an additional regular difference. Since the autocorrelations don't damp out that rapidly, especially from lags 5 through 12, the second possibility is probably the correct choice. However, let's estimate a model with one RD and one RAR parameter, as suggested in possibility (1), and see if our suspicions are correct.

The estimation results for this model are shown in Figure 16.26. The parameter diagnostics indicate that the RAR parameter is apparently significant, and the closeness-of-fit statistics indicate a close fit; but the Q-statistics indicate that the residuals are significantly autocorrelated. This result implies that the correct model has not been identified.

At this point the AC correlogram for the residuals might provide some clues for determining where the model is deficient. The AC correlogram for the residuals appears in Figure 16.27. The residual autocorrelation pattern uncovers some surprising information. The large autocorrelation spike at lag 4 indicates that a seasonal moving-average relationship of order 4 was not accounted for and is still present in the residuals. This result confirms our original supposition that the differenced series, after one RD, was not stationary because the seasonal moving-average AC pattern was covered up by the nonstationarity of the once-differenced series. Thus we will proceed to apply two regular differences to the series.

The AC correlogram for the twice-differenced series appears in Figure 16.28. This AC pattern now indicates that the differenced series is stationary, and the two spikes at lags 1 and 4 identify a model with one RMA and one SMA of order 4.

```
AUTOCORRELATIONS
----------------
 RD=0/SD=0/
 MEAN= 2.2685E+00   SDEV= 3.7745E-01
 (MEAN IS SIGNIFICANTLY NONZERO)

              5    10    15    20    25    30    35    40
         ++++:++++:++++:++++:++++:++++:++++:++++:
   1.00+
        :**
        :***                 --------------------
        :*****           ----
   0.80+******      ---
        :*******--
        :******X*
        :*****X***
   0.60+****X*****
        :***X*******
        :*************
        :**X***********
   0.40+***************
        :*X**************
        :*****************
        :******************
   0.20+X******************
        :**********************
        :************************
        :**************************
   0.00+++++:++++:++++:++++:++++:++++:++++:++++:
        :                              *********
        :                                *******
        :                                 ******
  -0.20+-                                   ****
        :                                     **
        :
        : -
  -0.40+
        :  -
        :
        :   -
  -0.60+     -
        :     -
        :      -
        :       --
  -0.80+         ---
        :           ----
        :               --------------------
        :
  -1.00+
         ++++:++++:++++:++++:++++:++++:++++:++++:
              5    10    15    20    25    30    35    40
```

Figure 16.24 AC Correlogram of Original Insurance Claims Series

Note, also, the two smaller spikes at lags 3 and 5, which imply a multiplicative model. (The spike at lag 1 was not present in the residual autocorrelations associated with the model containing one RD and one RAR, since the RAR parameter partially accounted for the regular moving-average relationship.)

The estimation results for this newly identified model containing two RDs and one RMA and one SMA parameter are given in Figure 16.29. All diagnostics

```
AUTOCORRELATIONS
----------------
 RD=1/SD=0/
 MEAN=-6.8046E-03   SDEV= 3.2471E-02


             5    10    15    20    25    30    35    40
        ++++:++++:++++:++++:++++:++++:++++:++++:
   1.00+
       :
       :
       :
   0.80+
       :
       :
       :*
   0.60+*
       :**
       :**
       :***         ------------------------
   0.40+***   ------
       :**X--
       :*X* **
       :*** *******
   0.20+X***********
       :***********
       :************     * ** *
       :************************ *
   0.00+++++:++++:++++:++++:++++:++++:++++:++++:
       :                                ******
       :                                * ** *
       :                                   *
  -0.20+-
       :
       : -
       :  ---
  -0.40+     ------
       :            ------------------------
       :
       :
  -0.60+
       :
       :
       :
  -0.80+
       :
       :
       :
  -1.00+
        ++++:++++:++++:++++:++++:++++:++++:++++:
             5    10    15    20    25    30    35    40
```

Figure 16.25 AC Correlogram for Insurance Claims Series After Applying One RD

indicate that the model is valid, with 99.7% of the variation in the series explained by the model (index of determination is .997). The final estimated mmodel for the insurance claim series is therefore

$$Z_t = -.304E_{t-1} - .729E_{t-4} + .222E_{t-5} + E_t$$

where $Z_t = X_t - 2X_{t-1} - X_{t-2}$ is the differenced series.

```
 1 REGULAR DIFFERENCES
 0 SEASONAL DIFFERENCES OF ORDER  0

PAR PARAMETER           PARAMETER  ESTIMATED         95 PERCENT
 #    TYPE                ORDER                  LOWER LIMIT   UPPER LIMIT

 1 REG AUTOREGRESS          1         0.683         0.524          0.841


----------------------------------------------------------------------


RESIDUAL MEAN              =        0.0023    NO. RESIDUALS       =  86
RESIDUAL SUM OF SQUARES =        0.0506
RESIDUAL MEAN SQUARE       =        0.0006    DEGREES OF FREEDOM =  85
RESIDUAL STANDARD ERROR =        0.0244

INDEX OF DETERMINATION   =    0.996
AVG ABSOULTE % ERROR     =    0.91%
MEAN % ERROR             =    0.12%


 LAGS       Q-STATISTIC    CHISQ AT 5%    DF
1 -  12        26.89          19.67        11
1 -  24        32.04          35.17        23
1 -  36        48.80          49.81        35
```

Figure 16.26 Model Estimation Results for Insurance Claim Series (One RD and One RAR)

Example 16.6: Sears Retail Sales Revenue

This series represents monthly national sales revenue, in millions of dollars, for Sears, Roebuck and Co. [26]. The series runs from July 1965 to December 1971 (78 periods). The graph of the series is shown in Figure 16.30.

The AC correlogram for the original undifferenced series is shown in Figure 16.31. The graph of the series and its AC correlogram both clearly indicate a strong seasonal pattern with 12 periods per season. The large, slowly decreasing spikes at lags 12, 24, 36, etc. suggest the need for at least one seasonal difference of order 12. The clumps of autocorrelations at lags 1 to 9, and later at lags 37 to 41, also indicate the possible need for one regular difference. We will first apply one SD of order 12.

The AC correlogram for the differenced series appears in Figure 16.32. This AC correlogram shows that the seasonal nonstationarity has been eliminated, while the need for regular differencing now becomes obvious with the appearance of large alternating clumps of autocorrelations.

After applying one regular difference and one seasonal difference of order 12, we obtain the AC correlogram shown in Figure 16.33. This AC pattern indicates that stationarity has finally been achieved, and the spike at lag 1 identifies a model with at least one RMA of order 1. Also, the barely significant spike at lag 3 might indicate the need for one RMA of order 3 (plausible since an autocorrelation at lag

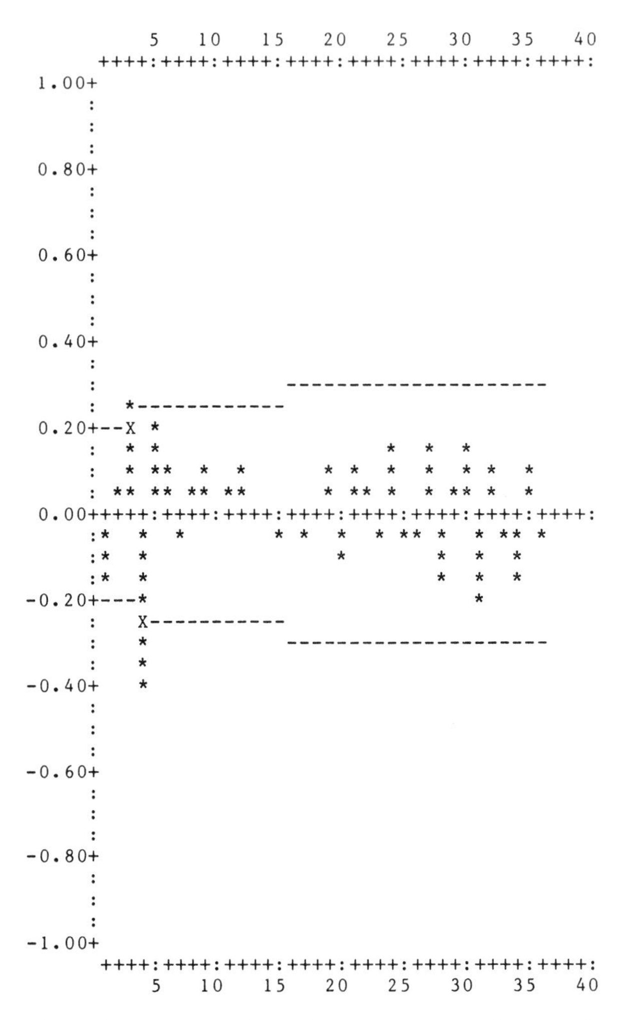

Figure 16.27 Residual AC Correlogram from Model of Figure 16.26

3 corresponds to a quarterly relationship). Thus a tentative model for this series is one RD, one SD of order 12, one RMA of order 1, and one RMA of order 3.

The estimation results for this model are given in Figure 16.34. These results indicate that the correlation between the two RMA parameters is fairly large (−.4271), which means that the RMA of order 3 may not really be needed. As further confirmation, note that the parameter confidence limits for the RMA of order 3 just barely include zero (upper limit = −.003). Thus even though all remaining statistics tend to validate the model, it would be reasonable to consider a model containing only the one RMA parameter of order 1, along with one RD and one SD of order 12.

The estimation results for this revised model are given in Figure 16.35. In these results all diagnostics indicate that the model is valid. Note also that excluding the

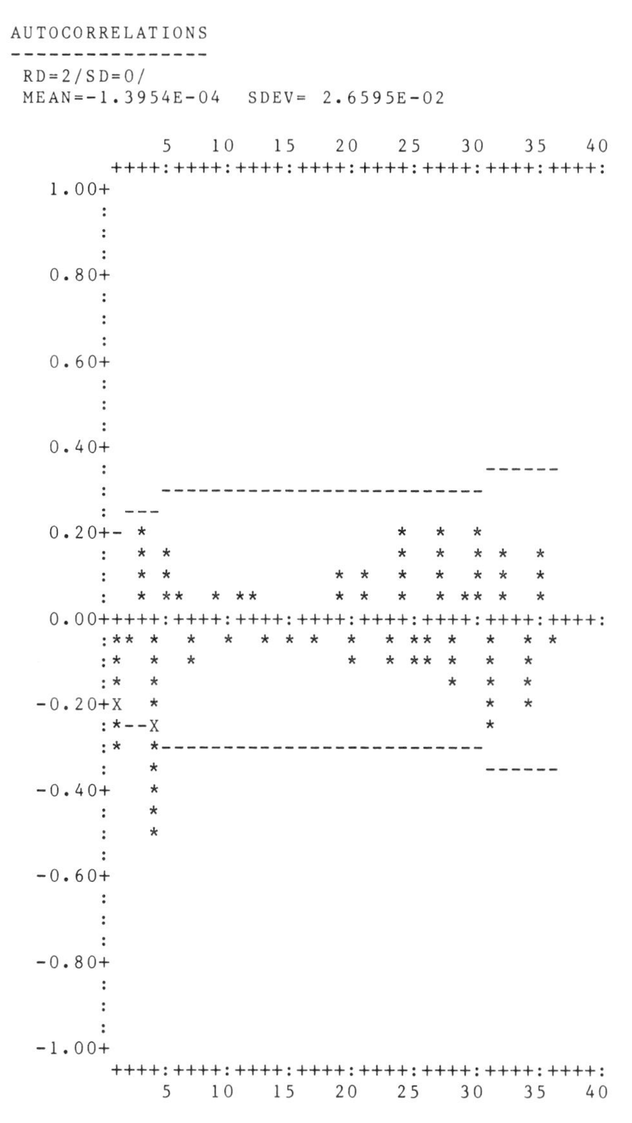

Figure 16.28 AC Correlogram for Insurance Claim Series After Applying Two RDs

RMA parameter of order 3 had no appreciable effect on the closeness-of-fit statistics, and the Q-statistics still remain significantly smaller than their test values. This model is therefore an acceptable and the preferable one for the series.

The final estimated model for the Sears sales series is

$$Z_t = -.7E_{t-1} + E_t$$

where $Z_t = X_t - X_{t-1} - X_{t-12} + X_{t-13}$ is the differenced series.

```
CORRELATION MATRIX OF THE PARAMETERS
------------------------------------

        1           2

 1   1.0000

 2  -0.0220      1.0000

----------------------------------------------------------------------

 2 REGULAR DIFFERENCES
 0 SEASONAL DIFFERENCES OF ORDER  4

PAR PARAMETER             PARAMETER  ESTIMATED          95 PERCENT
 #     TYPE                 ORDER               LOWER LIMIT   UPPER LIMIT

 1 REG MV   AVERAGE           1        0.304        0.097         0.512

 2 SEAS MV AVERAGE            4        0.729        0.557         0.901

----------------------------------------------------------------------

RESIDUAL MEAN            =          0.0012   NO. RESIDUALS        =  86
RESIDUAL SUM OF SQUARES  =          0.0342
RESIDUAL MEAN SQUARE     =          0.0004   DEGREES OF FREEDOM =  84
RESIDUAL STANDARD ERROR  =          0.0202

INDEX OF DETERMINATION   =     0.997
AVG ABSOULTE % ERROR     =     0.75%
MEAN % ERROR             =     0.03%

 LAGS        Q-STATISTIC     CHISQ AT 5%    DF
1 -  12          4.46          18.30        10
1 -  24          8.46          33.93        22
1 -  36         13.85          48.61        34
```

Figure 16.29 Model Estimation Results for Insurance Claim Series (Two RDs and One RMA and One SMA of Order 4)

Example 16.7: Airline Passengers

This series represents the monthly number, in thousands, of international airline passengers [2]. The series runs for 12 years, from January 1949 to December 1960. A graph of the last eight years of the series is shown in Figure 16.36.

This series is a classic Box-Jenkins example that appears in the original Box and Jenkins text [2] and much of the subsequent Box-Jenkins literature. One reason for its popularity is that the model for the series is one of the most common for seasonal economic series.

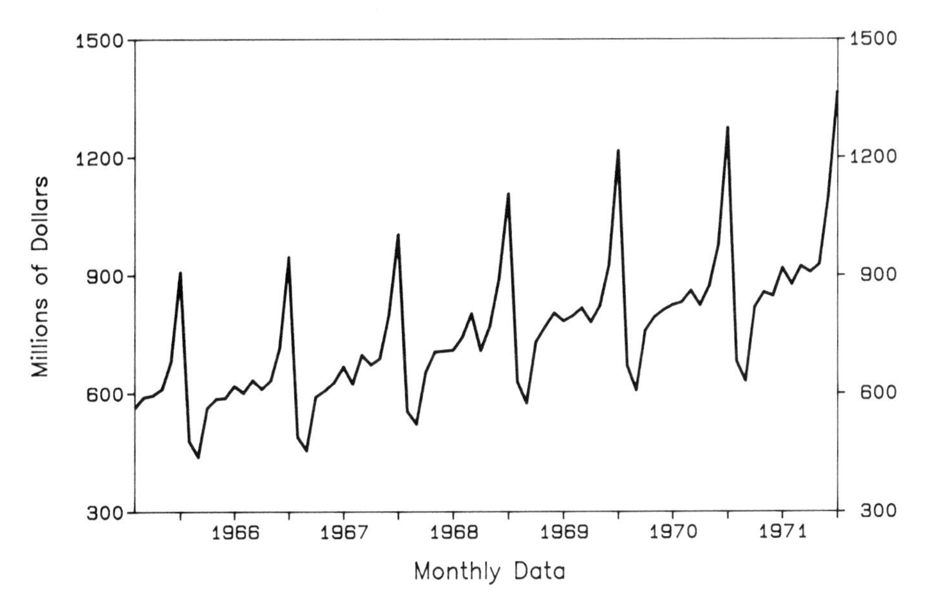

Figure 16.30 Sears Retail Sales Revenue

You will find that the model for this series is a familiar one (see Example 16.1) with one exception: namely, that instead of working with the original data, we will use logged data (i.e., the original data is transformed by computing the logarithm of each series value). This transformation is necessary, it turns out, because of a slight exponential trend in the series and the increasing magnitude of the seasonal variation. For details on how to identify the need for prior transformations, such as the logarithmic transformation, see Appendix C. (*Note:* When a log transformation is used, the forecasting results can be stated in terms of the original data as well as the logged data.)

The AC correlogram of the original undifferenced logged series is shown in Figure 16.37. This pattern indicates the obvious need for at least one regular difference, and the stair step pattern every 12 lags suggests the potential need for a seasonal difference of order 12.

Applying one regular difference first produces the AC correlogram shown in Figure 16.38. The need for at least one seasonal difference of order 12 is now clearly seen in this AC pattern.

After applying one RD and one SD of order 12, we obtain the AC correlogram shown in Figure 16.39. This AC pattern indicates that the differenced series is now stationary, and the two spikes at lags 1 and 12 indicate the need for one RMA and one SMA of order 12. (Although the spikes at lags 3 and 9 appear significant, they are barely so, and it is doubtful that they indicate the need for any more parameters.) A tentative model for this series is therefore one RD, one SD of order 12, one RMA, and one SMA of order 12.

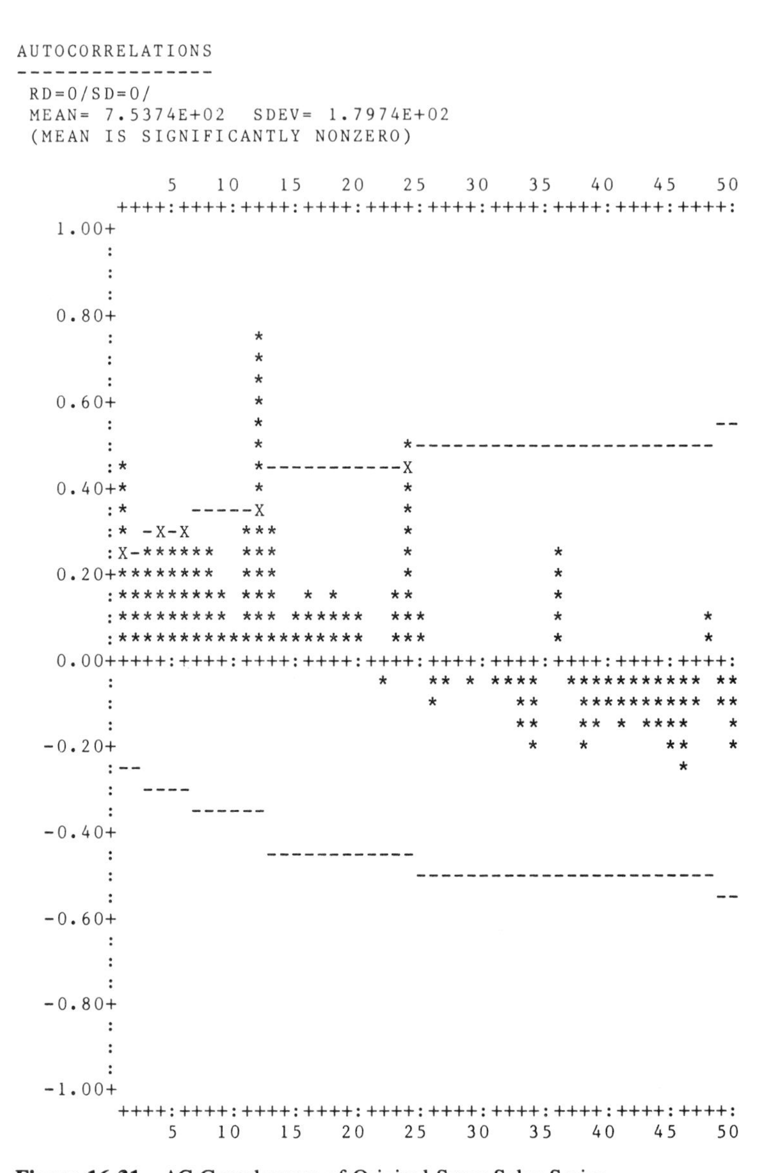

Figure 16.31 AC Correlogram of Original Sears Sales Series

The estimation results for this model are given in Figure 16.40. All diagnostics indicate that the model is valid. This model is thus represented as follows:

$$Z_t = -.377E_{t-1} - .571E_{t-12} + .215E_{t-13} + E_t$$

where $Z_t = {}^*X_t - {}^*X_{t-1} - {}^*X_{t-12} + {}^*X_{t-13}$ is the differenced series and the ${}^*X_t$ are logged values. This model is essentially the same as the model in Example 16.1. The only difference is the use of logged data.

```
AUTOCORRELATIONS
----------------
 RD=0/SD=1/PPS=12/
 MEAN= 5.5091E+01   SDEV= 2.8088E+01
 (MEAN IS SIGNIFICANTLY NONZERO)

              5    10   15   20   25   30   35   40   45   50
         ++++:++++:++++:++++:++++:++++:++++:++++:++++:++++:
   1.00+
        :
        :
        :
   0.80+
        :
        :
        :
   0.60+
        :
        :                             -----------------------
        :  *                    ------
   0.40+*  *                ----
        :*  *  -----------
        :**X-X*
        :XX****                                         *  *
   0.20+******                                       *   *  *
        :********                                  * *  ***** * *
        :*********                                 *********** *
        :*********                                ***************
   0.00+++++:++++:++++:++++:++++:++++:++++:++++:++++:++++:
        :         * ******************  *
        :           *****************
        :            * *************
  -0.20+             * ******** ****
        :--            ** *****  **
        :  ---         ** *****
        :     -----------   *  *
  -0.40+               ---- *
        :                   ------
        :                         -----------------------
        :
  -0.60+
        :
        :
        :
  -0.80+
        :
        :
        :
  -1.00+
         ++++:++++:++++:++++:++++:++++:++++:++++:++++:++++:
              5    10   15   20   25   30   35   40   45   50
```

Figure 16.32 AC Correlogram of Sears Sales Series After Applying One SD of Order 12

Example 16.8: New Car Registrations (United States)

This series represents the monthly number of new car registrations, in thousands, in the United States [27]. The complete series runs for 21 years, from January 1958 to December 1979 (252 periods). A graph of the series for the last 11 years is shown in Figure 16.41.

Like the airline series in the previous example, this series also requires a logarithmic transformation before proceeding with the modeling process. This transfor-

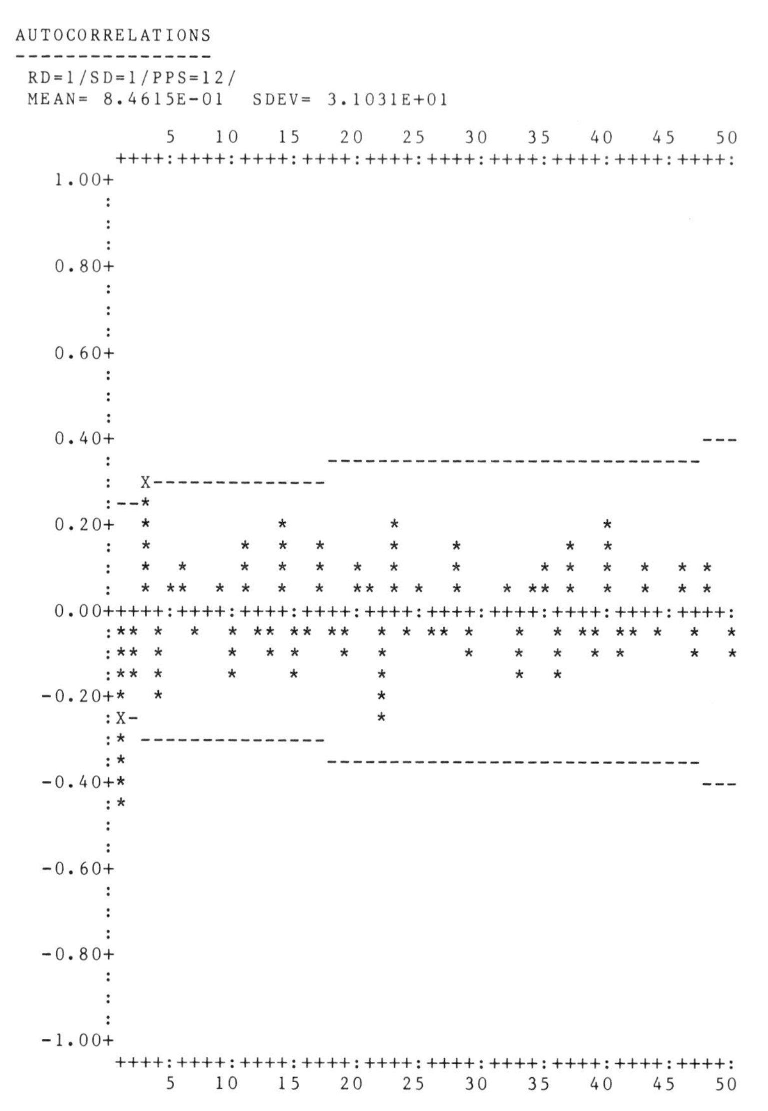

Figure 16.33 AC Correlogram of Sears Sales Series After Applying One RD and One SD of Order 12

mation is needed because of the increasing magnitude of the variation in the series. (For details on identifying the need for prior transformations, see Appendix C.) Note that a logarithmic transformation is usually needed when a monthly or quarterly series spans many years, as occurs in this example.

The AC correlogram for the original undifferenced logged series is shown in Figure 16.42. The need for regular differencing is immediately evident in this correlogram.

```
CORRELATION MATRIX OF THE PARAMETERS
------------------------------------

        1           2

  1  1.0000

  2 -0.4271      1.0000

----------------------------------------------------------------------

 1 REGULAR DIFFERENCES
 1 SEASONAL DIFFERENCES OF ORDER 12

PAR PARAMETER         PARAMETER  ESTIMATED         95 PERCENT
 #     TYPE             ORDER               LOWER LIMIT   UPPER LIMIT

 1 REG MV  AVERAGE        1        0.827        0.640         1.013

 2 REG MV  AVERAGE        3       -0.196       -0.388        -0.003

----------------------------------------------------------------------

RESIDUAL MEAN           =       -2.3075   NO. RESIDUALS        =  65
RESIDUAL SUM OF SQUARES =    36532.6289
RESIDUAL MEAN SQUARE    =      579.8831   DEGREES OF FREEDOM =  63
RESIDUAL STANDARD ERROR =       24.0808

INDEX OF DETERMINATION  =    0.982
AVG ABSOULTE % ERROR    =    2.39%
MEAN % ERROR            =   -0.06%


 LAGS      Q-STATISTIC   CHISQ AT 5%   DF
1 -  12        4.76         18.30      10
1 -  24       14.46         33.93      22
1 -  36       18.27         48.61      34
1 -  48       22.26         62.84      46
```

Figure 16.34 Model Estimation Results for Sears Sales Service (One RD, One SD of Order 12, and Two RMAs of Orders 1 and 3)

After applying one regular difference, we obtain the AC correlogram of Figure 16.43. Although the graph of the series did not indicate a strong seasonal pattern, the AC correlogram in Figure 16.43 does. The large, slowly decreasing spikes at lags 12, 24, 36, etc. leave no doubt about the presence of seasonality and the need for seasonal differencing.

The AC correlogram obtained after applying one RD and one SD of order 12 is shown in Figure 16.44. This correlogram indicates that the differenced logged series is definitely stationary. The only significant autocorrelations are at lags 2, 12, and 14. The large spike at lag 12, of course, indicates the need for one SMA of order 12, and the spike at lag 2 indicates the possible need for one RMA of order 2. The spike at lag 14, however, would appear to have no practical significance and is probably due to the interaction of the SMA of order 12 and RMA of order 2. We

```
 1 REGULAR DIFFERENCES
 1 SEASONAL DIFFERENCES OF ORDER 12

PAR PARAMETER            PARAMETER  ESTIMATED          95 PERCENT
 #     TYPE                ORDER                  LOWER LIMIT   UPPER LIMIT

 1 REG MV   AVERAGE          1          0.700          0.513          0.887

----------------------------------------------------------------------

RESIDUAL MEAN             =       -2.3572   NO. RESIDUALS       =  65
RESIDUAL SUM OF SQUARES =   38905.5781
RESIDUAL MEAN SQUARE      =     607.8997   DEGREES OF FREEDOM =  64
RESIDUAL STANDARD ERROR =      24.6556

INDEX OF DETERMINATION    =   0.980
AVG ABSOULTE % ERROR      =   2.53%
MEAN % ERROR              =  -0.07%


 LAGS      Q-STATISTIC   CHISQ AT 5%    DF
1 -  12        7.99          19.67       11
1 -  24       19.98          35.17       23
1 -  36       24.59          49.81       35
1 -  48       30.01          64.01       47
```

Figure 16.35 Model Estimation Results for Sears Sales Series (One RD, One SD of Order 12, and One RMA of Order 1)

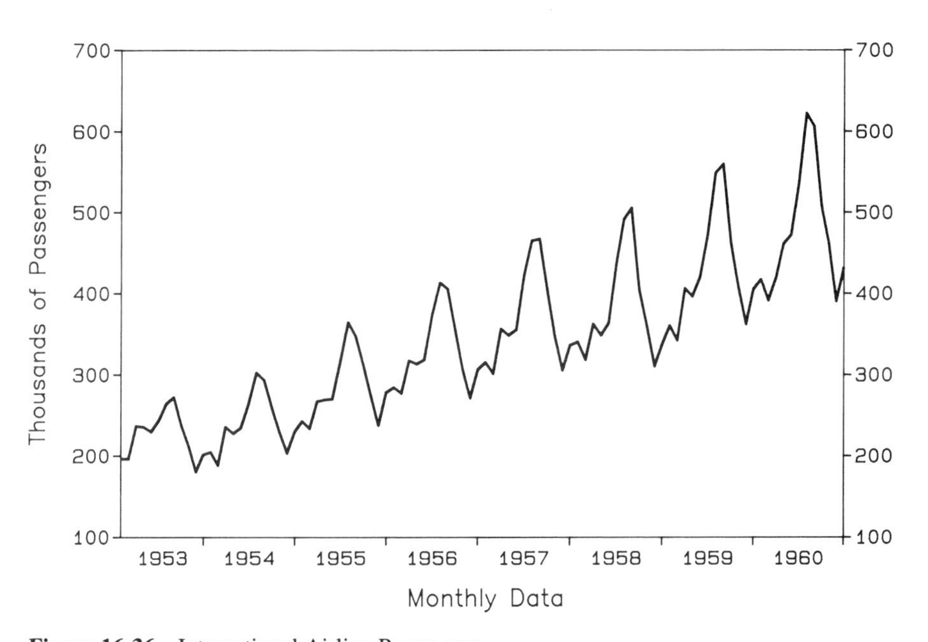

Figure 16.36 International Airline Passengers

```
AUTOCORRELATIONS
----------------
 RD=0/SD=0/LOGS/
 MEAN= 5.5421E+00   SDEV= 4.4145E-01
 (MEAN IS SIGNIFICANTLY NONZERO)

          5   10   15   20   25   30   35   40   45   50
      ++++:++++:++++:++++:++++:++++:++++:++++:++++:++++:
 1.00+
     :*
     :**
     :***                       ------------------------
 0.80+*****               ------
     :************    ----
     :*************---
     :***********XX*
 0.60+*********XX*****
     :*******XX********
     :*****XX******************
     :****X*********************
 0.40+***X***********************
     :**X**************************
     :*X***********************************
     :**************************************
 0.20+****************************************
     :X****************************************
     :*************************************************
     :**************************************************
 0.00+++++:++++:++++:++++:++++:++++:++++:++++:++++:++++:
     :
     :
     :-
-0.20+
     :
     : -
     :  -
-0.40+   -
     :    -
     :     --
     :       --
-0.60+         --
     :           --
     :             ---
     :                ----
-0.80+                    ------
     :                          ------------------------
     :
     :
     :
-1.00+
      ++++:++++:++++:++++:++++:++++:++++:++++:++++:++++:
          5   10   15   20   25   30   35   40   45   50
```

Figure 16.37 AC Correlogram of Original Logged Airline Passenger Series (Logged Values)

can therefore identify a tentative model of one RD, one SD of order 12, one RMA of order 2, and one SMA of order 12.

The estimation results for this model are shown in Figure 16.45. The parameter diagnostics indicate that the two parameters are uncorrelated and significant, so the model is not overspecified. The Q-statistics, however, indicate significant autocorrelation in the residuals within the first 12 lags. Thus another parameter probably needs to be in the model.

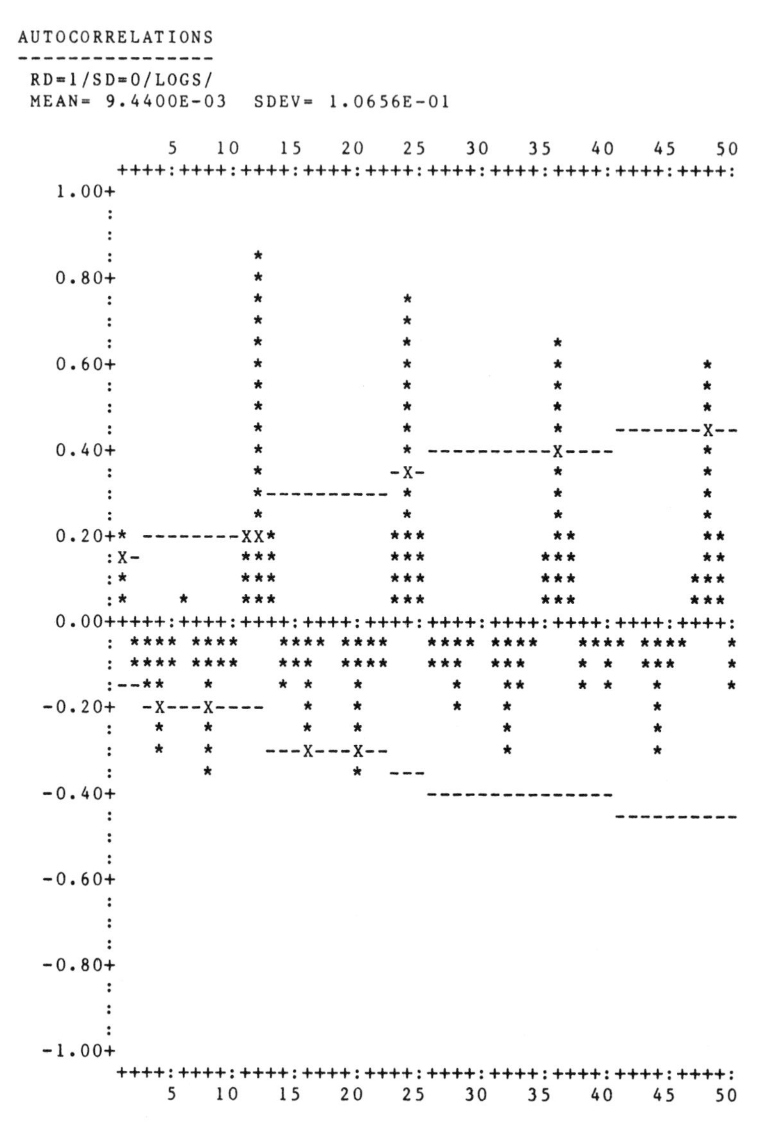

Figure 16.38 AC Correlogram of Logged Airline Passenger Series After Applying One RD

The AC correlogram of the residuals might give us a clue about what's missing. The residual AC correlogram is shown in Figure 16.46. This correlogram provides the clue to the missing parameter, namely, one RMA of order 1. Although the need for this parameter was not evident in the AC correlogram of the differenced series, it is unmasked after the other relationships in the series have been accounted for by the model containing one RMA of order 2 and one SMA of order 12. Recall, also,

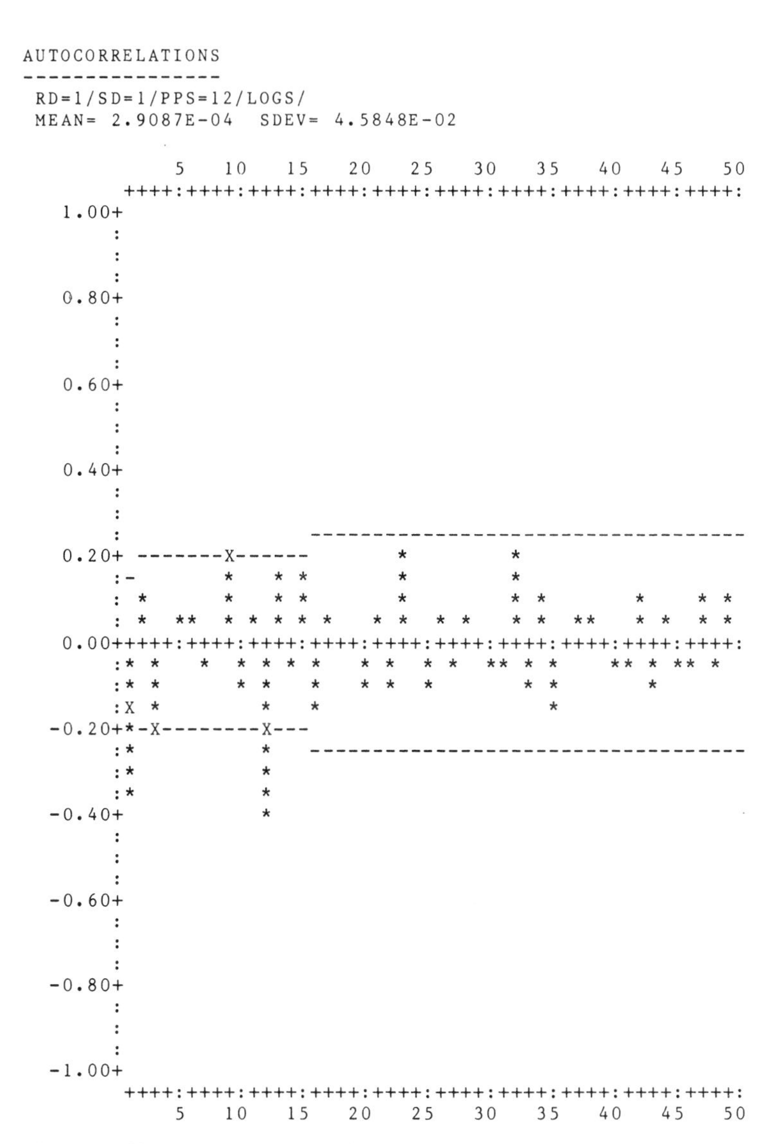

Figure 16.39 AC Correlogram of Logged Airline Passenger Series After Applying One RD and One SD of Order 12

that there is always a small chance that some insignificant sample autocorrelations should really be interpreted as significant (and vice versa).

After including the RMA of order 1 in the model, we obtain the estimation results shown in Figure 16.47. All diagnostics now indicate an acceptable model. There is a slight correlation between the two RMA parameters, but it is not enough to worry about.

```
CORRELATION MATRIX OF THE PARAMETERS
------------------------------------

        1            2

  1   1.0000

  2  -0.0913       1.0000


----------------------------------------------------------------------


 1 REGULAR DIFFERENCES
 1 SEASONAL DIFFERENCES OF ORDER 12

PAR PARAMETER              PARAMETER  ESTIMATED          95 PERCENT
 #     TYPE                  ORDER                  LOWER LIMIT   UPPER LIMIT

 1 REG MV   AVERAGE            1          0.377          0.213         0.541


 2 SEAS MV AVERAGE            12          0.571          0.416         0.726


----------------------------------------------------------------------


RESIDUAL MEAN             =      -0.0020    NO. RESIDUALS        = 131
RESIDUAL SUM OF SQUARES   =       0.1819
RESIDUAL MEAN SQUARE      =       0.0014    DEGREES OF FREEDOM = 129
RESIDUAL STANDARD ERROR   =       0.0376

INDEX OF DETERMINATION    =   0.991
AVG ABSOULTE % ERROR      =   0.52%
MEAN % ERROR              =  -0.04%


 LAGS      Q-STATISTIC    CHISQ AT 5%    DF
1 -  12        7.55          18.30        10
1 -  24       19.87          33.93        22
1 -  36       27.22          48.61        34
1 -  48       32.48          62.84        46
```

Figure 16.40 Model Estimation Results for Logged Airline Passenger Series One RD, One SD of Order 12, and One RMA and One SMA of Order 12

The final estimated model for the car registration series is therefore

$$Z_t = -.188E_{t-1} - .248E_{t-2} - .802E_{t-12} + .151E_{t-13} + .199E_{t-14} + E_t$$

where $Z_t = \dot{X}_t - \dot{X}_{t-1} - \dot{X}_{t-12} + \dot{X}_{t-13}$ is the differenced series and $\dot{X}_t$ is the logged series. (*Note:* $.151 = .802 \times .188$ and $.199 = .802 \times .248$.)

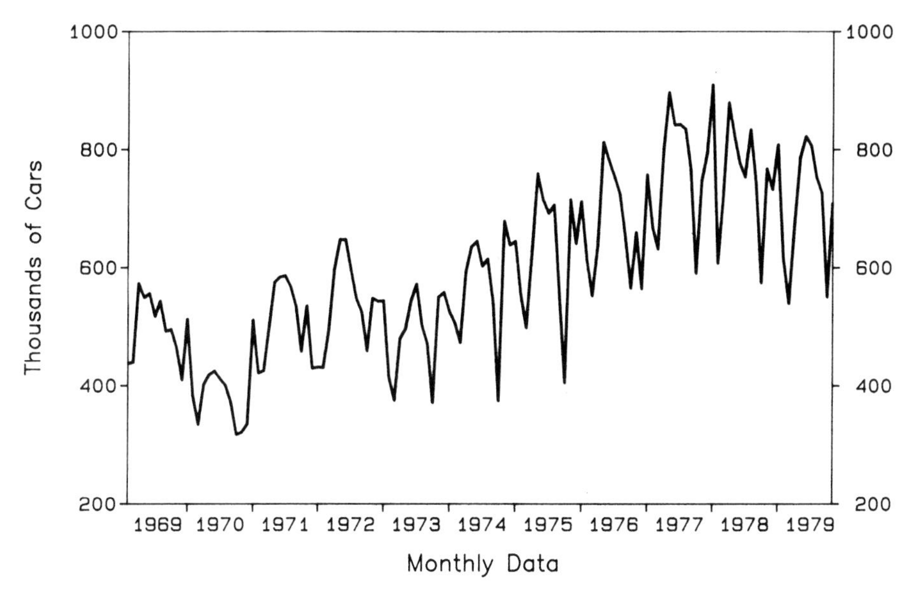

Figure 16.41 U.S. New Car Registrations

16.2 REVIEW OF KEY CONCEPTS

The Box-Jenkins modeling process for nonstationary series was illustrated in Figure 13.3. This process can now be updated to accommodate building models for seasonal series, as shown in Figure 16.48. This diagram, therefore, represents the complete Box-Jenkins modeling process. The added steps include the following:

- Apply a prior transformation if needed (see Appendix C).
- Apply seasonal differencing until stationarity is achieved in the seasonal pattern.
- Identify seasonal parameters from autocorrelations at lags that are multiples of the number of periods per season.

Again, the main principle to keep in mind while constructing Box-Jenkins models is to keep the model as simple as possible:

- Do not overdifference.
- Do not overparameterize.

In the literature this principle is alliteratively known as the "principle of parsimonious parameterization."

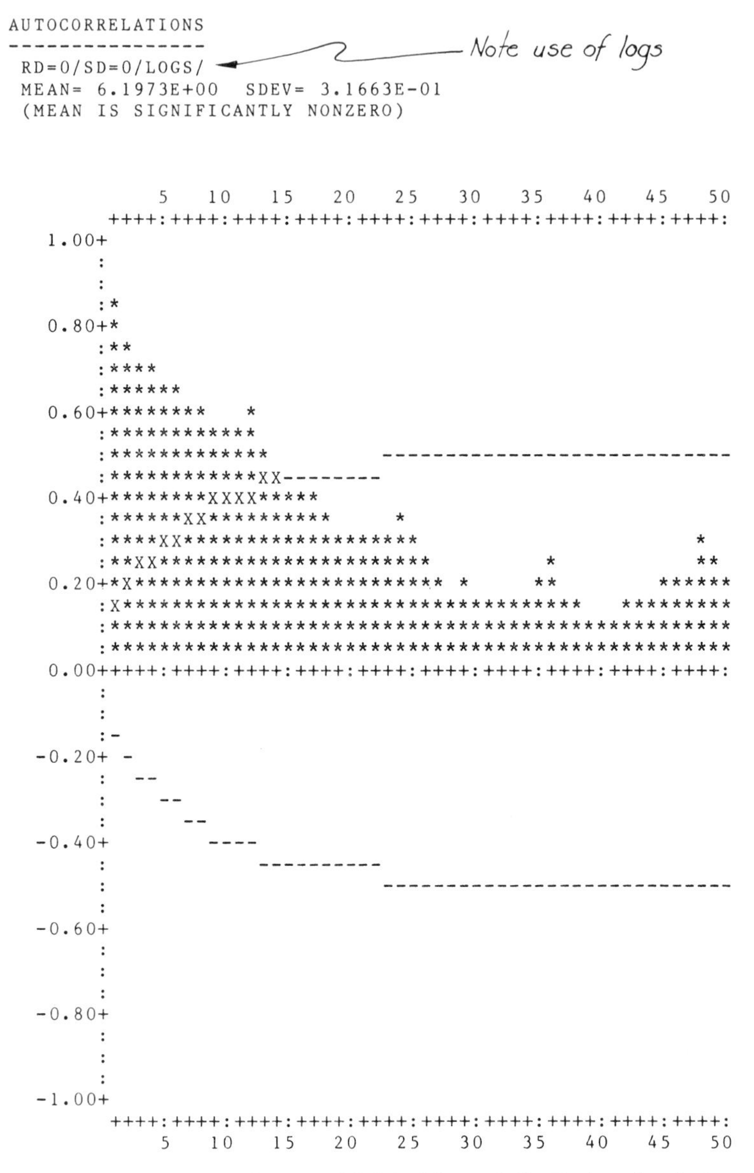

Figure 16.42 AC Correlogram of Original Logged Car Registration Series

```
AUTOCORRELATIONS
----------------
 RD=1/SD=0/LOGS/
 MEAN= 5.0263E-03   SDEV= 1.5938E-01
```

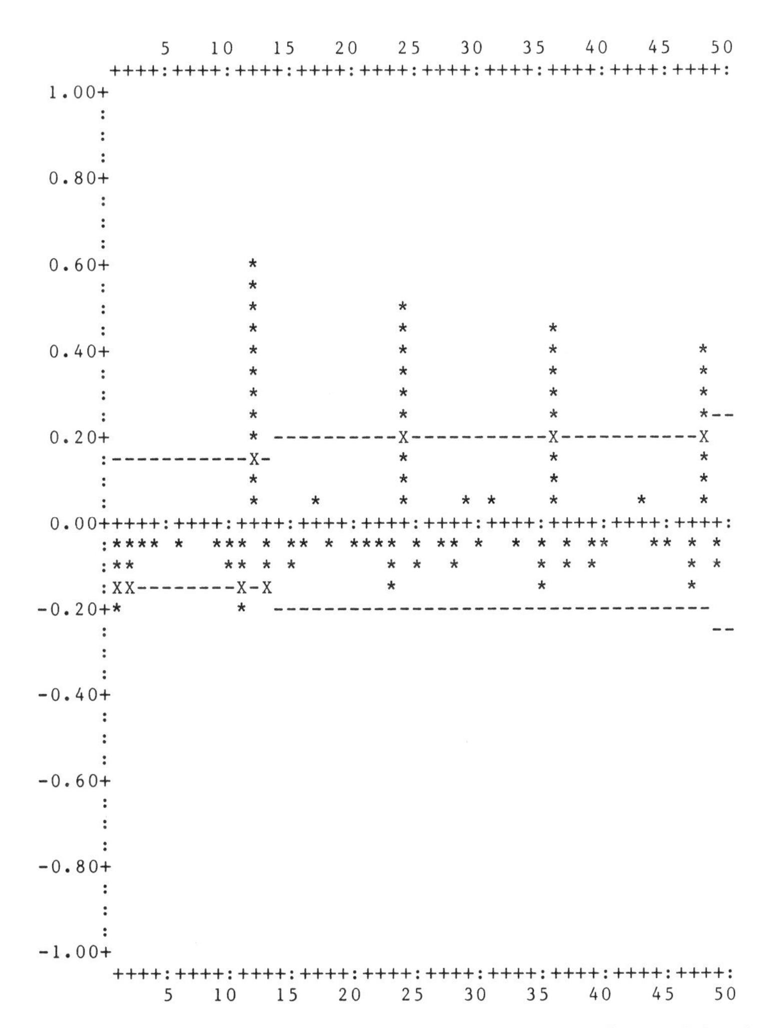

Figure 16.43 AC Correlogram of Logged Car Registration After Applying One RD

```
AUTOCORRELATIONS
----------------
 RD=1/SD=1/PPS=12/LOGS/
 MEAN=-1.5274E-03   SDEV= 1.4317E-01
```

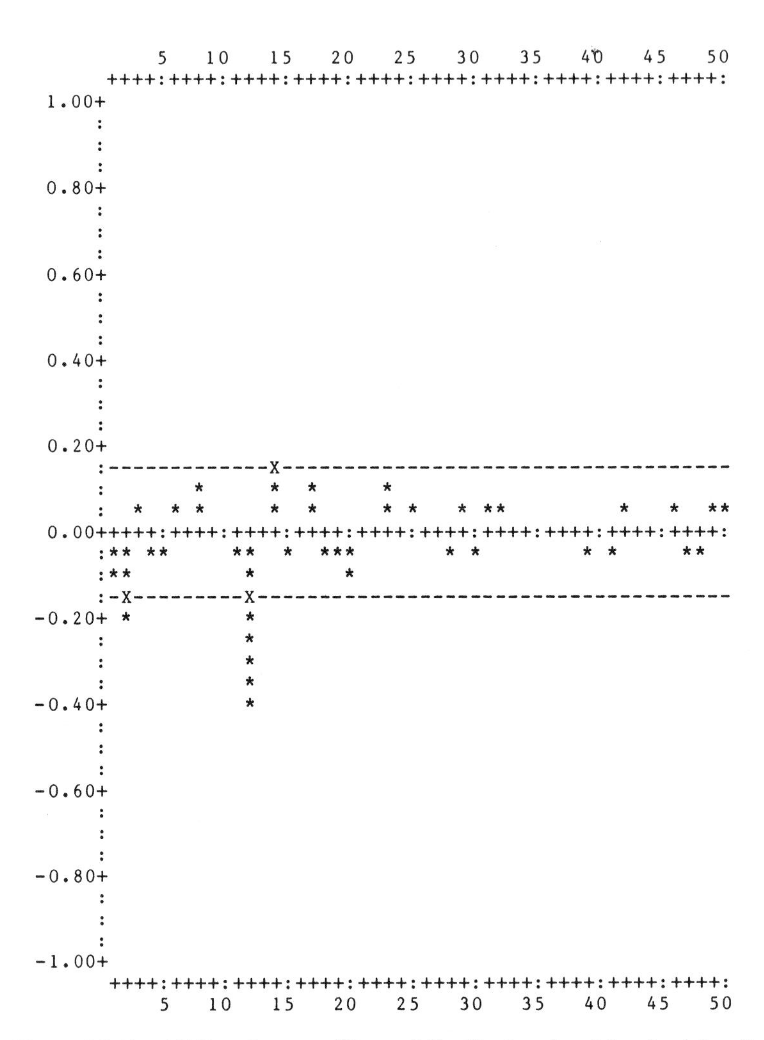

Figure 16.44 AC Correlogram of Logged Car Registration After Applying One RD and One SD of Order 12

```
CORRELATION MATRIX OF THE PARAMETERS
------------------------------------

        1            2

  1  1.0000

  2  0.0298       1.0000

----------------------------------------------------------------------

 1 REGULAR DIFFERENCES
 1 SEASONAL DIFFERENCES OF ORDER 12

PAR PARAMETER          PARAMETER  ESTIMATED         95 PERCENT
 #     TYPE              ORDER               LOWER LIMIT   UPPER LIMIT

 1 REG MV  AVERAGE         2         0.246        0.120         0.373


 2 SEAS MV AVERAGE        12         0.793        0.711         0.875


----------------------------------------------------------------------


RESIDUAL MEAN           =       0.0067   NO. RESIDUALS        = 239
RESIDUAL SUM OF SQUARES =       3.2061
RESIDUAL MEAN SQUARE    =       0.0135   DEGREES OF FREEDOM = 237
RESIDUAL STANDARD ERROR =       0.1163

INDEX OF DETERMINATION  =   0.837
AVG ABSOULTE % ERROR    =   1.35%
MEAN % ERROR            =   0.13%


 LAGS      Q-STATISTIC   CHISQ AT 5%    DF
1 -  12       20.71         18.30       10
1 -  24       30.10         33.93       22
1 -  36       34.64         48.61       34
1 -  48       41.24         62.84       46
```

Figure 16.45 Model Estimation Results for Logged Car Registration Series (One RD, One SD of Order 12, and One RMA of Order 2 and One SMA of Order 12)

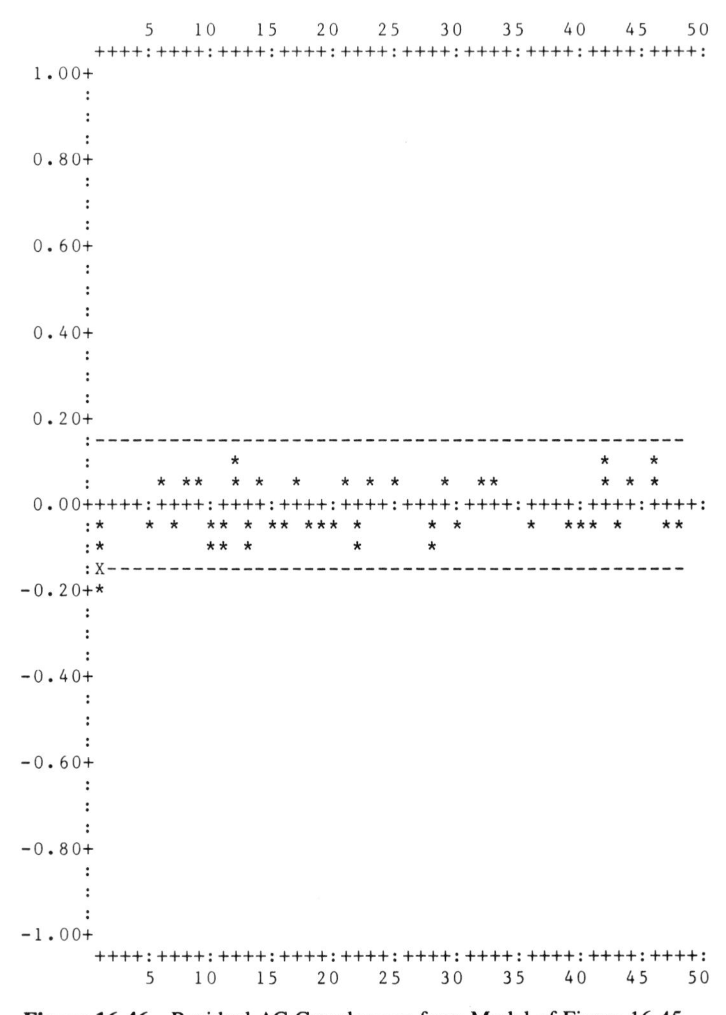

Figure 16.46 Residual AC Correlogram from Model of Figure 16.45

```
CORRELATION MATRIX OF THE PARAMETERS
------------------------------------

        1            2            3

  1  1.0000

  2 -0.2492       1.0000

  3  0.0044       0.0091       1.0000

---------------------------------------------------------------------------

 1 REGULAR DIFFERENCES
 1 SEASONAL DIFFERENCES OF ORDER 12

PAR PARAMETER          PARAMETER  ESTIMATED          95 PERCENT
 #     TYPE              ORDER                 LOWER LIMIT   UPPER LIMIT

 1 REG MV  AVERAGE          1        0.188         0.062         0.315

 2 REG MV  AVERAGE          2        0.248         0.122         0.375


 3 SEAS MV AVERAGE         12        0.802         0.722         0.883


---------------------------------------------------------------------------


RESIDUAL MEAN           =       0.0094   NO. RESIDUALS      = 239
RESIDUAL SUM OF SQUARES =       3.0884
RESIDUAL MEAN SQUARE    =       0.0131   DEGREES OF FREEDOM = 236
RESIDUAL STANDARD ERROR =       0.1144

INDEX OF DETERMINATION  =   0.843
AVG ABSOULTE % ERROR    =   1.36%
MEAN % ERROR            =   0.18%


 LAGS      Q-STATISTIC   CHISQ AT 5%   DF
1 -  12       11.01         16.91       9
1 -  24       19.31         32.67      21
1 -  36       23.37         47.41      33
1 -  48       29.82         61.66      45
```

Figure 16.47 Model Estimation Results for Logged Car Registration Series (One RD, One SD of Order 12, and Two RMAs and One SMA of Order 12)

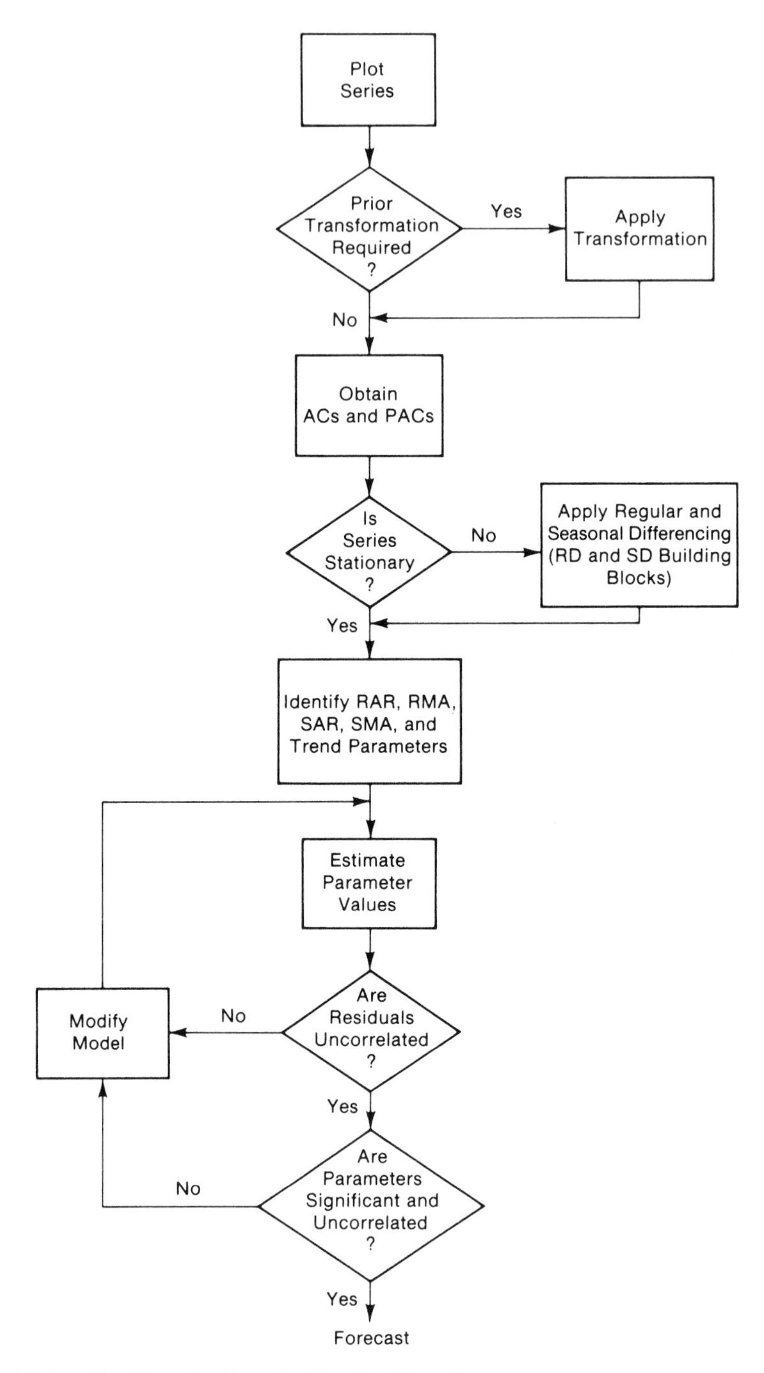

Figure 16.48 The Complete Box-Jenkins Modeling Process

Part 4

Putting the Model to Work

The major portion of this book, Parts 2 and 3, has been devoted to learning how to build Box-Jenkins forecasting models: the identification and estimation phases of the Box-Jenkins model-building process. We are now ready to put these models to work in the forecasting phase of Box-Jenkins.

Although it is an easy matter to generate forecasts of a time series from an estimated Box-Jenkins model, there is, as you learned in Part 1, much more to time series forecasting than simply generating numerical predictions for future time periods. An assessment of forecast accuracy, for example, is extremely important to the user of a forecast, and the tracking and updating of forecasts and forecasting models is an important part of the forecasting process.

In Part 4 you will learn about some forecasting tools and techniques that will help you:

- Measure forecast accuracy and uncertainty.
- Revise forecasts on the basis of new data.
- Update forecasting models.

CHAPTER **17**

Forecasting Tools and Techniques

After a Box-Jenkins model has been identified and estimated, it can immediately be used to generate forecasts of the time series for future time periods, and it can provide measurements of the uncertainty in the forecasts. In addition, when new data becomes available, the model may be used to generate revised forecasts on the basis of the new data, and the model itself may be updated.

In this chapter you will learn:

- How forecasts are generated from a Box-Jenkins model.
- How to measure the uncertainty in forecasts by using forecast confidence limits.
- How to judge the potential accuracy of forecasts by using after-the-fact forecasts.
- How to update forecasts and forecasting models on the basis of new data.

You will also get to see the forecasting results for each of the models developed in Chapter 16.

17.1 OBTAINING FORECASTS

As you learned in Part 1, once you have a forecasting model formula in which the parameter values are known, the formula may be used to generate forecasts. Thus once a Box-Jenkins model has been estimated (i.e., once the parameter values, fitted values, and residuals have been computed), forecasts may be easily generated by substituting the appropriate past series values, residual values, and previously generated forecasts into the model formula.

For example, suppose a Box-Jenkins model for a given series consisted of one RD, one AR, and one MA with parameter values $A_1 = .7$ and $B_1 = .5$. The model formula would then be written as follows:

$$\overbrace{X_t - X_{t-1}}^{Z_t} = .7(\overbrace{X_{t-1} - X_{t-2}}^{Z_{t-1}}) - .5E_{t-1} + E_t$$

or

$$X_t = \overbrace{1.7X_{t-1} - .7X_{t-2} - .5E_{t-1}}^{F_t} + E_t$$

Now if the series contained 50 values with $X_{49} = 292.0$ and $X_{50} = 319.0$, and the fitted value at period 50 was computed to be $F_{50} = 323.0$, then $E_{50} = -4.0$ and the first forecast for period 51 would be computed as follows:

$$\begin{aligned} F_{51} &= 1.7X_{50} - .7X_{49} - .5E_{50} \\ &= (1.7)(319.0) - (.7)(292.0) - (.5)(-4.0) \\ &= 339.9 \end{aligned}$$

The next forecast for period 52 would be

$$F_{52} = 1.7X_{51} - .7X_{50} - .5E_{51}$$

where $E_{51} = 0$ (since the random error component is assumed to be zero for all forecast periods) and $X_{51} = F_{51}$ (since $E_{51} = 0$). Thus

$$\begin{aligned} F_{52} &= (1.7)(339.9) - (.7)(319) \\ &= 354.53 \end{aligned}$$

Fortunately, you do not have to actually do these computations, since your Box-Jenkins program will do them for you automatically. All you have to do is indicate how many forecasts you want. For example, a table and a graph of the forecasts generated from the model for the paper industry sales series in Example 16.1 are shown in Figure 17.1. Twenty-four forecasts were generated, beginning with the first forecast period, January 1973.

17.2 ASSESSING THE ACCURACY OF FORECASTS

Since any forecast has a degree of uncertainty, or expected error, associated with it, the use of a forecast in any decision-making process makes sense only if you have some feeling for what the forecast error could possibly be, i.e., only if you can

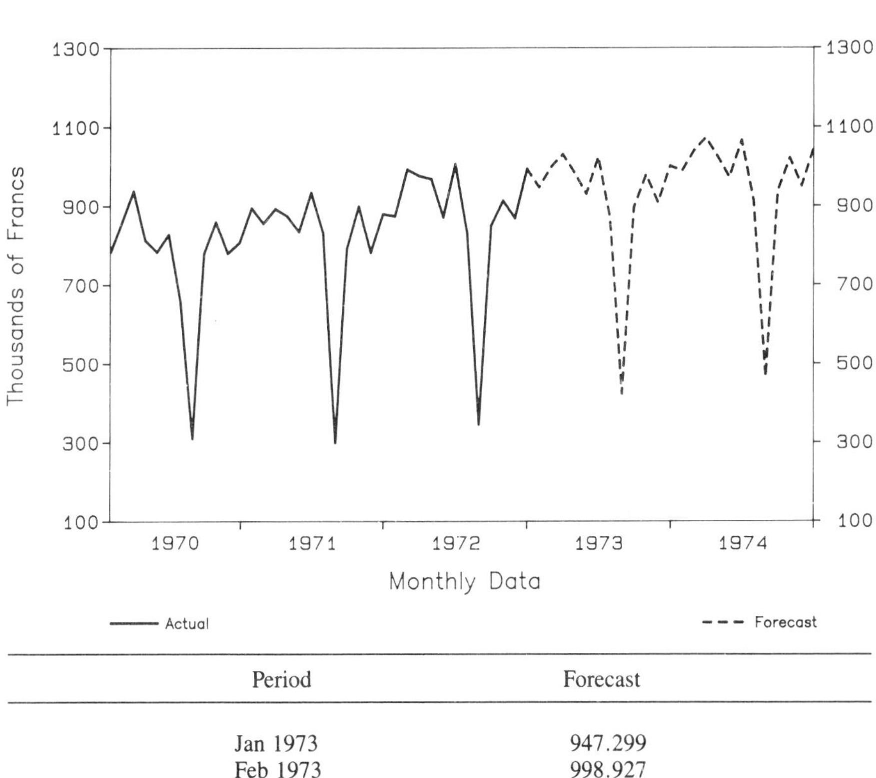

Period	Forecast
Jan 1973	947.299
Feb 1973	998.927
Mar 1973	1031.039
Apr 1973	983.977
May 1973	930.587
Jun 1973	1024.637
Jul 1973	870.783
Aug 1973	423.393
Sep 1973	896.050
Oct 1973	979.936
Nov 1973	908.651
Dec 1973	1001.658
Jan 1974	989.029
Feb 1974	1040.656
Mar 1974	1072.768
Apr 1974	1025.706
May 1974	972.317
Jun 1974	1066.367
Jul 1974	912.512
Aug 1974	465.122
Sep 1974	937.779
Oct 1974	1021.665
Nov 1974	950.381
Dec 1974	1043.387

Figure 17.1 Table and Graph of Forecasts Obtained from the Model for Paper Industry Sales

assess beforehand the accuracy of the forecast in some meaningful way. One way to do so is provided by forecast confidence limits that can be computed from a given forecasting model. Forecast confidence limits are essentially a high and a low forecast, or forecast range, about which statistical statements can be made concerning how much confidence you can have in the validity of the range. Another way to assess the accuracy of forecasts is to start generating forecasts from your model at some time period before the end of the series. These "after-the-fact" forecasts for the final periods of the series can then be compared with the actual data already known for these periods. Forecast confidence limits and after-the-fact forecasts are described in the following sections.

17.2.1 Forecast Confidence Limits

When a model has been estimated for a series, the magnitude of the random error component of the series can be measured by some of the closeness-of-fit statistics. For example, the average absolute error and the residual standard error provide two such yardsticks. It turns out that the residual standard error is also used to compute certain statistics that help you measure the expected error in a forecast. These statistics are called forecast confidence limits.

Specifically, *forecast confidence limits* are values placed above and below a forecast, indicating a range, or *confidence interval,* in which the future actual series value is expected to fall. The forecast itself is located at the midpoint of the confidence interval.* Obviously, the larger the interval, the more confidence we will have that the future actual value will actually fall in the confidence interval. Thus forecast confidence limits are computed relative to a certain degree, or level, of confidence. A level of confidence is usually expressed as some percentage. For example, a 95% level of confidence indicates that for 95 out of every 100 forecasts, the corresponding actual series values (when they occur) should actually fall between the forecast confidence limits. A 95% level of confidence is the level most commonly used, although many Box-Jenkins programs can compute confidence limits at other confidence levels, such as 50%, 75%, 90%, and 99%.

Figure 17.2 shows the forecasts and forecast confidence limits produced from the Box-Jenkins model for the paper industry sales series. For example, the forecast confidence limits for January 1973 are 859.99 and 1034.609. Thus the future actual value for January 1973 can be expected to fall between the range of 859,990 and 1,034,609 francs with 95% confidence. To put it another way, we can say that the forecast of 947,299 francs for this period will not be in error by more than 87,310 francs (87,310 = 1,034,609 − 947,299 = 947,299 − 859,990). Note that the forecast confidence interval widens as the forecast horizon lengthens. This result expresses, statistically, the intuitive idea that your confidence in a forecast decreases the further into the future the forecast is made.

*When a prior transformation is used, as in Examples 16.7 and 16.8, the confidence limits will no longer be symmetric about the forecast value (see Figures 17.9 and 17.10 and Appendix C).

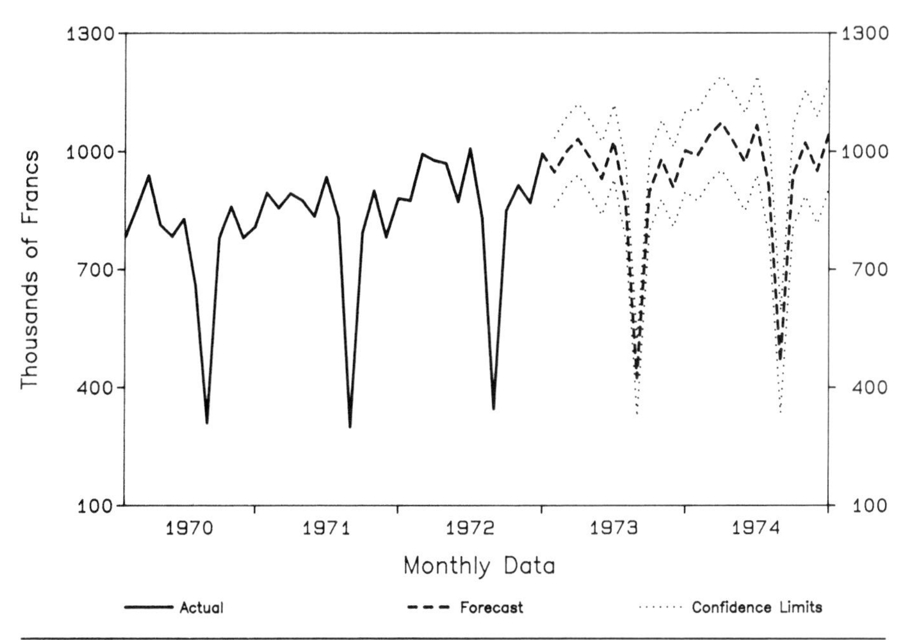

Period	Lower Limit	Forecast	Upper Limit	Confidence Interval
Jan 1973	859.990	947.299	1034.609	174.620
Feb 1973	910.081	998.927	1087.772	177.689
Mar 1973	940.683	1031.039	1121.395	180.707
Apr 1973	892.136	983.977	1075.818	183.684
May 1973	837.285	930.587	1023.890	186.605
Jun 1973	929.896	1024.637	1119.379	189.484
Jul 1973	774.624	870.783	966.942	192.318
Aug 1973	325.836	423.393	520.948	195.112
Sep 1973	797.116	896.050	994.983	197.867
Oct 1973	879.644	979.936	1080.227	200.586
Nov 1973	807.019	908.651	1010.283	203.261
Dec 1973	898.703	1001.658	1104.612	205.907
Jan 1974	873.985	989.029	1104.072	230.085
Feb 1974	923.331	1040.656	1157.980	234.649
Mar 1974	953.206	1072.768	1192.330	239.124
Apr 1974	903.948	1025.706	1147.464	243.512
May 1974	848.401	972.317	1096.232	247.829
Jun 1974	940.330	1066.367	1192.403	252.070
Jul 1974	784.390	912.512	1040.634	256.240
Aug 1974	334.948	465.122	595.296	260.348
Sep 1974	805.584	937.779	1069.973	264.386
Oct 1974	887.481	1021.665	1155.849	268.369
Nov 1974	814.235	950.381	1086.526	272.295
Dec 1974	905.309	1043.387	1181.465	276.161

Figure 17.2 Table and Graph of Forecasts and 95% Forecast Confidence Limits Obtained from the Model for Paper Industry Sales

Note: When interpreting the width of forecast confidence intervals from a graph, you must be careful to measure the distance from the lower to the upper confidence limit vertically. For example, the confidence band between July 1973 and August 1973 in the graph in Figure 17.2 looks deceivingly narrow because of your eyes' tendency to measure the width of the band horizontally on steep slopes. Actually, the confidence interval in this region is wider than the confidence interval for previous time periods, as can be verified by looking at the table.

What happens when we choose a different confidence level? Figure 17.3 shows a graph of the same forecasts depicted in Figure 17.2 but with confidence limits computed at a 75% level of confidence. Note that these confidence limits provide narrower confidence intervals than those at a 95% level of confidence. Thus when using the 75% forecast confidence limits, we can say future actual values can possibly differ from the forecasts by a lesser amount than when using 95% forecast confidence limits, but we say so with less confidence.

17.2.2 After-the-Fact Forecasting

Another technique you can use to assess the accuracy of the forecasts produced from a Box-Jenkins model is to see how well the model generates forecasts for some time periods prior to the last period in the series. That is, after estimating a model by using all the series data, pretend that the last few periods of data are not there, and begin forecasting from an earlier period. This technique allows you to compare forecasts computed from the current model with actual series values that are already known. Such forecasts are called *after-the-fact forecasts*. They provide another way to measure the accuracy that can be expected in a forecast for the "real" future. The period at which forecasting begins is called the *forecast origin*. Your Box-Jenkins program will allow you to indicate a forecast origin other than the normal forecast origin.

For example, using the same model we used before for the paper industry sales series, we can specify a new forecast origin 12 periods before the end of the series (January 1972), and after-the-fact forecasts and their confidence limits will be produced for January 1972 through December 1972. These forecasts may then be compared with the actual known values for those periods, as shown in Table 17.1.

Note that forecasts were also produced for January 1973 and beyond (i.e., for the real future). Since these forecasts are based on the forecast origin of January 1972, they are different from those produced by the same model using the normal forecast origin of January 1973 (see Figure 17.1). That is, when the forecast origin is moved to an earlier period, the actual series values following the forecast origin do not take part in producing forecasts for those periods or beyond.

Note also that the actual values for January 1972 through December 1972 fall well within the 95% confidence limits, and the largest after-the-fact forecast error (95,608 francs) occurs in February 1972. From these results we can get a good feeling for how accurate forecasts of the real future generated by our model are expected to be.

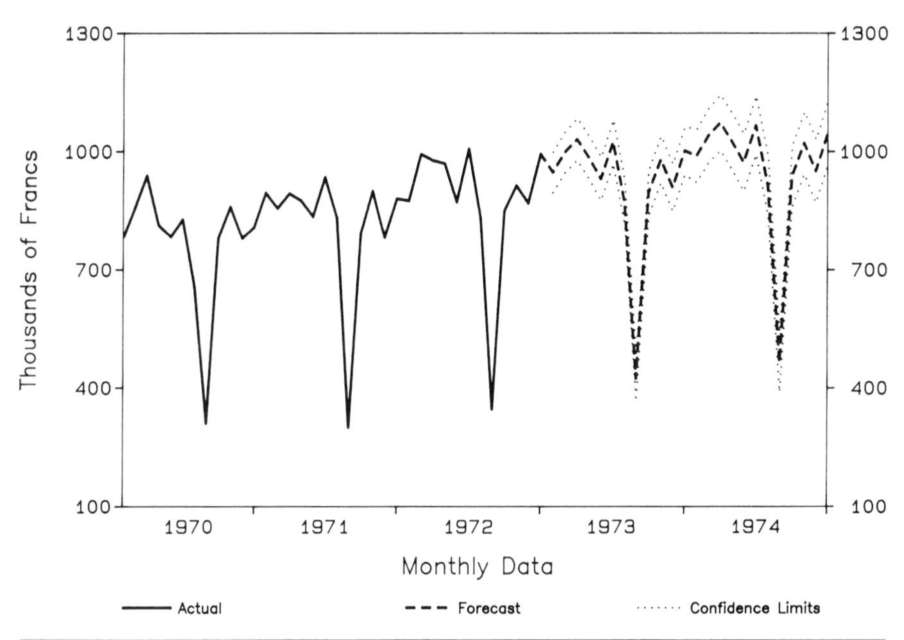

Period	Lower Limit	Forecast	Upper Limit	Confidence Interval
Jan 1973	896.072	947.299	998.527	102.455
Feb 1973	946.798	998.927	1051.055	104.262
Mar 1973	978.024	1031.039	1084.054	106.026
Apr 1973	930.090	983.977	1037.863	107.770
May 1973	875.843	930.587	985.331	109.488
Jun 1973	969.049	1024.637	1080.226	111.181
Jul 1973	814.363	870.783	927.203	112.840
Aug 1973	366.153	423.393	480.632	114.479
Sep 1973	838.002	896.050	954.097	116.095
Oct 1973	921.091	979.936	1038.780	117.689
Nov 1973	849.020	908.651	968.282	119.262
Dec 1973	941.251	1001.658	1062.065	120.809
Jan 1974	921.529	989.029	1056.528	135.001
Feb 1974	971.819	1040.656	1109.494	137.673
Mar 1974	1002.617	1072.768	1142.919	140.300
Apr 1974	954.266	1025.706	1097.146	142.884
May 1974	899.611	972.317	1045.022	145.409
Jun 1974	992.417	1066.367	1140.317	147.903
Jul 1974	837.338	912.512	987.686	150.348
Aug 1974	388.744	465.122	541.500	152.756
Sep 1974	860.216	937.779	1015.342	155.124
Oct 1974	942.934	1021.665	1100.396	157.466
Nov 1974	870.500	950.381	1030.261	159.760
Dec 1974	962.372	1043.387	1124.402	162.028

Figure 17.3 Table and Graph of Forecasts and 75% Forecast Confidence Limits Obtained from the Model for Paper Industry Sales

You can also compute closeness-of-fit statistics for the after-the-fact forecast errors in the same way as is done for the residuals. For example, the 12 forecast errors shown in Figure 17.1 have a mean percent error of 3.39% and an average absolute percent error of 4.91%. These statistics compare with the residual mean percent error of 1.76% and residual average absolute percent error of 5.76% (see estimation results in Figure 16.7). The relatively large forecast mean percent error (3.39% versus 1.76%) suggests that there might be a tendency for forecasts of the real future to underestimate, rather than overestimate, the actual value, while the slightly smaller forecast average absolute percent error (4.91% versus 5.76%) suggests that the forecasts might be slightly closer to the actual values than the fitted values were.

17.2.3 Some Further Examples

Figures 17.4 through 17.10 show the forecasting results generated by the models constructed for each of the remaining series in Chapter 16 (Examples 16.2 through 16.8). For each example a graph and table of the forecasts and their 95% forecast confidence limits are shown. The figure associated with each series is listed below:

Figure 17.4: Postal volume
Figure 17.5: Lawn chair sales revenue
Figure 17.6: Champagne sales
Figure 17.7: Insurance claims
Figure 17.8: Sears retail sales revenue
Figure 17.9: Airline passengers
Figure 17.10: Car registrations

Note that the forecast results for the airline passenger series and the car registration series (Figures 17.9 and 17.10) show confidence limits that are not symmetric about the forecast values. This result is due to the use of logged data instead of the original data when estimating the models.

17.3 UPDATING FORECASTS AND MODELS

As additional data for a time series is made available, the same model estimated for the original series can also be used to generate new (revised) forecasts based on this additional data. (Many Box-Jenkins programs allow you to save your estimated model so that you can use it for forecasting over and over again without having to reestimate it.) Table 17.2 compares the old and new forecasts (using the same model for the paper industry sales series) after three additional periods of actual data (values 946.23, 1013.38, and 1051.97) were added to the end of the series for January through March 1973. The new data is not used to reeestimate the model; it only

Table 17.1 After-the-Fact Forecasts and 95% Forecast Confidence Limits Obtained from the Model for the Paper Industry Sales Series, Using an Earlier Forecast Origin

Period	Lower Limit	Forecast	Upper Limit	Actual	Forecast Error
Jan 1972	802.552	889.862	977.171	875.024	−14.838
Feb 1972	808.514	897.360	986.206	992.968	95.608
Mar 1972	871.241	961.597	1051.953	976.804	15.207
Apr 1972	796.772	888.613	980.454	968.697	80.084
May 1972	770.956	864.259	957.561	871.675	7.416
Jun 1972	836.200	930.941	1025.683	1006.852	75.911
Jul 1972	694.875	791.035	887.194	832.037	41.002
Aug 1972	272.080	369.636	467.192	345.587	−24.049
Sep 1972	722.542	821.476	920.409	849.528	28.052
Oct 1972	818.075	918.367	1018.658	913.871	−4.496
Nov 1972	728.043	829.675	931.307	868.746	39.071
Dec 1972	798.447	901.401	1004.355	993.733	92.332
Jan 1973	802.151	917.195	1032.238		
Feb 1973	807.368	924.692	1042.017		
Mar 1973	869.368	988.930	1108.492		
Apr 1973	794.187	915.946	1037.704		
May 1973	767.675	891.591	1015.507		
Jun 1973	832.237	958.274	1084.310		
Jul 1973	690.245	818.367	946.489		
Aug 1973	266.795	396.969	527.143		
Sep 1973	716.614	848.808	981.003		
Oct 1973	811.515	945.699	1079.883		
Nov 1973	720.862	857.008	993.153		
Dec 1973	790.656	928.734	1066.812		

takes part in recomputing the forecasts beginning in April 1973. Note that all forecasts were revised upward on the basis of the new data.

This process of adding new data and using the same model to compute revised forecasts can be continued as long as desired. In practice, however, it is usually desirable to update the model as well after a significant amount of new data has been added to the series. When the model is updated, either the current model is reestimated by using the new data (which simply recomputes the parameter values) or a change is made in the model structure (i.e., a new model must be identified and estimated).

A new model will seldom need to be identified since it is unlikely that the basic relationships that existed in the original series will change drastically because of the new data. At any rate, you will know if a change in model structure is required when you reestimate the current model. If the resulting parameter and residual

Table 17.2 Revised Versus Original Forecasts After Adding New Data

Period	Original Forecast	Revised Forecast
Apr 1973	983.977	989.998
May 1973	930.587	936.608
Jun 1973	1024.637	1030.658
Jul 1973	870.783	876.804
Aug 1973	423.393	429.414
Sep 1973	896.050	902.071
Oct 1973	979.936	985.957
Nov 1973	908.651	914.672
Dec 1973	1001.658	1007.679
Jan 1974	989.029	994.622
Feb 1974	1040.656	1052.454
Mar 1974	1072.768	1087.152
Apr 1974	1025.706	1034.133
May 1974	972.317	980.743
Jun 1974	1066.367	1074.793
Jul 1974	912.512	920.939
Aug 1974	465.122	473.549
Sep 1974	937.779	946.206
Oct 1974	1021.665	1030.092
Nov 1974	950.381	958.807
Dec 1974	1043.387	1051.814

diagnostics indicate an inadequate model, you will need to go through the model identification phase again.

The need for a change in parameter values, however, is a much more common situation. You can be alerted to cases when you should reestimate your model by tracking the accuracy of your forecasts. If your forecasts begin to consistently vary further than expected (see forecast confidence limits) from what actually happens, then it is probably time to reestimate the model. However, it is advisable to reestimate the model on a regular basis anyway. This process assures you that your model is always up to date.

17.4 REVIEW OF KEY CONCEPTS

Assessing the accuracy of forecasts is an important function in the forecasting process. One-number forecasts simply do not provide sufficient information on which to make informed decisions: The uncertainty in the forecasts must also be visible and

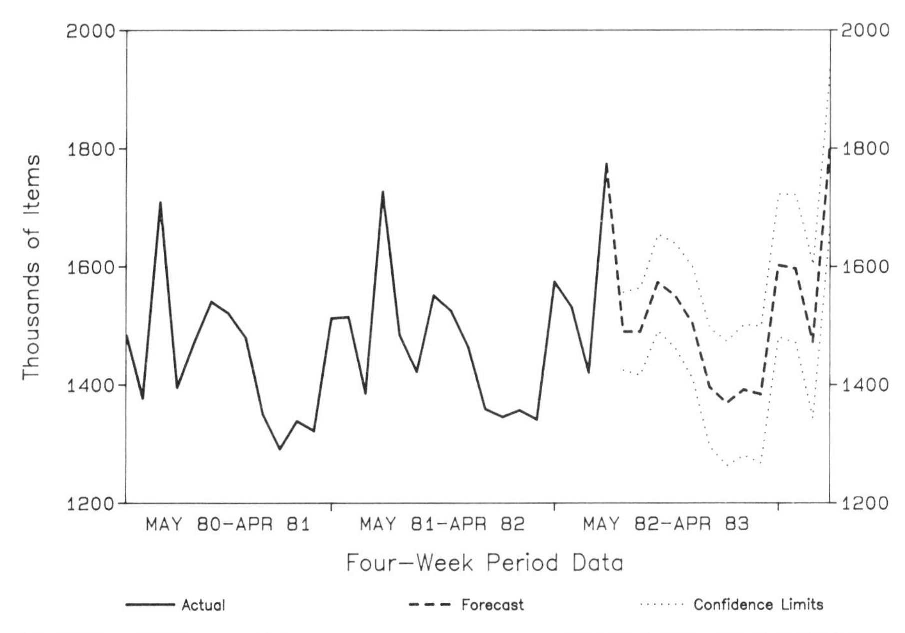

Period	Lower Limit	Forecast	Upper Limit	Confidence Interval
43	1424.061	1490.134	1556.206	132.150
44	1415.897	1490.018	1564.139	148.240
45	1491.631	1573.009	1654.386	162.760
46	1462.781	1550.818	1638.856	176.080
47	1410.026	1504.254	1598.483	188.450
48	1295.610	1395.647	1495.684	200.070
49	1263.250	1368.776	1474.302	211.050
50	1280.653	1391.396	1502.139	221.490
51	1267.926	1383.652	1499.377	231.450
52	1481.458	1601.960	1722.462	241.000
53	1471.364	1596.460	1721.557	250.200
54	1343.118	1472.646	1602.174	259.050
55	1667.708	1801.521	1935.333	267.620

Figure 17.4 Forecast Results for Postal Volume Series

understood. Two ways for quantifying the uncertainty in a forecast are provided by the following:

- *Forecast confidence limits*. Forecast confidence limits can be computed from your forecasting model and are used to indicate a range of values, or interval, in which the future actual value is expected to fall. Forecast confidence limits are computed relative to a stated level of confidence; so the larger the confidence interval, the greater is the confidence you will have that the future actual

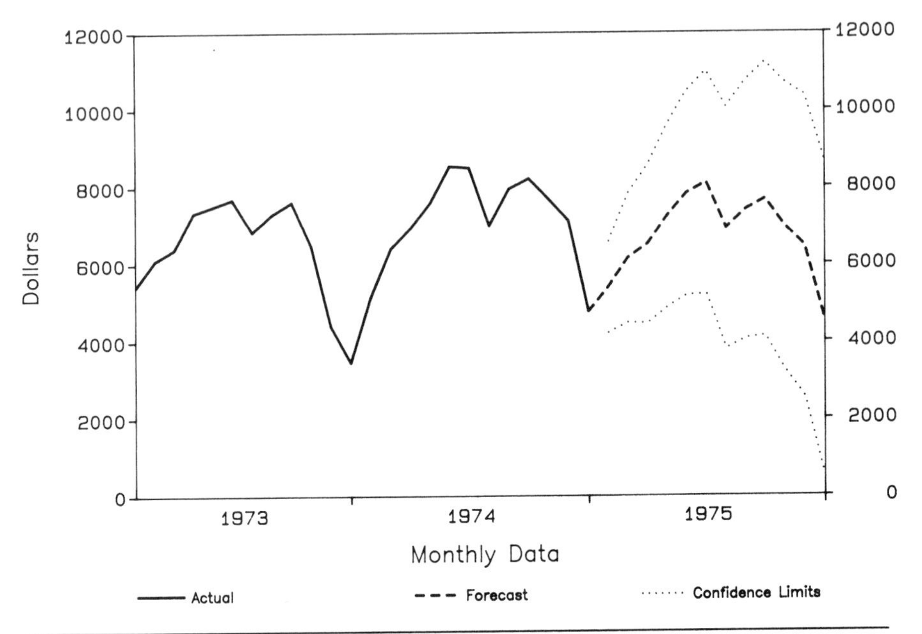

Period	Lower Limit	Forecast	Upper Limit	Confidence Interval
Jan 1975	4220.102	5401.523	6582.941	2362.840
Feb 1975	4483.305	6154.086	7824.863	3341.560
Mar 1975	4469.449	6515.730	8562.008	4092.560
Apr 1975	4879.758	7242.598	9605.434	4725.670
May 1975	5188.203	7829.941	10471.680	5283.500
Jun 1975	5213.773	8107.652	11001.530	5787.730
Jul 1975	3800.168	6925.910	10051.650	6251.530
Aug 1975	4064.300	7405.859	10747.420	6683.100
Sep 1975	4134.641	7678.902	11223.160	7088.560
Oct 1975	3233.907	6969.883	10705.860	7471.990
Nov 1975	2531.371	6449.695	10368.020	7836.630
Dec 1975	473.030	4565.586	8658.141	8185.110

Figure 17.5 Forecast Results for Lawn Chair Sales Series

value will actually fall within the interval. A level of confidence is expressed as some percentage. The most common confidence level is 95%. At a 95% level of confidence you can expect that for 95 out of every 100 forecasts, the corresponding realized actual value will fall within the forecast confidence limits.

- *After-the-fact forecasts*. After-the-fact forecasts provide a test of how well your model forecasts for time periods for which actual series values are already known. That is, the model is used to generate forecasts starting at some period

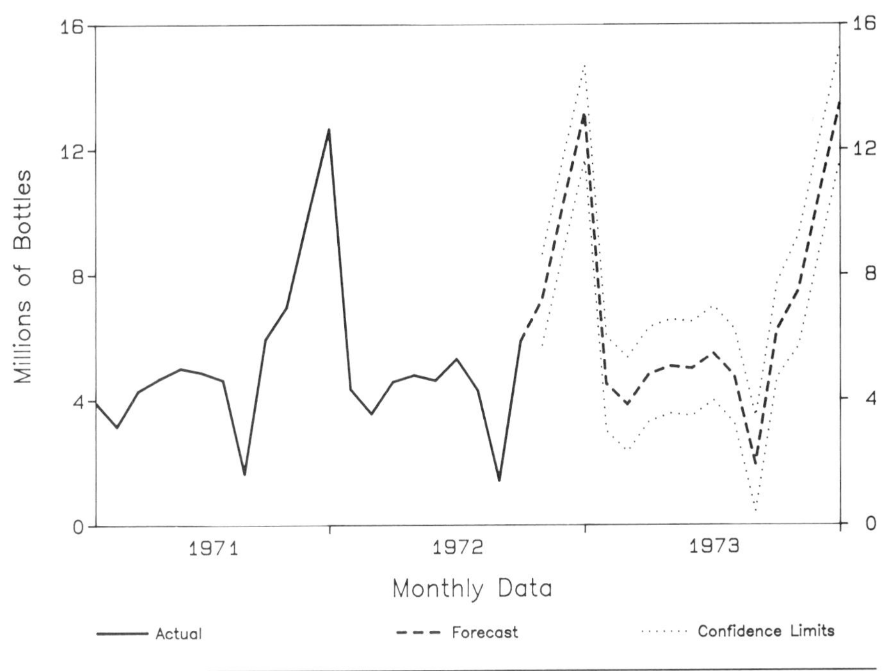

Period	Lower Limit	Forecast	Upper Limit	Confidence Interval
Oct 1972	5.757	7.210	8.663	2.906
Nov 1972	8.802	10.311	11.820	3.018
Dec 1972	11.689	13.198	14.707	3.018
Jan 1973	3.002	4.511	6.021	3.018
Feb 1973	2.339	3.848	5.357	3.018
Mar 1973	3.306	4.815	6.324	3.018
Apr 1973	3.556	5.065	6.575	3.018
May 1973	3.484	4.993	6.502	3.018
Jun 1973	3.978	5.487	6.996	3.018
Jul 1973	3.243	4.752	6.261	3.018
Aug 1973	0.430	1.939	3.448	3.018
Sep 1973	4.709	6.218	7.727	3.018
Oct 1973	5.709	7.484	9.259	3.550
Nov 1973	8.783	10.578	12.372	3.589
Dec 1973	11.671	13.465	15.259	3.589

Figure 17.6 Forecast Results for Champagne Sales Series

(the forecast origin) prior to the last period in the series, and the forecasts produced from the forecast origin to the end of the series (after-the-fact forecasts) are compared with the known series values for these periods. After-the-fact forecasts can therefore give you some idea of the potential accuracy or inaccuracy of forecasts generated from all the data for the real future.

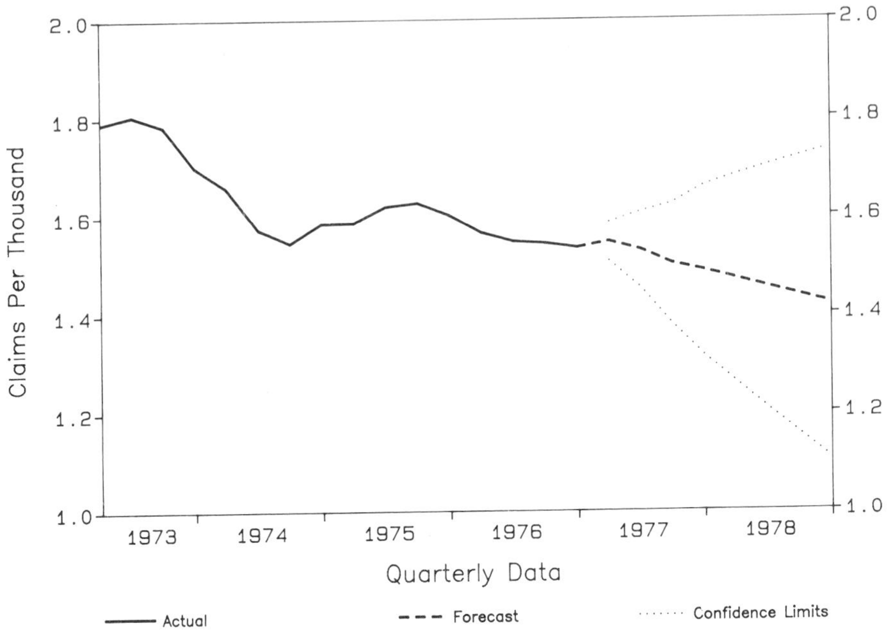

Period	Lower Limit	Forecast	Upper Limit	Confidence Interval
Q1 1977	1.508	1.547	1.587	0.079
Q2 1977	1.452	1.530	1.608	0.156
Q3 1977	1.380	1.502	1.625	0.245
Q4 1977	1.315	1.488	1.661	0.346
Q1 1978	1.260	1.471	1.682	0.422
Q2 1978	1.207	1.454	1.700	0.494
Q3 1978	1.155	1.437	1.718	0.563
Q4 1978	1.104	1.420	1.736	0.632

Figure 17.7 Forecast Results for Insurance Claim Frequency Series

Revising and updating forecasts and forecasting models is also an important step in the forecasting process. As time moves on and new data becomes available, you should use this new data to accomplish the following steps:

- *Revise previously generated forecasts* for time periods further in the future. Many Box-Jenkins programs will allow you to do this step without reestimating the model.
- *Recompute the optimum values of the model parameters,* i.e., reestimate the model on the basis of both the new and old data. If computer costs are not an issue, this step should be done after each new period of data is added to the

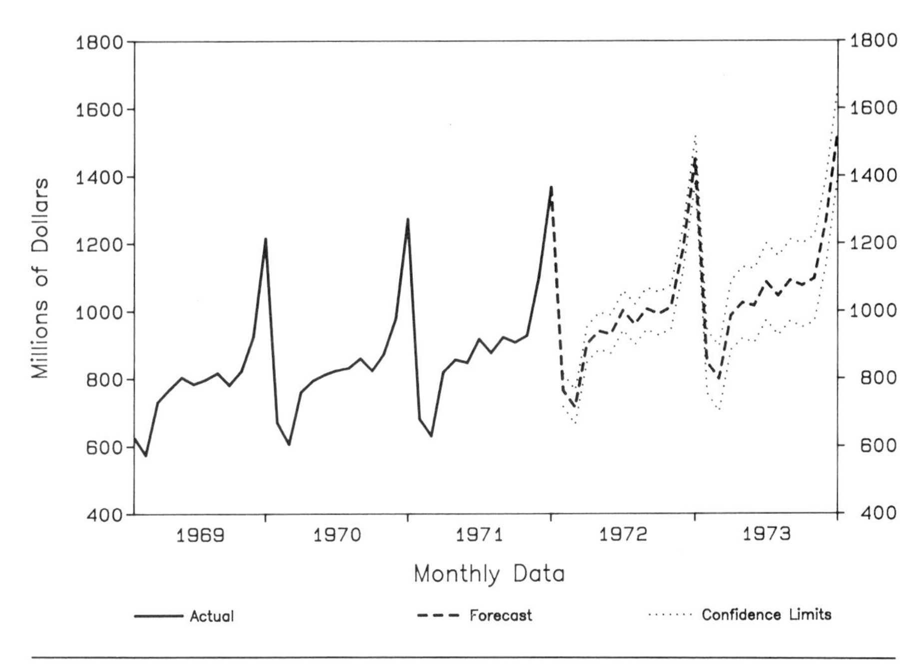

Period	Lower Limit	Forecast	Upper Limit	Confidence Interval
Jan 1972	716.653	764.978	813.303	96.650
Feb 1972	664.523	714.978	765.433	100.910
Mar 1972	850.480	902.978	955.476	104.996
Apr 1972	886.513	940.978	995.443	108.930
May 1972	875.615	931.978	988.341	112.726
Jun 1972	943.779	1001.978	1060.177	116.401
Jul 1972	900.999	960.978	1020.957	119.961
Aug 1972	945.270	1006.978	1068.686	123.420
Sep 1972	928.588	991.978	1055.367	126.782
Oct 1972	946.950	1011.978	1077.006	130.060
Nov 1972	1118.353	1184.978	1251.604	133.250
Dec 1972	1382.792	1450.978	1519.164	136.370
Jan 1973	756.237	848.956	941.675	185.438
Feb 1973	701.805	798.956	896.107	194.302
Mar 1973	885.567	986.956	1088.345	202.773
Apr 1973	919.499	1024.956	1130.413	210.911
May 1973	906.582	1015.956	1125.330	218.748
Jun 1973	972.801	1085.956	1199.111	226.309
Jul 1973	928.142	1044.956	1161.770	233.628
Aug 1973	970.594	1090.956	1211.318	240.726
Sep 1973	952.148	1075.956	1199.764	247.612
Oct 1973	968.795	1095.956	1223.117	254.325
Nov 1973	1138.528	1268.956	1399.384	260.850
Dec 1973	1401.341	1534.956	1668.571	267.230

Figure 17.8 Forecast Results for Sears Retail Sales Series

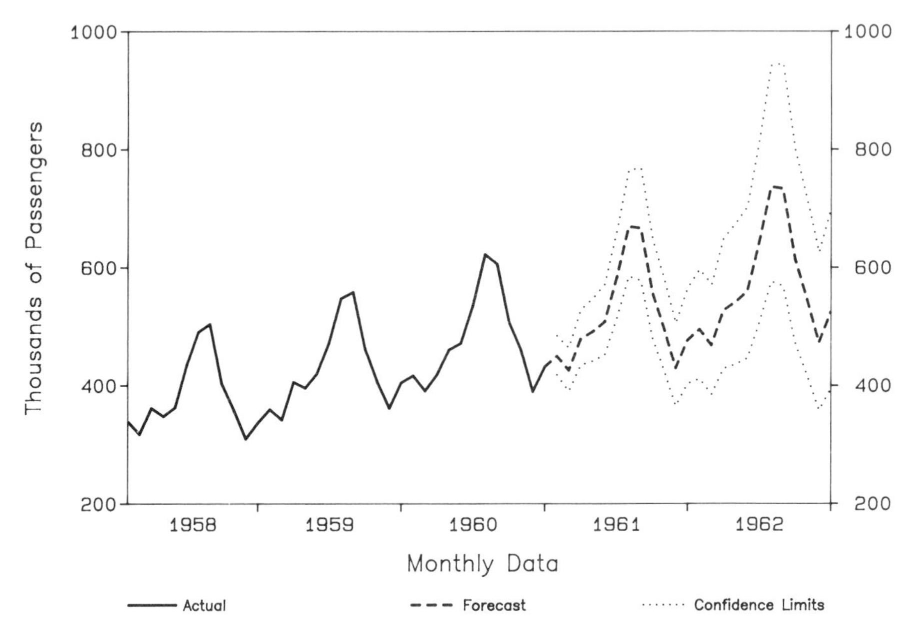

Period	Lower Limit	Forecast	Upper Limit	Confidence Interval
Jan 1961	418.175	450.116	484.497	66.322
Feb 1961	390.284	425.642	464.202	73.918
Mar 1961	434.653	479.455	528.874	94.221
Apr 1961	441.541	492.041	548.316	106.775
May 1961	452.133	508.557	572.023	119.890
Jun 1961	513.871	583.008	661.445	147.574
Jul 1961	585.082	669.179	765.364	180.282
Aug 1961	578.256	666.427	768.042	189.786
Sep 1961	480.668	557.977	647.719	167.051
Oct 1961	424.993	496.759	580.642	155.649
Nov 1961	365.188	429.682	505.565	140.377
Dec 1961	402.949	477.127	564.961	162.012
Jan 1962	411.508	495.563	596.788	185.280
Feb 1962	384.794	468.617	570.701	185.907
Mar 1962	428.869	527.864	649.709	220.840
Apr 1962	435.708	541.721	673.528	237.820
May 1962	446.011	559.905	702.883	256.872
Jun 1962	506.596	641.873	813.271	306.675
Jul 1962	576.320	736.745	941.825	365.505
Aug 1962	569.039	733.714	946.044	377.005
Sep 1962	472.495	614.315	798.701	326.206
Oct 1962	417.280	546.915	716.823	299.543
Nov 1962	358.123	473.066	624.901	266.778
Dec 1962	394.652	525.302	699.202	304.550

Figure 17.9 Forecast Results for Airline Passenger Series

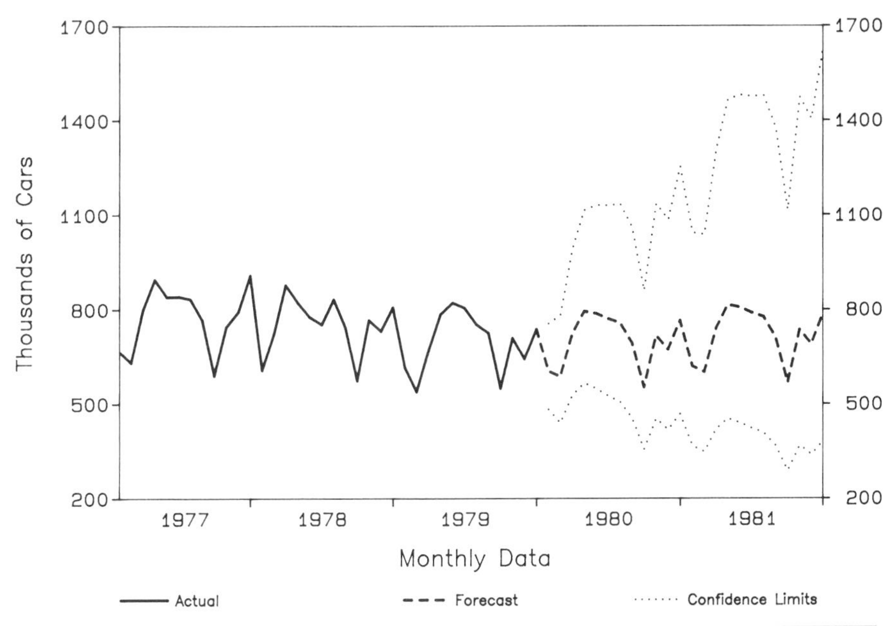

Period	Lower Limit	Forecast	Upper Limit	Confidence Interval
Jan 1970	482.668	603.981	755.785	273.117
Feb 1970	439.363	586.449	782.776	343.413
Mar 1970	527.490	722.915	990.740	463.250
Apr 1970	566.084	794.933	1116.298	550.216
May 1970	548.431	787.842	1131.762	583.329
Jun 1970	525.409	771.078	1131.614	606.201
Jul 1970	506.388	758.363	1135.719	629.332
Aug 1970	454.578	694.024	1059.596	605.022
Sep 1970	355.670	553.123	860.191	504.521
Oct 1970	453.658	718.106	1136.708	683.052
Nov 1970	418.151	673.279	1084.067	665.919
Dec 1970	468.133	766.260	1254.245	786.107
Jan 1971	367.835	619.608	1043.712	675.875
Feb 1971	348.359	601.446	1038.402	690.041
Mar 1971	420.687	741.401	1306.611	885.923
Apr 1971	453.514	815.261	1465.554	1012.040
May 1971	440.930	807.988	1480.608	1039.680
Jun 1971	423.599	790.795	1476.295	1052.690
Jul 1971	409.160	777.756	1478.404	1069.240
Aug 1971	367.929	711.771	1376.944	1009.010
Sep 1971	288.259	567.267	1116.327	828.071
Oct 1971	368.048	736.469	1473.683	1105.630
Nov 1971	339.497	690.496	1404.383	1064.880
Dec 1971	380.278	785.855	1623.990	1243.710

Figure 17.10 Forecast Results for Car Registration Series

series; otherwise, it can be done on a less frequent basis. If a model is not reestimated, revised forecasts should still be computed by using the new data (and old model parameters). Also, by tracking the actual accuracy of forecasts generated by your model, you can be alerted to situations in which it is necessary to reestimate your model. However, reestimation should still be done on a regular basis, if possible, to assure yourself that your model is up to date.

- *Reidentify and reestimate a new model.* Sometimes, but rarely, the addition of enough new data can cause the basic structure of the model to change because of a change in the behavior patterns of the series. You can be alerted to the need to identify a new model for your series if you recompute the optimum values of the model parameters and examine the residual and parameter diagnostics on a regular basis.

GLOSSARY

AC Autocorrelation

Additive Seasonal Model A model for a seasonal series containing both regular and seasonal parameters where the regular and seasonal terms in the model are just added together. For example, the model $X_t = A_1 X_{t-1} + A_1^* X_{t-12} + E_t$ is an additive, seasonal autoregressive model.

After-the-Fact Forecast A Box-Jenkins model can be used to start generating forecasts at some time period prior to the last period of the series. The forecasts then produced for time periods for which actual time series values already exist are called after-the-fact forecasts. After-the-fact forecasts provide one means of assessing the potential accuracy of forecasts produced by the model for the real future. Forecast errors and error statistics, like residual statistics, can also be computed for these time periods.

AR Autoregressive

ARI Autoregressive integrated

ARIMA Autoregressive, integrated moving-average

ARMA Autoregressive moving-average

Autocorrelation confidence limits Autocorrelation (and partial-autocorrelation) confidence limits provide the means for testing the significance of a sample autocorrelation. Autocorrelation confidence limits are upper and lower bounds between which the true value of the autocorrelation would be expected to fall with a certain level of confidence (usually 95% confidence). The interval between the lower and upper confidence limits is called the autocorrelation confidence interval, and its midpoint is the value of the sample autocorrelation. If this interval is translated so that it is centered on zero, the upper and lower limits of the translated interval then provide an easy way for testing whether the sample autocorrelation value is significantly nonzero. If the sample autocorrelation value exceeds the interval limits, then we can conclude that the autocorrelation is significantly nonzero. If the autocorrelation falls within the limits, we cannot conclude it is not significantly different from zero, and in practice we assume it is zero. The translated confidence interval is what is usually shown in an autocorrelation correlogram, and its upper and lower limits are usually two times the autocorrelation standard error.

Autocorrelation Patterns The means for identifying the appropriate Box-Jenkins model for a stationary series, i.e., for identifying the appropriate autoregressive and moving-average parameters to include in the model.

Autocorrelation Standard Error An autocorrelation computed from a real-life series is expected to be in error since the series is only one sample of the true underlying process that generated the series. The standard error of an autocorrelation is a statistical measure of the magnitude of this error in the computed, or sample, autocorrelation. The standard error of an autocorrelation is used in computing confidence limits for an autocorrelation.

Autocorrelations If there is a consistent relationship throughout a time series between what happens in a given period to what happened k periods earlier (e.g., if a higher-than-average value is usually preceded k periods earlier by a higher-than-average value), then the series is said to be autocorrelated at lag k, and the extent of that relationship is measured by an autocorrelation value computed from the data. An autocorrelation value ranges from -1 to $+1$, where $+1$ indicates high positive correlation, -1 indicates high negative, or opposite, correlation, and 0 indicates no correlation. Autocorrelation values for a series may be computed for consecutive lags of 1, 2, 3, ... The autocorrelations computed from a real-life series are called sample autocorrelations since the series from which they are computed is only one sample of the true underlying process (model) that generated the series. In the identification of a Box-Jenkins model for a given stationary series, the sample autocorrelation pattern of the series is compared with the known theoretical autocorrelation pattern associated with the Box-Jenkins models.

Autoregressive Integrated (ARI) Model A model for a nonstationary, nonseasonal series containing only regular autoregressive parameters, where the series has been differenced to achieve stationarity.

Autoregressive, Integrated, Moving-Average (ARIMA) Model A model for a nonstationary, nonseasonal series containing both regular autoregressive and regular moving-average parameters, where the series has been differenced to achieve stationarity.

Autoregressive Model If a series is stationary and each value of the series X_t is related to one or more past series values, the model for the series is called an autoregressive model and is written as follows:

$$X_t = A_1X_{t-1} + A_2X_{t-2} + \cdots + A_pX_{t-p} + E_t$$

The A_k in the above model are called autoregressive parameters, and E_t is the random error term. The order of each A_k is k, and the order of the model is p. (The above model representation assumes that the mean of the series is not significantly nonzero—see *mean parameter.*)

Autoregressive, Moving-Average (ARMA) Model A model for a stationary, nonseasonal series containing both regular autoregressive and regular moving-average parameters.

Autoregressive Parameter See *autoregressive model*

Average Absolute Error The average of all residuals (or forecast errors) without regard to sign; i.e., all residuals are treated as positive values. The average absolute error is simply the average distance the fit is from the original series values.

Average Absolute Percent Error The average of all the ratios $100E_t/X_t$ without regard to the sign of E_t/X_t, where X_t is the series value and E_t is the residual (or forecast error) at period t. The average absolute percent error measures the distance the fit is from the original series values on an average percentage basis.

Basic Model In this book a basic model is a model for a stationary series containing only regular autoregressive and regular moving-average parameters.

Cause-and-Effect Model Models for time series forecasting constructed on the basis of establishing relationships between the time series to be forecasted and one or more other series that are assumed to influence or cause the behavior of the first time series. A time series regression model for product sales is one example of a cause-and-effect forecasting model.

Chi-square Statistic In estimation for Box-Jenkins models the chi-square statistic is used to test whether a set of residual autocorrelations up to some lag k reveal any evidence that the residuals are significantly correlated. The appropriate chi-square statistic, which depends on the number of autocorrelations being tested, can usually be obtained from a chi-square distribution table in any statistics textbook. Many Box-Jenkins programs automatically provide the appropriate chi-square statistic for the given set of residual autocorrelations being tested.

Closeness-of-Fit Statistics After a model has been estimated and validated, the closeness-of-fit statistics provide a number of different measures of the magnitude of the residual, or random error, component of the series; i.e., they measure how close the fit is to the original series. There are a number of different ways to provide this measurement. The residual statistics generated by a Box-Jenkins program may include some or all of the following: residual standard error, average absolute percent error, mean error, index of determination, average absolute percent error, mean percent error, mean square error, and residual sum of squares.

Combined Model A model containing both regular and seasonal parameters and/or a model for which both regular and seasonal differencing has been applied to the series to achieve stationarity.

Confidence Level Statistical conclusions are not generally made with absolute certainty. When a statistical hypothesis is concluded to be true on the basis of some statistical test, the conclusion is usually stated in terms of a degree, or level, of confidence expressed as some percentage, such as 95% confidence. The higher the level of confidence desired, the more difficult it is to pass the test of acceptance. Thus such statistics as forecast confidence limits, parameter confidence limits, and autocorrelation confidence limits are computed in terms of a level of confidence (usually 95%). Alternatively, the conclusion can be stated in terms of a risk level; e.g., a 95% confidence level (95% chance of being right) is the same as a 5% risk level (5% risk of being wrong).

Confidence Limits See *parameter confidence limits* and *forecast confidence limits.*

Correlation See *autocorrelations, partial-autocorrelations,* and *parameter correlation matrix.*

Correlogram A graphical representation of a set of autocorrelations (or partial-autocorrelations) where the autocorrelation values are plotted against the lag values 1, 2, 3, ... When the autocorrelation values are on the vertical axis and the lag values are on the horizontal axis, the magnitude of each autocorrelation is often represented as a bar or spike that extends vertically from zero to the value of the autocorrelation (positive or negative). Correlograms are useful for visually identifying the autocorrelation and partial-autocorrelation patterns of a series.

Cyclic Component/Pattern A series that exhibits a more or less consistently rising and falling pattern over an extended period of time contains a cyclic component or pattern. The occurrence of peaks and valleys is generally not as consistent or regular as in seasonal patterns and the distance between peaks and valleys is usually long.

Decomposition Model A model that is a combination (usually multiplicative) of several models that separately model the distinct pattern components of the series. For example, a trend, seasonal, and smoothing model may be determined by decomposing the series into its various subpatterns and then combined to produce a composite model that fits all patterns simultaneously.

Diagnostics See *residual diagnostics* and *parameter diagnostics.*

Difference A transformation applied to a time series whereby the differences between successive time series values, $X_t - X_{t-1}$, are computed (called a regular difference or a difference of order 1), or the differences between successive s period-apart time series values, $X_t - X_{t-s}$, are computed (called a *seasonal difference* or a difference of order s). The computed series of differences is called the differenced series. Successive differences may also be applied; i.e., a series of differences may also be differenced. In this book the resultant differenced series is denoted by Z_t. Differences are used to transform a nonstationary series into a stationary series.

Differencing The process of applying one or more differences to a time series. Differencing is used to transform a nonstationary series into a stationary one.

Estimation Phase In the Box-Jenkins method the estimation phase refers to the process of computing, or estimating, the parameters of a model from the time series data. The parameter values are estimated so that the resulting sum of squared residuals is minimized. After the parameters have been estimated, the residual and parameter diagnostics are examined in order to determine the validity, or adequacy, of the estimated model.

Estimation Results See *residual diagnostics* and *parameter diagnostics.*

Extreme Value Extreme values, or outliers, in a time series are usually caused by unusual circumstances such as bad weather, strikes, or recording errors. Extreme values should be adjusted before attempting to build a Box-Jenkins forecasting model.

Fit Statistics See *closeness-of-fit statistics.*

Fitted Values The values generated by a forecasting model formula for periods 1 through N, where N is the number of time periods of the series. If the model is a valid one for the series, then the fitted values will represent the pattern component of the series.

Forecast A numerical prediction for a time series of what will happen at some future period, along with an assessment or measurement of the uncertainty in the prediction. To be of value, the forecast should also include a description of how, and under what assumptions, it was derived.

Forecast Confidence Limits A high and a low value that bracket a given forecast and between which the actual value of the time series, when it is realized, is expected to fall with a certain level of confidence (usually 95% confidence). The interval between the upper and lower confidence limit is called the confidence level, and the forecast value falls at the midpoint of this interval (assuming no prior transformations have been applied to the series). The width of the interval is based on the confidence level desired and the magnitude of the residual standard error. The wider the interval, the more confidence we have that the actual value will fall in the interval. Forecast confidence limits provide a means for measuring the degree of uncertainty in a forecast, or the potential inaccuracy of a forecast.

Forecast Error The difference between the forecast and the actual time series value, when it is realized. An after-the-fact forecast error is the difference between a forecast generated by a model at a time period for which actual data is already known.

Forecast Error Statistics Statistics similar to residual statistics (e.g., mean error, average absolute error) computed for a set of forecast errors. These statistics are particularly useful for analyzing after-the-fact forecast errors.

Forecast Horizon A point of time in the future up to which forecasts are required or desired. A forecast horizon is usually referred to as a short-term, medium-term, or long-term forecast horizon. The meaning of *short, medium,* and *long* is subjective, but a common interpretation in the forecast of monthly time series, for example, is to consider short term as 1 to 3 months, medium term as 4 to 24 months, and long term as over 2 years. Time series forecasting methods are often characterized as short-, medium-, or long-term forecasting methods.

Forecast Origin The first time period for which a model begins generating forecasts (versus fitted values) of a time series.

Forecasting Model See *model.*

Forecasting Phase In the Box-Jenkins method the forecasting phase refers to the process of generating forecasts and forecast confidence limits from an estimated model. Revised forecasts may also be generated from the model on the basis of new data, and after-the-fact forecasts can be obtained by forecasting from an earlier forecast origin.

Identification Phase In the Box-Jenkins method the identification phase refers to the process of determining what differencing is required (if any) and identifying the model parameters. The principal tools used in the identification phase are the autocorrelations and partial-autocorrelations of the (stationary) series and its differences.

IMA Integrated moving-average

Index of Determination Measures how much of the original variation in the series is accounted for by the fitted values (pattern component) generated from a model. One measure of the variation in the original series is given by the sum of the squared deviations of the series values from their mean. The sum of the squared residuals (error component) is then compared with this sum, producing a ratio representing the proportion of the variation in the original series not accounted for by the fit. This ratio is then subtracted from 1 to obtain the proportion of the variation in the original series accounted for by the fitted values, i.e., the index of determination. Thus the larger the index of determination, the smaller is the error component in a series.

Integrated Model A model for a nonstationary series where differencing has been applied to the series to achieve stationarity.

Integrated Moving-Average (IMA) Model A model for a nonstationary, nonseasonal series containing only regular moving-average parameters, where the series has been differenced to achieve stationarity.

Lag The number of time periods between two time periods in a series. For example, the lag between period t and $t - k$ is k. A series is said to be lagged k periods (relative to itself or another series) if its $(t - k)$th period is made to correspond to the tth period.

Lagged Series See *lag.*

Least Squares Method A common technique for determining (estimating) appropriate values for the parameters in a forecasting model. In this technique parameter values are computed such that the sum of the squared residuals is minimized. The Box-Jenkins method employs a type of least squares technique, called nonlinear least squares estimation, to estimate model parameters.

MA Moving average

Mean Absolute Error See *average absolute error.*

Mean Error The average value of all residuals (or forecast errors). If the residuals are truly random error, the mean error should not be significantly nonzero; otherwise, the fit is biased above or below the actual series values.

Mean Parameter If a series is stationary, but has a significantly nonzero mean value, the Box-Jenkins model for the series must contain a constant parameter called the mean parameter. This parameter is estimated along with any other parameters during the estimation phase. For example, a model containing the mean parameter and one AR and one MA parameter is written as follows:

$$X_t - M = A_1(X_{t-1} - M) - B_1E_{t-1} + E_t$$

where M is the mean parameter.

Mean Percent Error The average value of all the ratios $100E_t/X_t$, where X_t is the series value and E_t is the residual (or forecast error) at period t. The mean percent error provides another way to measure bias on a relative or percentage basis.

Missing Value In a time series a missing value is where no data could be obtained for a given time period. Missing values should be filled in, or the time series truncated to eliminate periods with missing values, before one attempts to build a Box-Jenkins forecasting model.

Model A time series forecasting model is simply a mathematical representation of a time series. Since a series is assumed to have two main components, a pattern component and a random error component, the most general representation of a time series is written as follows:

$$X_t = F_t + E_t$$

where X_t is the time series, F_t represents the pattern component in the series, and E_t represents the random error component. In the construction of a model for a series, the object is to find some mathematical formula (see *model formula*) to reproduce the pattern component. The random error component by definition cannot be reproduced, but it is often assumed to have certain statistical properties. There are a wide variety of model formulas that have been developed over the years that attempt to reproduce the many patterns demonstrated in time series.

Model-Building Process Building models for time series forecasting is an iterative process usually consisting of four steps: (1) identification of the model type and model elements on the basis of some analysis of the data, (2) estimation of the parameters in the model from the time series data, followed by a check of the adequacy of the model, (3) a forecast of the time series from the estimated model, and (4) a tracking of the accuracy of forecasts generated by the model and an updating of the forecasts and the model on the basis of new data.

Model Formula A mathematical expression that is used to generate the pattern component F_t in a series, where the series is represented by the model $X_t = F_t + E_t$. The formula consists of three elements: a particular form that indicates how certain numerical values (time-dependent data and parameters) are to be combined arithmetically, an indication of what data values are to be substituted into the formula at each period, and a set of parameters that remain constant for all periods. The values computed for F_t are called fitted values when t is less than the last period number in the series and are called forecast values when t is greater than the last period number. Two series may have models whose model formula has the same form but whose parameter values are different. For example, the form of a straight-line trend model formula is $A + Bt$, where A and B are parameters and t is the period number that is substituted into the formula each period.

Model Validity After a model has been estimated, the Box-Jenkins method provides a variety of statistical diagnostics for checking the validity, or adequacy, of the estimated model. See *residual diagnostics* and *parameter diagnostics*.

Moving-Average Model If a series is stationary, and each value of the series X_t is related to only one or more past random errors, the model for the series is called a moving-average model and is written as follows:

$$X_t = -B_1E_{t-1} - B_2E_{t-2} - \cdots - B_qE_{t-q} + E_t$$

The B_k are called (regular) moving-average parameters, and the E_i are the random errors. The order of each B_k is k, and the order of the model is q. (The above model representation assumes that the mean of the series is not significantly nonzero—see *mean parameter*.)

Moving-Average Parameter See *moving-average model*.

Multiplicative Seasonal Model A model for a seasonal series containing both regular and seasonal parameters, where the regular and seasonal terms in the model also produce additional terms that are combinations of the regular and seasonal terms. For example, the model

$$X_t = A_1X_{t-1} + A_1^*X_{t-12} + A_1A_1^*X_{t-13} + E_t$$

is a multiplicative, seasonal autoregressive model.

Multivariate Model A model in which more than one time series is forecasted simultaneously in order to forecast the time series of interest (multivariate models are cause-and-effect models, but not necessarily vice versa). The general Box-Jenkins method also encompasses multivariate models. Multivariate Box-Jenkins models are not addressed in this book.

Nonstationary Series A series that appears to have no fixed level. Examples of nonstationary behavior include series that exhibit overall trends, random changes in established levels, or random changes in established levels and slopes. A nonstationary series must be transformed into a stationary series before Box-Jenkins model parameters can be identified. Most nonstationary behavior can be transformed into stationary behavior by differencing the series. Both regular and seasonal differencing may be required since nonstationarity may also be present in the seasonal pattern of a series. Differencing alone may not be sufficient for some nonstationary series that exhibit exponentiallike trends or changes in the variation of the series. In this case a prior transformation may have to be applied before differencing can be used.

Order See *parameter order* and *order of differencing*.

Order of Differencing When a series is differenced, the number of time periods separating each pair of time series values whose difference is computed is called the order of differencing. For example, the set of values $X_t - X_{t-k}$ represents a difference of order k.

Overall Trend See *trend component/pattern*.

PAC Partial-autocorrelation

Parameter A constant value in a model formula for a particular time series. The actual numerical value used for the parameter is computed from the time series data itself. For example, the model formula $F_t = At$, where A represents a parameter, may be used for more than one time series, but the value of A will be different for each time series.

Parameter Confidence Limits A means for testing the statistical significance of an estimated parameter. In general it is concluded, with a certain level of confidence (usually 95% confidence), that the true parameter value would fall within the upper and lower parameter confidence limits. Thus if the value 0 falls between these limits, we cannot conclude that the true parameter value is significantly different from zero, and in practice we assume that it really is zero and therefore not needed. Parameter confidence limits are useful for detecting overspecified models.

Parameter Correlation Matrix The degree of dependence of an estimated parameter on another estimated parameter can be measured by computing a correlation value that measures the correlation between the two estimated parameters (theoretically, parameters should not be correlated). The correlation value ranges from -1 to $+1$, where $+1$ indicates high positive correlation and -1 indicates high negative, or opposite, correlation. If two parameters are highly correlated, then the model is probably overspecified and one of the parameters is not needed. The correlation value for every unique pair of estimated parameters is usually computed and displayed as a matrix of correlation values after the parameters have been estimated.

Parameter Diagnostics Certain statistics that are used to test the validity of the estimated parameters in a model. They can indicate whether two estimated parameters are highly correlated (parameter correlation matrix) or are statistically insignificant (parameter confidence limits). If the parameters are highly correlated or insignificant, then the model is either overspecified or incorrectly specified.

Parameter Order The order of a parameter in a Box-Jenkins model is the number of periods separating the current period from the past period with which the parameter is associated. For example, in the model $X_t = AX_{t-k}$ the parameter A is of order k since it is associated with the period that is k periods earlier than period t.

Parameter Standard Error A parameter estimated from a real-life series is expected to be in error since the series is only one sample of the true underlying process that generated the series. The standard error of an estimated parameter is a statistical measure of the magnitude of this error in the estimated parameter. The standard error of a parameter is used in computing parameter confidence limits.

Partial-Autocorrelations A set of statistical measurements, similar to autocorrelations, that reveal how time series values are related to each other at specified lags. Just like autocorrelations, partial-autocorrelations range in value from -1 to $+1$ and are computed for successive lags of 1, 2, 3, ... It turns out that the partial-autocorrelation patterns associated with autoregressive relationships in a stationary series are the same as the autocorrelation patterns associated with moving-average relationships, and vice versa. Thus partial-autocorrelations are useful for identifying the number of autoregressive parameters in a Box-Jenkins model since those patterns are easier to recognize (as are the autocorrelation patterns for moving-average parameters). Technically, a partial-autocorrelation at lag k is an estimate of the last autoregressive parameter in an autoregressive model of order k if it were fitted to the series; see Appendix B.

Pattern Component That part of a series that demonstrates certain relationships over time. The pattern component can consist of many different types of patterns, some of which may be visually recognizable in a graph of a time series (e.g., seasonal, trend, and cyclic patterns) or only recognizable by some statistical measurement (e.g., autoregressive and moving-average patterns).

Patterns See *autocorrelation patterns* and *time series patterns*.

Periods Per Season See *seasonal component pattern*.

Principle of Parsimonious Parameterization This principle simply states that when one is constructing a Box-Jenkins model, as little differencing and as few parameters as possible should be used to obtain an adequate model.

Prior Transformation See *transformations*.

Purely Seasonal Model A model containing only seasonal autoregressive and/or seasonal moving-average parameters.

Q-Statistics A statistic computed from a set of residual autocorrelations up to some lag k. It is used to test whether this set of autocorrelations, as a whole, is significant, i.e., whether the residuals are significantly autocorrelated over k lags. The test is performed by comparing the computed Q-statistic with a test value, called the chi-square statistic. If the Q-statistic is greater than the test value, then it can be concluded with a certain level of confidence that the residuals are significantly autocorrelated over the k lags and hence that the model is underspecified or incorrectly specified. The appropriate chi-square statistic depends on the number of autocorrelations being tested and the number of parameters in the model, and it can be found in a chi-square distribution table in any statistical text. Usually, the appropriate chi-square test value is provided as part of the estimation results of a Box-Jenkins program.

Random Error Component That part of a series that has no explanation outside of chance alone. That is, any data collected to represent some process or activity will always be in error owing to measurement errors or outside random influences. Each data value in a time series, therefore, is expected to be in error from the "true" value for that period. The manner in which these errors occur is generally assumed to be random. (If they weren't, they would demonstrate some relationship and therefore be part of the pattern component.)

Range-Mean Plot A plot for a series obtained by dividing the series into a group of subseries and computing the range (maximum minus minimum value) and the mean of each subseries. The range values are then plotted against the mean values. The range-mean plot can give a rough indication of the type of transformation that should be applied to some types of nonstationary series (see Appendix C).

RAR Regular autoregressive

RD Regular difference

Regular Autoregressive Parameter See *autoregressive model.*

Regular Moving-Average Parameter See *moving-average model.*

Residual Autocorrelations The autocorrelations computed from a series of residuals. Since the residuals computed from a valid estimated model should behave as random errors, they should be uncorrelated, and hence the residual autocorrelations should not be significantly nonzero. Thus residual autocorrelations are useful for testing whether an estimated model is a valid one and for suggesting alternative models.

Residual Diagnostics Certain statistics computed from the residuals of an estimated model. Some residual diagnostics (residual autocorrelations, residual mean error, residual mean percent error, Q-statistics) are used to check the validity of the estimated model by determining whether the residuals are uncorrelated and unbiased. If the residuals are correlated or biased, then the model is either underspecified or incorrectly specified. Other residual diagnostics (closeness-of-fit statistics) are used to measure how well the model fits the original data, i.e., how large the residuals (random error component) of the series is.

Residual Mean Square Error The average of the squared residuals (the residual standard error is the square root of this value).

Residual Standard Error The square root of the average of the squared residuals, i.e., the square root of the residual mean square error. The residual standard error is the most common measurement of the magnitude of the residuals in statistical work, and it is used in deriving other common statistics such as forecast confidence limits.

Residual Statistics See *closeness-of-fit statistics.*

Residual Sum of Squares The sum of all the squared residuals. The minimization of the residual sum of squares is the criterion by which the appropriate parameter values are estimated for a given Box-Jenkins model (least squares method).

Residuals The differences between the actual time series values X_t and the fitted values generated by some forecasting model formula. If F_t represents the fitted values and E_t represents the computed residuals, then $E_t = X_t - F_t$ for $t = 1, 2, \ldots, N$, where N is the number of time periods of the series.

Revised Forecast After a Box-Jenkins model has been used to generate a series of forecasts and new (actual) time series values become available for one or more of the original forecast periods, revised forecasts can be computed for the remaining forecast periods by using the same model and the new data.

RMA Regular moving average

Sample Autocorrelations See *autocorrelations.*

SAR Seasonal autoregressive

Seasonal, Autoregressive, Integrated, Moving-Average (Seasonal ARIMA) Model A model for a nonstationary, seasonal series containing both autoregressive and moving-average parameters, where the series has been differenced to achieve stationarity. The model may contain both regular and seasonal parameters and both regular and seasonal differencing may be used.

Seasonal Autoregressive Model If a series is stationary and seasonal, and each value of the series X_t is related to only one or more past series values that are separated by the number of periods per season, the model for the series is called a purely seasonal autoregressive model and is written as follows:

$$X_t = A_1^* X_{t-s} + A_2^* X_{t-2s} + \cdots + A_P^* X_{t-Ps} + E_t$$

The A_k^* are called seasonal autoregressive parameters, and E_t is the random error term. The order of each A_k^* is ks, where s is the number of periods per season, and the order of the model is Ps. A seasonal autoregressive model may also contain regular autoregressive parameters, in which case it is called a combined regular and seasonal model.

Seasonal Autoregressive Parameter See *seasonal autoregressive model.*

Seasonal Component/Pattern A series that exhibits a repeating pattern of behavior on a precise regular basis contains a seasonal component or pattern. If the pattern occurs every s periods, then the seasonal pattern is said to have s periods per season. The length of time it takes the pattern to run its course is called the seasonal period. For example, a monthly time series might have 12 periods (months) per season with a seasonal period of one year; or it might have 3 periods (months) per season with a seasonal period of one quarter.

Seasonal Difference See *difference.*

Seasonal Moving-Average Model If a seasonal series is stationary and seasonal, and a current value of the series X_t is related to only one or more past random errors that are separated by the number of periods per season, the model for the series is called a purely seasonal moving-average model and is written as follows:

$$X_t = -B_1^*E_{t-s} - B_2^*E_{t-2s} - \cdots - B_Q^*E_{t-Qs} + E_t$$

The B_k^* are called seasonal moving-average parameters, and the E_i are the random errors. The order of each B_k^* is ks, and the order of the model is Qs. A seasonal moving-average model may also contain regular moving-average parameters, in which case it is called a combined regular and seasonal model.

Seasonal Moving-Average Parameter See *seasonal moving-average model.*

Seasonal Period See *seasonal component/pattern.*

Self-Projecting Model A model for time series forecasting constructed by using only the values of the time series to be forecasted. Thus the historical patterns of the time series play central roles in constructing self-projecting models.

Series See *time series.*

SMA Seasonal moving-average

Smoothing Model A model usually based on the concept of computing averages or weighted averages of past time series values to produce fitted and forecast values. Smoothing models represent some of the more common, traditional time series forecasting models (e.g., simple moving averages and exponential smoothing). In general, smoothing models are more responsive to, and more dependent on, what has happened in the most recent past, so that they are better able to handle intermediate behavioral patterns than are overall trend models. Often smoothing models are combined with other types of traditional models, such as trend and seasonal models, to construct a decomposition model.

Standard Error See *autocorrelation standard error, residual standard error,* and *parameter standard error.*

Stationary Series A series that varies more or less uniformly about some fixed level. The series may stray from this fixed level, but it will always return to it. Also, the magnitude of the variation in a stationary series does change significantly throughout the series. When one is building Box-Jenkins models, the series must be stationary or transformed to a stationary series before the autocorrelation and moving-average parameters can be identified.

Theoretical Autocorrelation Pattern See *autocorrelations.*

Time Period An interval of time, such as a day, a week, a month, a year. Time series are recorded over a continuous sequence of equal-length time periods. In time series, time periods are usually numbered from 1 to N, where N is the number of time periods in the series.

Time Series A set of numerical values that represents the level, or status, of some ongoing activity over time. Each value is associated with a particular time period, such as a month, and the time periods for all values are assumed to be of equal (or approximately equal) length. In this book an original (untransformed) time series is represented as X_t, where t represents the time period number, $t = 1, 2, \ldots, N$.

Time Series Forecasting Model See *model* and *model-building process.*

Time Series Patterns Any consistent historical behavior or relationship in a time series. Traditional patterns include seasonal, trend, and cyclic patterns. These patterns are generally recognizable when the series is plotted in a graph. Other patterns, or relationships, may also exist that are not visually detectable but may be recognized through statistical analyses and techniques.

Trading-Day Variation Variations that can occur in some business-oriented, monthly and quarterly time series. These variations arise from the uneven distribution of days in the week in different months and from unequal-length months. When possible, these variations should be identified and the series adjusted to eliminate the effects of trading-day variations.

Transformations Differencing alone may not be sufficient to achieve stationarity for a nonstationary series when the series exhibits exponentiallike growth or when the variation in the series changes over time. In this case certain "instantaneous," or prior, mathematical transformations may be applied to the series to eliminate this type of behavior. Differencing can then be applied to the transformed series to achieve stationarity. The most common transformation used is the logarithmic transformation, which converts exponential behavior into straight-line behavior. Other transformations, such as taking square roots or reciprocals of the data, may also be appropriate. A range-mean plot of successive groups of time series values is useful in determining the appropriate transformation to use (see Appendix C).

Trend Component/Pattern A series that exhibits an identifiable overall movement from beginning to end contains a trend component or pattern. The overall trend may be as simple as a straight-line or exponential growth pattern or may be more complex as in certain S-curve patterns.

Trend Parameter If a series is nonstationary, but the differenced (stationary) series has a significantly nonzero mean value, the integrated Box-Jenkins model for this series must contain a constant parameter called the trend parameter. An integrated model containing the trend parameter is written as follows:

$$Z_t = B_0 + (\text{AR and MA terms}) + E_t$$

where Z_t is the differenced series and B_0 is the trend parameter.

Univariate Model A model in which only one time series is forecasted (self-projecting models are univariate models, but not necessarily vice versa). The Box-Jenkins models described in this book are univariate models.

Appendix A

Autocorrelation Computations

A.1 SIMPLE CORRELATION

Measuring the autocorrelation of a time series is based on the general concept of measuring the correlation between two sets of ordered data. In measuring the correlation between two sets of data, we are interested in comparing how corresponding data values from each set tend to behave relative to the mean value of their respective data sets. For example, if greater-than-average values in one set tend to correspond to greater-than-average values in the second set and less-than-average values in one correspond to less-than-average values in the other, then we would conclude that the two sets are highly correlated (positively correlated, in this case). On the other hand, if greater-than-average values in one set tend to correspond to less-than-average values in the other set, then we would conclude that the two sets are also highly correlated but in an opposite, or negative, sense. If there does not appear to be any consistent correspondence, then we conclude there is no correlation.

Mathematically, this correspondence is measured by using the procedure illustrated in the following example with two sets of data X and Y:

Set X	Set Y
15	100
13	130
10	80
18	150
24	140

The mean of X is 16 and of Y is 120. Subtracting these means from each value in their respective data sets gives the following sets of values:

$(X - 16)$	$(Y - 120)$
-1	-20
-3	10
-6	-40
2	30
8	20

Now data values in the original data sets that are larger than their mean yield positive values in the new data sets, and values that are less than their mean yield negative values. Thus *positive correlation* exists if corresponding data values in the new data sets tend to have the same sign (positive or negative), and *negative correlation* exists if corresponding data values tend to have opposite signs.

Whether corresponding data values have the same or opposite signs can be conveniently determined by multiplying the two values together:

$(X - 16) \times (X - 120)$
$+20$
-30
$+240$
$+60$
$+160$

If a product is positive, then the corresponding original values were either both higher or both lower than their respective means; and if a product is negative, then one value was higher and the other lower than its respective mean. If all, or most, of the products are positive, there is a high positive correlation between X and Y; and if all, or most, of the products are negative, there is a high negative correlation between X and Y.

A single measure of the magnitude or degree of correlation between X and Y can now be obtained by simply finding the average of all the products. This average value is called the *covariance of X and Y* and is denoted as $\text{COV}(X, Y)$. A large positive covariance value indicates high positive correlation, and a large negative covariance value indicates a high negative correlation between X and Y. Note that if the covariance is close to zero, then there were both positive and negative products, which tended to cancel each other; i.e., there was no consistent correspondence between the original data values and there is no correlation between X and Y.

Continuing with our example, the covariance of X and Y is computed as

$$\text{COV}(X, Y) = (20 - 30 + 240 + 60 + 160) \div 5 = 90$$

We now have a measure of the correlation between X and Y, but how do we interpret it; i.e., does the value 90 mean there is a high correlation between X and Y or not? Actually, we really can't tell yet, since the value of the covariance depends on the magnitude of the values in $X - 16$ and $Y - 120$: If the values in $X - 16$ and $Y - 120$ are large (small) in an absolute sense, then the covariance of X and Y will tend to be large (small) in an absolute sense. We must therefore "normalize" the covariance of X and Y so that no matter how large or small the values are in $X - 16$ and $Y - 120$, we will be able to tell if the covariance is large on a relative scale.

This "normalization" is accomplished by comparing the covariance of X and Y to some measure of the magnitude of the values in $X - 16$, called the *standard deviation* of X, and some measure of the magnitude of the values in $Y - 120$, called the standard deviation of Y. The standard deviation of X, for example, is simply the square root of the average of the squared values of $X - 16$, and it is denoted as STD(X). Thus the standard deviation of X in our example is

$$\text{STD}(X) = (1 + 9 + 36 + 4 + 64) \div 5 = 4.8$$

And similarly, the standard deviation of Y is

$$\text{STD}(Y) = (400 + 100 + 1600 + 900 + 400) \div 5 = 26.1$$

The square of the standard deviation is called the *variance* and is denoted as VAR(X); i.e., $\text{VAR}(X) = [\text{STD}(X)]^2$.

Now to normalize the covariance of X and Y, we simply divide the covariance by the standard deviation of both X and Y. The result is called the *correlation* of X and Y, and it is denoted as COR(X, Y). Thus the correlation of X and Y in our example is computed as follows:

$$\text{COR}(X, Y) = \frac{\text{COV}(X, Y)}{\text{STD}(X)\,\text{STD}(Y)} = \frac{90}{(4.8)(26.1)} = .72$$

It turns out that the correlation between any two data sets will always be some value between -1 and $+1$. A correlation value close to $+1$ means high positive correlation, and a value close to -1 means high negative correlation. A value of zero means no correlation. In our example the correlation .72 indicates a fairly high degree of correlation between X and Y.

A.2 AUTOCORRELATION

In computing an autocorrelation at a given lag for a series, we are really computing the correlation between two distinct data sets, namely, the original series and the same series moved forward in time a specified number of periods (lag). For example, the autocorrelation at lag 1 is the correlation between the original series X_t and the same series moved forward one period (represented as X_{t-1}), as illustrated below:

Period t	Set X (X_t)	Set Y (X_{t-1})
1	15	—
2	13	15
3	10	13
4	18	10
5	24	18
6	—	24

Note that when the series is moved forward one period, there is no corresponding data value in X_{t-1} for the first data value in X_t and no corresponding data value in X_t for the last data value in X_{t-1}. Thus we can only compare the data values that are shown in the box. The mean values of X_t and X_{t-1}, however, are the same, namely, $(15 + 13 + 10 + 18 + 24) \div 5 = 16$ (note that we use all the values to compute the means).

To compute the covariance of X_t and X_{t-1}, we first compute the product of $(X_t - 16)$ and $(X_{t-1} - 16)$, as shown below:

Period	$(X_t - 16)$	$(X_{t-1} - 16)$	$(X_t - 16) \times (X_{t-1} - 16)$
1	−1	—	—
2	−3	−1	3
3	−6	−3	18
4	2	−6	−12
5	8	2	−16
6	—	8	—

The covariance of X_t and X_{t-1} is then given by

$$\text{COV}(X_t, X_{t-1}) = (3 + 18 - 12 + 16) \div 5 = 5$$

Note that the covariance of X_t and X_{t-1} is computed by dividing the sum of the products $(X_t - 16) \times (X_{t-1} - 16)$ by the original number of data values (5 in this case), not the number of products used in the sum (4 in this case).

Now since the data values in X_t and X_{t-1} are the same, then the standard deviations of X_t and X_{t-1} are the same (4.8), and the correlation between X_t and X_{t-1} is therefore

$$\text{COR}(X_t, X_{t-1}) = \frac{\text{COV}(X_t, X_{t-1})}{\text{STD}(X_t)\,\text{STD}(X_{t-1})} = \frac{5}{(4.8)(4.8)} = .22$$

Note that $\text{STD}(X_t)\,\text{STD}(X_{t-1}) = \text{VAR}(X_t)$, i.e. the variance of X_t.

But the correlation of X_t and X_{t-1} is actually the autocorrelation of X_t at lag 1. Thus if we denote the autocorrelation of X_t at lag 1 by R_1, then

$$R_1 = \text{COR}(X_t, X_{t-1}) = .22$$

The autocorrelation of X_t at lag 2 is computed similarly, except that the second data set is now X_t moved forward two periods (represented as X_{t-2}), as illustrated below:

Period t	X_t	X_{t-2}	$(X_t - 16)$	$(X_{t-2} - 16)$	$(X_t - 16) \times (X_{t-2} - 16)$
1	15	—	−1	—	—
2	13	—	−3	—	—
3	10	15	−6	−1	6
4	18	13	2	−3	−6
5	24	10	8	−6	−48
6	—	18	—	2	—
7	—	24	—	8	—

The mean values of X_t and X_{t-2} are still the same (16); and the standard deviations of X_t and X_{t-2} are also the same (4.8), i.e., the variance of X_t = 23.04. Therefore

$$\text{COV}(X_t, X_{t-2}) = (6 - 6 - 48) \div 5 = -9.6$$

and

$$R_2 = \text{COR}(X_t, X_{t-2}) = \frac{\text{COV}(X_t, X_{t-2})}{\text{VAR}(X_t)} = \frac{-9.6}{23.04} = -.42$$

The generalization of the preceding arguments to autocorrelations at higher lags is obvious, and we have

$$R_j = \text{COR}(X_t, X_{t-j}) = \frac{\text{COV}(X_t, X_{t-j})}{\text{VAR}(X_t)}, \qquad j = 1, \ldots, N-2$$

Autocorrelations for lags greater than $N - 2$ are not meaningful since data is no longer available to compare.

Appendix B

Interpretation of Partial-Autocorrelations

From a computational point of view, partial-autocorrelations are computed directly from the autocorrelations of a series through a given formula. Conceptually, however, partial-autocorrelations arise from the consideration of autoregressive models. In this appendix we will show how these two ideas come together and provide a helpful interpretation of partial-autocorrelations.

In the following discussion we will use the notation developed in Appendix A for variance, standard deviation, covariance, correlation, and autocorrelation. Thus for data sets X and Y and a time series X_t, we have

$$\text{Variance of } X = \text{VAR}(X)$$

$$\text{Standard deviation of } X = \text{STD}(X) = \sqrt{\text{VAR}(X)}$$

$$\text{Covariance of } X \text{ and } Y = \text{COV}(X, Y)$$

$$\text{Correlation of } X \text{ and } Y = \text{COR}(X, Y) = \frac{\text{COV}(X, Y)}{\text{STD}(X)\,\text{STD}(Y)}$$

$$\text{Autocorrelation of } X_t \text{ at lag } j = R_j = \text{COR}(X_t, X_{t-j}) = \frac{\text{COV}(X_t, X_{t-j})}{\text{VAR}(X_t)}$$

We will also need to use the following properties of the covariance function:

1. Given three data sets X, Y, and Z, then

   $$\text{COV}(X + Y, Z) = \text{COV}(X, Z) + \text{COV}(Y, Z).$$

   Thus the covariance of the sum $X + Y$ and Z is the covariance of X and Z plus the covariance of Y and Z.
2. Given a constant value A, then

   $$\text{COV}(AX, Z) = \text{A COV}(X, Z)$$

   Thus the covariance of AX and Z is just A times the covariance of X and Z.

Now let's consider an autoregressive model with only one AR parameter of order 1:

$$X_t = A_1 X_{t-1} + E_t$$

The autocorrelation of X_t at lag 1 is given by

$$R_1 = \text{COR}(X_t, X_{t-1}) = \frac{\text{COV}(X_t, X_{t-1})}{\text{VAR}(X_t)}$$

Substituting the autoregressive expression for X_t into $\text{COV}(X_t, X_{t-1})$ yields

$$R_1 = \frac{\text{COV}(A_1X_{t-1} + E_t, X_{t-1})}{\text{VAR}(X_t)}$$

and from the properties of covariance

$$R_1 = \frac{A_1 \text{COV}(X_{t-1}, X_{t-1}) + \text{COV}(E_t, X_{t-1})}{\text{VAR}(X_t)}$$

But

$$\frac{\text{COV}(X_{t-1}, X_{t-1})}{\text{VAR}(X_t)} = \text{COR}(X_{t-1}, X_{t-1}) = 1$$

and $\text{COV}(E_t, X_{t-1}) = 0$ (since E_t is independent of X_{t-1}). Therefore

$$R_1 = A_1$$

Hence the autocorrelation at lag 1 associated with an AR model with one AR parameter of order 1 is equal to the value of the AR parameter.

Remembering this fact, consider next an autoregressive model with two AR parameters of orders 1 and 2:

$$X_t = A_1X_{t-1} + A_2X_{t-2} + E_t$$

The autocorrelation at lag 1 for this model is given by

$$\begin{aligned} R_1 &= \text{COR}(X_t, X_{t-1}) \\ &= \frac{\text{COV}(X_t, X_{t-1})}{\text{VAR}(X_t)} = \frac{\text{COV}(A_1X_{t-1} + A_2X_{t-2} + E_t, X_{t-1})}{\text{VAR}(X_t)} \\ &= A_1\left[\frac{\text{COV}(X_{t-1}, X_{t-1})}{\text{VAR}(X_t)}\right] + A_2\left[\frac{\text{COV}(X_{t-2}, X_{t-1})}{\text{VAR}(X_t)}\right] + \frac{\text{COV}(E_t, X_{t-1})}{\text{VAR}(X_t)} \end{aligned}$$

The first term above simplifies to A_1 as before, and the third term is again zero. In the second term, since X_{t-2} and X_{t-1} are offset from each other by only one period, then

$$\frac{\text{COV}(X_{t-2}, X_{t-1})}{\text{VAR}(X_t)} = R_1 = \text{Autocorrelation of } X_t \text{ at lag } 1$$

Thus

$$R_1 = A_1 + A_2R_1$$

Similarly the autocorrelation at lag 2 for the AR model of order 2 is computed as follows:

$$
\begin{aligned}
R_2 &= \mathrm{COR}(X_t, X_{t-2}) \\
&= A_1\left[\frac{\mathrm{COV}(X_{t-1}, X_{t-2})}{\mathrm{VAR}(X_t)}\right] + A_2\left[\frac{\mathrm{COV}(X_{t-2}, X_{t-2})}{\mathrm{VAR}(X_t)}\right] + \frac{\mathrm{COV}(E_t, X_{t-2})}{\mathrm{VAR}(X_t)} \\
&= A_1R_1 + A_2
\end{aligned}
$$

Thus for the AR model of order 2 we have two equations involving R_1, R_2, A_1, and A_2:

$$
\begin{aligned}
R_1 &= A_1 + A_2R_1 \\
R_2 &= A_1R_1 + A_2
\end{aligned}
$$

Note that we can solve for the parameter values A_1 and A_2 in terms of the autocorrelation R_1 and R_2.

In general, if we consider an autoregressive model with p AR parameters, we can obtain a set of p equations involving $R_1, R_2, \ldots, R_p$ and $A_1, A_2, \ldots, A_p$, as follows:*

$$
\begin{aligned}
R_1 &= A_1 \quad + A_2R_1 \quad + A_3R_2 \quad + \cdots + A_pR_{p-1} \\
R_2 &= A_1R_1 \quad + A_2 \quad + A_3R_1 \quad + \cdots + A_pR_{p-2} \\
R_3 &= A_1R_2 \quad + A_2R_1 \quad + A_3 \quad + \cdots + A_pR_{p-3} \\
&\vdots \\
R_p &= A_1R_{p-1} + A_2R_{p-2} + A_3R_{p-3} + \cdots + A_p
\end{aligned}
$$

From these equations $A_1, \ldots, A_p$ can be solved for in terms of $R_1, \ldots, R_p$.

For example, if $p = 4$, the set of four equations associated with an AR model of order 4 is written as follows:

$$
\begin{aligned}
R_1 &= A_1 \quad + A_2R_1 + A_3R_2 + A_4R_3 \\
R_2 &= A_1R_1 + A_2 \quad + A_3R_1 + A_4R_2 \\
R_3 &= A_1R_2 + A_2R_1 + A_3 \quad + A_4R_1 \\
R_4 &= A_1R_3 + A_2R_2 + A_3R_1 + A_4
\end{aligned}
$$

Now let's see what all of this has to do with partial-autocorrelations. Suppose we have an arbitrary stationary series X_t generated by some basic Box-Jenkins model (not necessarily an AR model), and suppose we know the autocorrelations R_1, R_2, . . . of the series. Even though we don't know whether or not the model for the series is autoregressive, we can still set up the previous set of equations for any p,

*This set of equations is known as the Yule-Walker equations.

and proceed to solve for the A_i in terms of the autocorrelations R_i. The values computed for the A_i, however, may not have anything to do with autoregressive parameters (since the model for the series may not be autoregressive). Let's denote these A_i values by the symbol $\hat{A}_i$ to distinguish them from the autoregressive parameter notation A_i. With these preliminaries we are now in a position to describe what partial-autocorrelations are.

First, let $p = 1$ and solve for $\hat{A}_1$ from the one equation corresponding to $p = 1$ ($\hat{A}_1$ is simply equal to R_1 in this case); $\hat{A}_1$ is called the partial-autocorrelation of the series at lag 1. Since we computed this $\hat{A}_1$ from one equation, we will denote it by $\hat{A}_1^1$. Next, let $p = 2$ and solve for $\hat{A}_1$ and $\hat{A}_2$. Since $\hat{A}_1$ and $\hat{A}_2$ were computed from a set of two equations, we will denote them by $\hat{A}_1^2$ and $\hat{A}_2^2$, respectively. The term $\hat{A}_2^2$ is called the partial-autocorrelation at lag 2. In general, for any p, form the required set of equations and solve for $\hat{A}_1, \hat{A}_2, \ldots, \hat{A}_p$. These values are denoted by $\hat{A}_1^p, \hat{A}_2^p, \ldots, \hat{A}_p^p$, and $\hat{A}_p^p$ is called the partial-autocorrelation at lag p. The partial-autocorrelations of the series are thus the values $\hat{A}_1^1, \hat{A}_2^2, \hat{A}_3^3, \ldots,$ *where each one was obtained from a different* set of equations.

Now let's see how the partial-autocorrelation patterns discussed in Chapter 8 for AR and MA models are derived. Suppose the model for the series really was an AR model of order 1, i.e., $X_t = A_1 X_{t-1} + E_t$. Then the partial-autocorrelation for the series X_t at lag 1, $\hat{A}_1^1$, is equivalent to the autoregressive parameter A^1. This equivalence results because the equations set up to compute partial-autocorrelations were originally set up to compute the the autoregressive parameters for a model that was strictly autoregressive. Moreover, since A_1 is nonzero, $\hat{A}_1^1$ must also be nonzero. Now what about the partial-autocorrelation at lag 2 for the series generated by the one-AR model? Well, $\hat{A}_2^2 = A_2$ since the model is purely autoregressive; but $A_2 = 0$ since the model is only of order 1. Thus, $\hat{A}_2^2 = 0$ for a one-AR model. Similarly, it can be argued that $\hat{A}_k^k$ is zero for all $k \geq 2$, if the model for the series is an AR model of order 1. Thus corresponding to an AR model of order 1, the partial-autocorrelations are nonzero for lag 1 and zero for all other lags. This result, of course, is the partial-autocorrelation pattern described in Chapter 8 for a model with only one AR parameter.

In a similar manner, we can show that for an AR model of order p (i.e., $A_p = 0$ and $A_k = 0$ for $k > p$), the partial-autocorrelation at lag p is nonzero, and all partial-autocorrelations at lags greater than p are zero. Also, if the model contains an AR parameter of order $j < p$, then the partial-autocorrelation at lag j will be nonzero. This result corresponds to the general partial-autocorrelation patterns described in Chapter 8 for autoregressive models.

On the other hand, if we have a series generated by a model containing MA parameters, then the partial-autocorrelations must be nonzero at all lags. That is, if the partial-autocorrelations were zero for all lags after some lag p, then the model would have to be purely autoregressive of order p. It can also be shown that these partial-autocorrelations decrease rapidly in magnitude at an exponential rate. This is the type of partial-autocorrelation pattern that was described in Chapter 8 for MA models.

Appendix C

The Use of Prior Transformations to Achieve Stationarity

As discussed in Chapter 12, the process of differencing is the main tool for converting a nonstationary series into a stationary one. Differencing will generally produce stationarity for many of the series you will encounter. For example, nonstationary behavior in the form of certain types of overall trends can be reduced to stationary behavior through differencing. The types of trends that can be handled in this way are straight-line trends, where the rate of change is constant, or parabolic (curvilinear) trends, where the rate of change itself is changing but at a constant rate.* In the case of straight-line trends a first-order difference will produce stationarity, while for parabolic trends two first-order differences will do so (analogous statements can be made about seasonal differencing and trending behavior in the seasonal pattern).

There are, however, certain types of nonstationary behavior that differencing alone cannot convert into stationary behavior. Exponential trends are examples of such behavior, because the rate of change in exponential trends is itself exponential. Thus successive differences will only produce another series that is exponential. You may often encounter this situation in the economic environment where exponential growth is a common phenomenon. Figure C.1 illustrates this type of behavior.

Another type of nonstationary behavior that differencing by itself cannot reduce to stationary behavior occurs when the variation of the series about its overall trend (horizontal or otherwise) changes over time. (Recall that a series is stationary when the pattern in the series is independent of time, i.e., when the level of the series is fixed and the series varies more or less uniformly about this level.) Differencing in this case may be able to produce a series that varies about a fixed level (i.e., the mean is "stationary"), but the magnitude of the variation about that level will be changing (i.e. the variance is not "stationary"). Figure C.2 illustrates this type of nonstationary behavior.

Fortunately, in most cases these problems can be overcome by applying an "instantaneous," or *prior,* transformation to the series before applying any differencing. The most common transformation used is the logarithmic transformation, $ln(X)$. The logarithmic transformation can convert exponential behavior into straight-line behavior and can convert certain nonconstant variation into constant variation. Model identification and estimation is then applied to the logged series $\dot{X}_t = ln(X_t)$. Forecasts can be expressed in terms of the logged series or in terms of the original

*The straight-line trend and the parabolic trend are actually called polynomials of the first degree and second degree, respectively. Higher-degree polynomial trends (trends that become increasingly "wavy" as the degree increases) may also be modeled through the differencing process (e.g., a third-degree polynomial trend is a cubic trend that may be modeled by applying three regular differences). Such trends, however, are rarely present in real-life time series.

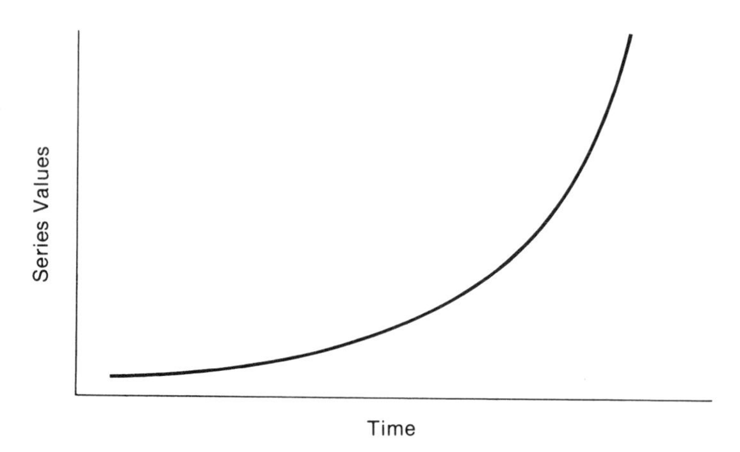

Figure C.1 Exponential Growth

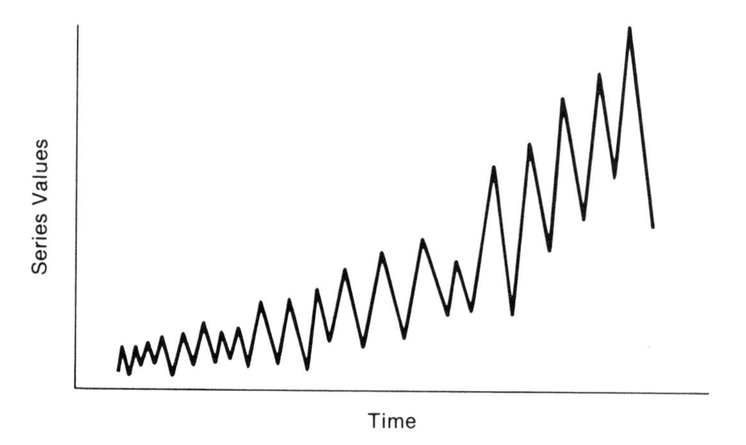

Figure C.2 Increasing Variation

series by applying a "reverse" transformation [the exponential transformation $exp(X)$] to the forecasts of the logged series.

Although the logarithmic transformation is the most often used prior transformation, other transformations, such as $\sqrt{X_t}$ or $1/X_t$, may be appropriate for other types of exponential trend growth. Most Box-Jenkins computer programs will automatically apply the logarithmic transformation for you. All you have to do is indicate that it is to be used. Some programs also allow the other common types of transformations to be applied automatically.

How do you know when you need to use a prior transformation? Unfortunately, the autocorrelation patterns of a series in many cases do not give you any clues, particularly in the case of nonconstant variation. Thus you may have to rely on a

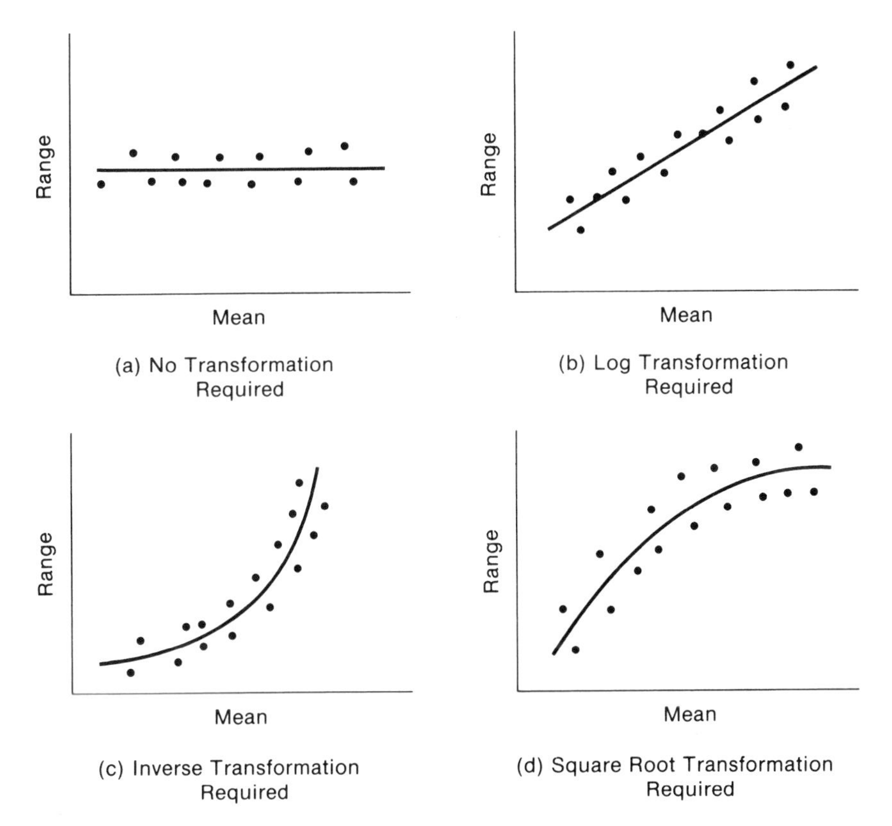

Figure C.3 Mean-Versus-Range Curves for Identifying Prior Transformations

visual inspection of a graph of the series for detecting any exponential trends or pronounced changes in variation (see the examples at the end of this appendix). Another technique is to plot the mean value of successive groups of periods (e.g., every 12 periods) against the range (maximum minus minimum) of the series values in each group. The resulting plot can suggest the most appropriate transformation to use as identified by the shape of the curve produced. Figure C.3 shows some of the possibilities.

Now let's look at some examples.

Example C.1

Consider the series shown in Figure C.4. This series follows an obvious exponential trend.

The means and ranges for successive groups of 20 data values were then computed. The mean-versus-range plot for this data is shown in Figure C.5. The plot follows an upward straight-line trend, which indicates the need for a logarithmic transformation.

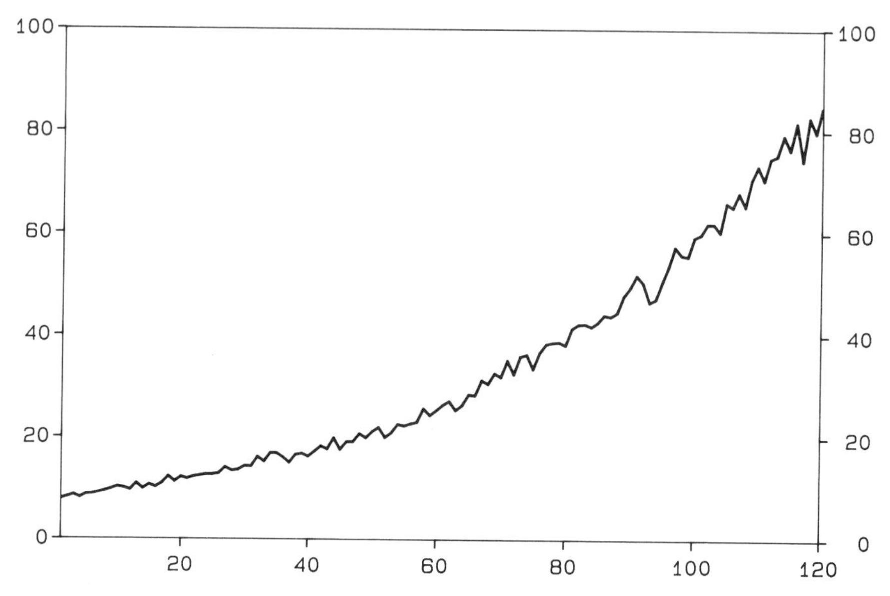

Figure C.4 Time Series with Exponential Trend

The next two examples are the series from Examples 16.7 and 16.8 of Chapter 16. In both cases a logarithmic transformation is used.

Example C.2

The airline series (Example 16.7) is shown in Figure C.6. Although it may not be immediately obvious, the graph of this series does suggest a slight exponential growth rate superimposed on the seasonal pattern. The most obvious pattern, however, is the increasing variation in the seasonal pattern; i.e., the distance between peaks and valleys is constantly increasing as time goes on.

Plotting the average value of each year in the series against the range of values in each year yields the mean-versus-range plot shown in Figure C.7. The resulting straight-line plot corroborates the initial need for a logarithmic transformation. Thus before proceeding with the Box-Jenkins modeling process, we should first transform the series by taking the logarithm of each series value.

Example C.3

The car registration series (Example 16.8) is shown in Figure C.8. In this case there appears to be no exponential trend, but the variation of the series increases in magnitude as time goes on. Again, this inspection suggests the need for a logarithmic transformation.

A mean-versus-range plot for this series, based on successive groups of 12 months, is shown in Figure C.9. Although the plotted points don't all fall as nicely

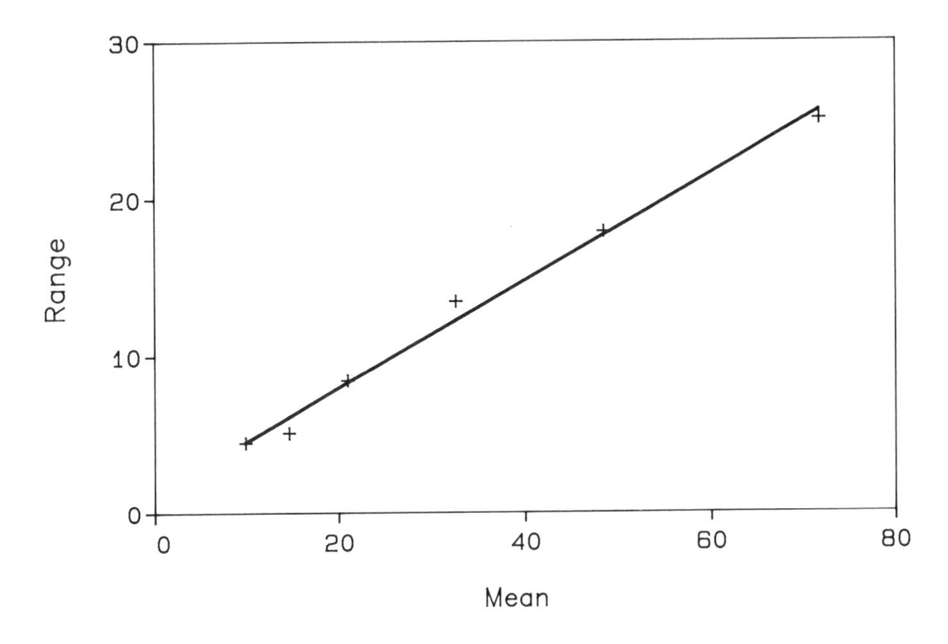

Periods	Mean	Range
1 to 20	9.84	4.48
21 to 40	14.55	5.09
41 to 60	21.00	8.43
61 to 80	32.60	13.46
81 to 100	48.47	17.89
101 to 120	71.73	25.03

Figure C.5 Mean-Versus-Range Plot for Exponential Trend Series

on a straight-line trend as the airline series points did, the general pattern is still upward, as indicated. Thus a logarithmic transformation should be applied.

In general, if you suspect, but aren't quite sure, that a series requires a prior transformation, you should model the series both with and without the transformation and compare the results. A common problem in not using a transformation when it is really needed is the inability to identify seasonal parameters. The AC correlograms in Figure C.10 illustrate this situation. The correlogram in part (a) corresponds to the unlogged airline series (see Example 16.7) after one RD and one SD of order 12; the correlogram in part (b) corresponds to the logged airline series after the same amount of differencing. Note the autocorrelation spike at lag 12 in the part (b) correlogram, indicating an SMA parameter; this spike is missing in the part (a) correlogram for the unlogged series.

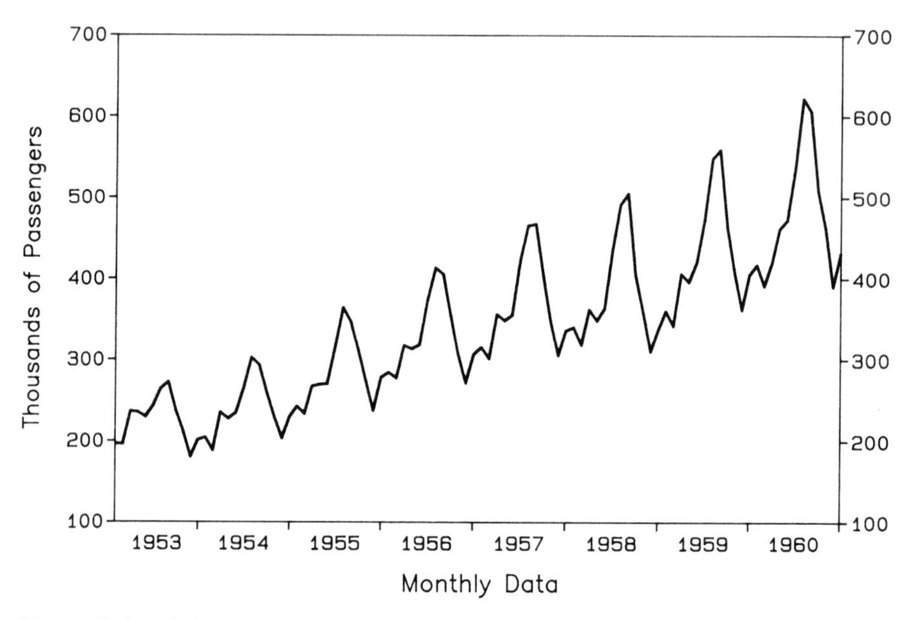

Figure C.6 Airline Series Showing Slight Exponential Growth and Increasing Variation

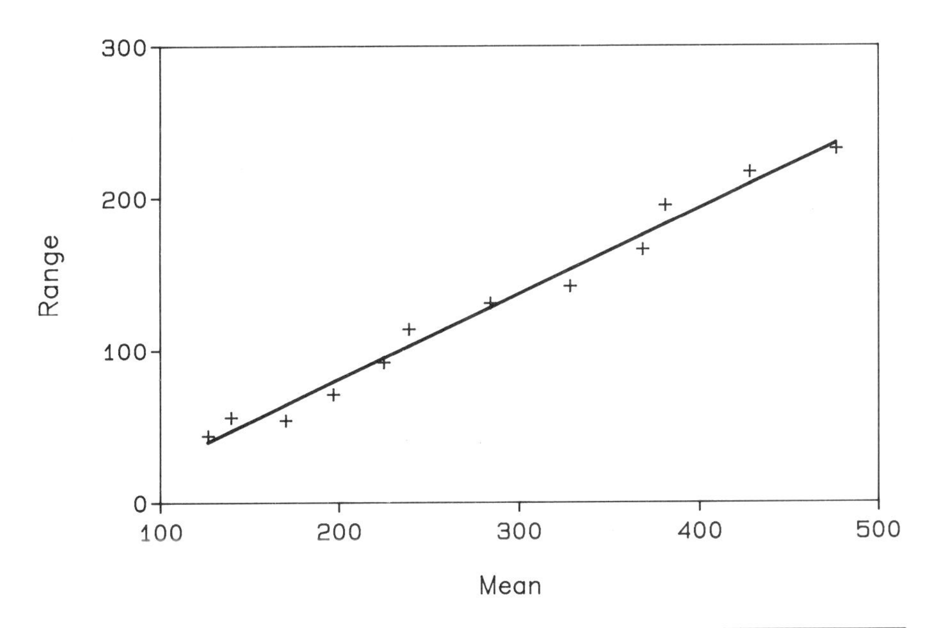

Periods	Mean	Range
1 to 12	126.67	44.00
13 to 24	139.67	56.00
25 to 36	170.17	54.00
37 to 48	197.00	71.00
49 to 60	225.00	92.00
61 to 72	238.92	114.00
73 to 84	284.00	131.00
85 to 96	328.25	142.00
97 to 108	368.42	166.00
109 to 120	381.00	195.00
121 to 132	428.33	217.00
133 to 144	476.17	232.00

Figure C.7 Mean-Versus-Range Plot for Airline Series

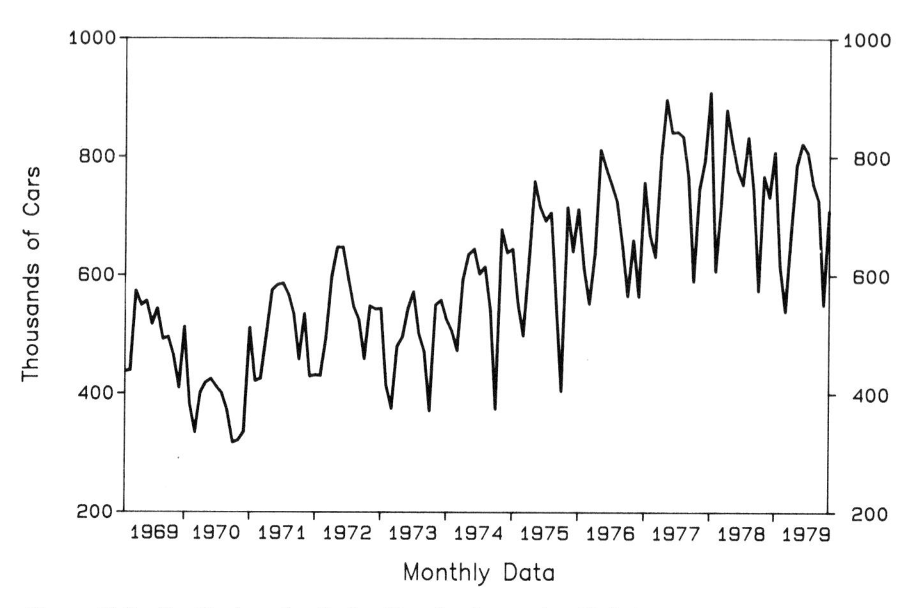

Figure C.8 Car Registration Series Showing Increasing Variation

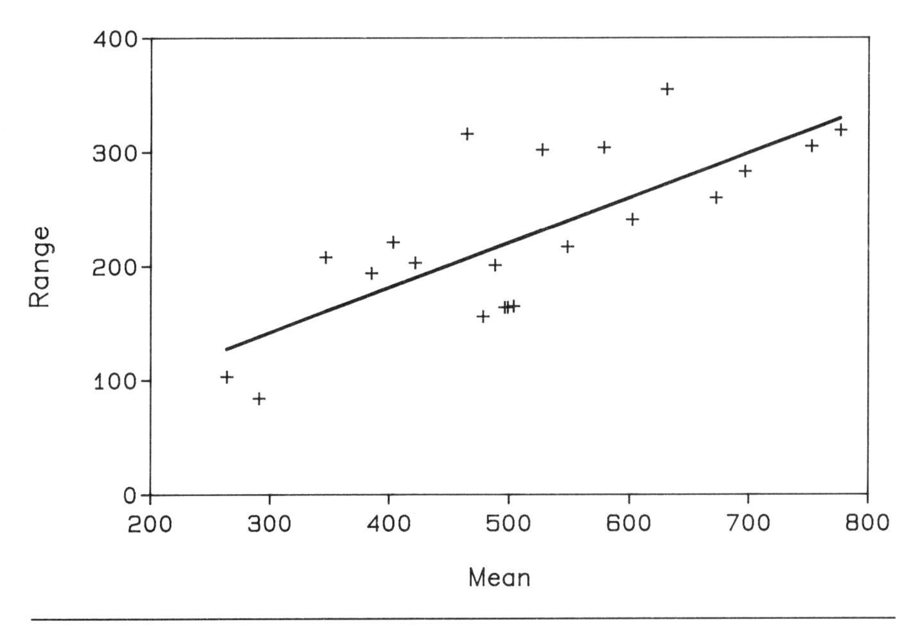

Periods	Mean	Range
1 to 12	263.92	103.00
13 to 24	290.92	84.00
25 to 36	403.42	221.00
37 to 48	527.17	302.00
49 to 60	421.67	203.00
61 to 72	347.42	208.00
73 to 84	478.25	156.00
85 to 96	464.50	316.00
97 to 108	602.08	241.00
109 to 120	496.25	164.00
121 to 132	498.83	164.00
133 to 144	385.42	194.00
145 to 156	503.58	165.00
157 to 168	548.08	217.00
169 to 180	488.17	201.00
181 to 192	578.33	304.00
193 to 204	631.00	355.00
205 to 216	672.17	260.00
217 to 228	776.33	319.00
229 to 240	751.58	305.00
241 to 252	696.67	283.00

Figure C.9 Mean-Versus-Range Plot for Car Registration Series

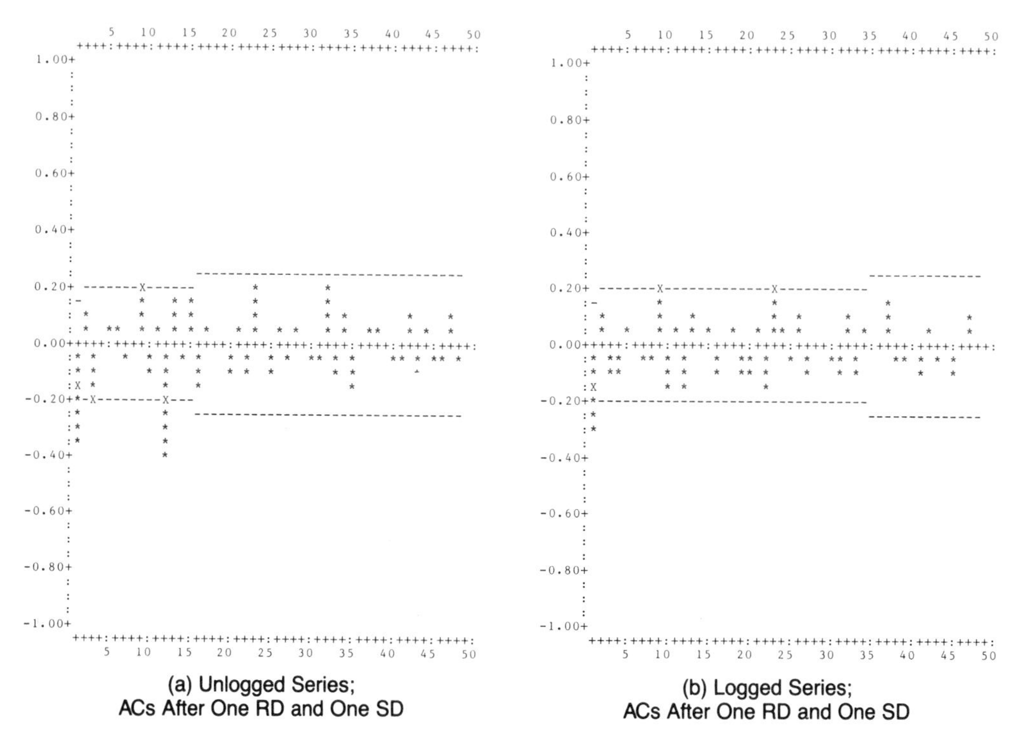

Figure C.10 AC Correlograms for the Unlogged and Logged Airline Series After Differencing

Appendix D

The Mathematical Representation of Box-Jenkins Models

For consistency with most literature on Box-Jenkins, the following notation will be used to describe the general Box-Jenkins model:

t = Time period, $t = 1, 2, \ldots$
s = Periods per season
x_t = Series values, $t = 1, 2, \ldots, N$
e_t = Random error, $t = 1, 2, \ldots, N$
(assumed to be normally distributed with zero mean)
ϕ_i = Regular autoregressive parameter, $i = 1, \ldots, p$
ϕ_i^* = Seasonal autoregressive parameter, $i = 1, \ldots, P$
θ_i = Regular moving-average parameter, $i = 1, \ldots, q$
θ_i^* = Seasonal moving-average parameter, $i = 1, \ldots, Q$
θ_0 = Trend parameter
d = Number of regular differences
D = Number of seasonal differences

Let B denote an "operator" called the backward-difference operator. This operator has the following meaning: Given a series value x_t, Bx_t represents the preceding series value x_{t-1}; i.e.,

$$Bx_t = x_{t-1}$$

Also, B can be applied more than once. For example,

$$B^2x_t = B(Bx_t) = Bx_{t-1} = x_{t-2}$$

In general,

$$B^kx_t = x_{t-k}$$

With the B operator the following composite operators can be constructed:

Differencing operator: $D(B) = (1 - B)^d(1 - B^s)^D$
RAR operator: $\Phi(B) = (1 - \phi_1 B - \cdots - \phi_p B^p)$

SAR operator: $\Phi^*(B) = (1 - \phi_1^* B^s - \cdots - \phi_p^* B^{sp})$

RMA operator: $\Theta(B) = (1 - \theta_i B - \cdots - \theta_q B^q)$

SMA operator: $\Theta^*(B) = 1 - \theta_1^* B^s - \cdots - \theta_Q^* B^{sQ})$

The complete Box-Jenkins model may then be represented as follows:

$$\Phi(B)\Phi^*(B)D(B)x_t = \theta_0 + \Theta(B)\Theta^*(B)e_t$$

To illustrate how the above representation works, we'll consider a stationary model with one RAR and one RMA parameter. Then $p = 1$ and $q = 1$, so that $\mathbf{\Phi}(B) = (1 - \phi_1 B)$ and $\Theta(B) = (1 - \theta_1 B)$. Since there are no differencing or seasonal parameters, then $d = 0$, $D = 0$, $P = 0$, $Q = 0$, and $D(B) = 1$, $\Phi^*(B) = 1$, and $\Theta^*(B) = 1$. Also, $\theta_0 = 0$ since there is no trend parameter. The one-RAR and one-RMA model may thus be represented as follows:

$$(1 - \phi_1 B)x_t = (1 - \theta_1 B)e_t$$

Multiplying both sides yields

$$x_t - \phi_1 Bx_t = e_t - \theta_1 Be_t$$

and by definition of the B operator, we have

$$x_t - \phi_1 x_{t-1} = e_t - \theta_1 e_{t-1}$$

Rearranging terms, then, gives the following familiar representation:

$$x_t = \phi_1 x_{t-1} - \theta_1 e_{t-1} + e_t$$

To illustrate a multiplicative seasonal model, we'll consider a nonstationary model with one regular difference, one seasonal difference of order 12, and one RMA and one SMA of order 12. Then

$$D(B) = (1 - B)(1 - B^{12})$$
$$\Phi(B) = 1$$
$$\Phi^*(B) = 1$$
$$\Theta(B) = (1 - \theta_1 B)$$
$$\Theta^*(B) = (1 - \theta_1^* B)$$
$$\theta_0 = 0$$

Therefore

$$(1 - B)(1 - B^{12})x_t = (1 - \theta_1 B)(1 - \theta_1^* B^{12})e_t$$

i.e.,

$$x_t - Bx_t - B^{12}x_t + B^{13}x_t = e_t - \theta_1 Be_t - \theta_1^* B^{12}e_t + \theta_1\theta_1^* B^{13}e_t$$

This expression reduces to

$$x_t - x_{t-1} - x_{t-12} + x_{t-13} = -(\theta_1 e_{t-1} + \theta_1 e_{t-12} - \theta_1\theta_1^* e_{t-13}) + e_t$$

The left-hand side above is just the differenced series, and the right-hand side is the multiplicative seasonal MA model with one RMA and one SMA of order 12.

Appendix E

Box-Jenkins Computer Programs

As indicated in Chapter 3, Box-Jenkins programs may be accessed through commercial computer time-sharing services or purchased directly from computer software distributors (commercial and noncommercial) and installed on your own computer facilities.

Most programs provide both univariate and multivariate Box-Jenkins modeling capabilities. Table E.1 lists some of the companies that offer a Box-Jenkins program on their time-sharing service. If you know of, or deal with, a time-sharing company not included in this list, chances are they offer a Box-Jenkins program too.

Most of these Box-Jenkins programs are interactive, or *conversational*. That is, you are able to operate the program in real time while sitting at a computer terminal hooked up to the time-sharing computer, and the program will ask you for instructions on what you want to do (e.g., estimate a model, generate forecasts, or plot a correlogram, etc.) and prompt you for the required input (e.g., parameters to include in a model or forecasts required, etc.). The conversational procedure is a very convenient way to use a Box-Jenkins program because of the iterative nature of the Box-Jenkins modeling process.

In most cases the Box-Jenkins modeling and forecasting capability on a time-sharing service is part of a more general forecasting or modeling system which can also prove to be convenient since you can perform other types of analyses and computations that you may wish to do with your data under a single umbrella. (The computer output shown in this book was obtained from the TIMEPACK forecasting program on the Control Data Business Information Services time-sharing system.)

Most Box-Jenkins programs produce similar output since they were developed from the same set of procedures, the ones outlined in the original book by Box and Jenkins [2]. In addition, many of the programs are based on a program written by David J. Pack at the University of Ohio [9, 10].* This program was acquired by a number of time-sharing companies or software developers who then wrote conversational-user interfaces for the program. The conversational-user interfaces, therefore, vary widely, and you should compare the program documentation from the various time-sharing services to determine the type of program operation you like the best.

The cost of time-sharing services can also vary widely, depending on the type of service, contractual arrangements, etc. Although only the address of the corporate headquarters is given for the companies listed in Table E.1, many of the companies may have local sales offices in your city from which you can obtain cost information.

*Pack is currently at Oak Ridge National Laboratory, Computer Sciences Division, Oak Ridge, Tennessee.

Table E.1 Commercial Time-Sharing Services with Access to Box-Jenkins Programs

Company	Headquarters Address	Program Names	Type of Models
CompuServe, Inc.	5000 Arlington Center Boulevard Columbus, OH 43220	AUTOBOX* and SYBIL/RUNNER*	Univariate and Multivariate
Control Data Corporation Business Information Services (CALL/370)	500 West Putnam Avenue P.O. Box 7100 Greenwich, CT 06830	TIMEPACK*	Univariate and Multivariate (two variables only)
Computer Sciences Corporation (INFONET)	650 N. Sepulveda Boulevard El Segundo, CA 90245	AUTOBJ*	Univariate and Multivariate
Comshare, Inc. (COMMANDER II)	300 S. State Street P.O. Box 1588 Ann Arbor, MI 48106	ORION	Univariate and Multivariate (two variables only)
General Electric Information Services Co. (MARK III)	401 N. Washington Street Rockville, MD 20850	LABJ* and MAPBJ*	Univariate and Multivariate
Chase Econometrics/ Interactive Data Corporation, Dynamics Associates Division	1033 Massachusetts Avenue, Cambridge, MA 02138	XSIM*/ AUTOBJ	Univariate and Multivariate
National CSS	187 Danbury Road Wilton, CT 06897	SPX/TIME	Univariate and Multivariate
On Line Business Systems, Inc.	115 Sansome San Francisco, CA 94104	MODEL	Univariate and Multivariate
National Data Corporation, RAPIDATA Division	20 New Dutch Lane P.O. Box 1049 Fairfield, NJ 07006	PROBE	Univariate and Multivariate
Tymshare, Inc.	20705 Valley Green Drive Cupertino, CA	EXPRESS*	Univariate

Table E.2 lists some of the companies or organizations that will sell you a Box-Jenkins program for installation on your own computer. Many of these programs

*TIMEPACK is a registered trademark of TimeWare Corporation, Palo Alto, California. SYBIL/RUNNER is a registered trademark of Applied Decisions Systems, Inc., Lexington, Massachusetts. AUTOBJ and AUTOBOX are proprietary products of Automatic Forecasting Systems, Inc., Hatboro, Pennsylvania. XSIM is a servicemark of Interactive Data Corporation, Cambridge, Massachusetts. LABJ and MAPBJ are products of Lochrie and Associates, Milwaukee, Wisconsin. EXPRESS is a product of Management Decision Systems, Inc., Waltham, Massachusetts.

Table E.2 Software Vendors of Box-Jenkins Programs

Organization	Address	Program Name(s)	Type of Models	Batch/ Conversational
Applied Decisions Systems, Inc.	33 Hayden Avenue Lexington, MA 02173	SIBYL/RUNNER	Univariate and Multivariate	Conversational
Automatic Forecasting Systems, Inc.	P.O. Box 563 Hatboro, PA 19040	PACK System and AUTOBJ	Univariate and Multivariate	Conversational and Batch
Gwilym Jenkins & Partners, Ltd.	1700 Echo Trail Norman, OK 73069	GENISIS	Univariate and Multivariate	Batch
IBM Corporation	Data Processing Division 1133 Westchester Avenue White Plains, NY 10604	APL Forecasting and Time Series Analysis	Univariate and Multivariate	
Charles R. Nelson Associates, Inc.	4921 N.E. 39th St. Seattle, WA	PDQ, et al.	Univariate and Multivariate	Conversational and Batch
SAS Institute, Inc.	Box 8000 Cory, NC	SAS	Univariate and Multivariate	Conversational and Batch
Scientific Computing Associates, Inc.	P.O. Box 625 DeKalb, IL 60115	The SCA System	Univariate Multivariate	Conversational and Batch
Statistical Laboratory, Iowa State University	c/o Bill Meeker Route 1 Ames, IA 50010	TSERIES	Univariate	Batch
BMDP Statistical Software, Inc.	1964 Westwood Blvd. Suite 202 Los Angeles, CA 90025	BMDP	Univariate Multivariate	Conversational and Batch

are *batch*-oriented. That is, you must prepare all your instructions and input to the program beforehand (e.g., on punched cards or in a file stored on the computer) and then submit a request to execute the program using this prepared input. This method of operation usually means you will have to wait a period of time before getting your results, so the process of building a model is usually a more time-consuming process than it is when you are using a conversational program. Most of the programs, however, are conversational, which you can use if your computer facilities have an appropriate time-sharing operating system.

Most programs are compatible with IBM mainframe computers and operating systems, and many are hardware independent or are provided in a form compatible with non-IBM computers. You should make sure that the system requirements for a program (size and type of computer, operating system, programming language, etc.) are compatible with your computing facilities.

Some of these programs are actually large statistical systems (e.g., SAS and BMDP) or general forecasting systems (e.g., SYBIL/RUNNER and SCA) that include the Box-Jenkins method as one forecasting/statistical technique.

Some of the organizations that sell Box-Jenkins programs also provide installation support and offer ongoing maintenance, updating, and consulting services. These services, along with ease-of-use considerations and program capabilities, can be important factors to you depending on your requirements and circumstances.

Prices vary widely, ranging from a purchase price of several hundred dollars to several thousand dollars. Some programs may also be obtained by paying a monthly or annual license fee. Again, the services provided with the program, as well as the quality of the user interface, program capabilities, and your requirements, should be considered when making price comparisons.

Appendix F
Bibliography

References Devoted to the Box-Jenkins Method Exclusively

The starred references contain material on only multivariate Box-Jenkins and related topics. Most of the other references also contain material on multivariate Box-Jenkins as well as univariate Box-Jenkins.

1. O. D. Anderson. 1976. *Time Series Analysis and Forecasting—The Box-Jenkins Approach*. London: Butterworths.
2. G. E. P. Box and G. M. Jenkins. 1970. *Time Series Analysis Forecasting and Control*. San Francisco: Holden-Day.
3. ★ G. E. P. Box and G. C. Tiao. 1973. "Intervention Analysis with Application to Economic and Environmental Problems." Technical Report No. 335. Madison, Wis.: Department of Statistics, University of Wisconsin.
4. J. C. Hoff. 1977. *TIMEPACK II Concepts—Part II (Box-Jenkins)*. Control Data Business Information Services, Form No. 65-2699-1.
5. G. M. Jenkins. 1979. *Practical Experiences with Modelling and Forecasting Time Series*. Jersey, Channel Islands: Gwilyn Jenkins and Partners (Overseas Ltd).
6. ★ L. Liu and D. M. Hanssens. "Identification of Transfer Function Models Via Least Squares." Technical Report No. 68. Los Angeles, Calif.: BMDP Statistical Software, Department of Biomathematics, UCLA.
7. V. A. Mabert. 1975. *An Introduction to Short Term Forecasting Using the Box-Jenkins Methodology.* Publication No. 2 in the Monograph Series of Production Planning and Control Division. Norcross, Ga.: American Institute of Industrial Engineers.
8. C. R. Nelson. 1973. *Applied Time Series Analysis For Managerial Forecasting*. San Francisco: Holden-Day.
9. D. J. Pack. 1977. *"A Computer Program for the Analysis of Time Series Models Using the Box-Jenkins Philosophy."* Hatboro, Pa.: Automatic Forecasting Systems, Inc.
10. D. J. Pack. 1974. *Computer Programs for the Analysis of Univariate Time Series Models and Single Input Transfer Functions Models Using the Methods of Box and Jenkins*. Columbus: The Data Center, College of Administrative Science, Ohio State University.

11. D. J. Pack. 1978. "The Pitfalls in Combining Regression Analysis with a Time Series Model." Working Paper Series 78-49. Columbus: College of Administration Science, Ohio State University.

★ 12. D. J. Pack. April 8, 1977. "Revealing Time Series Interrelationships." *Decision Sciences,* pp. 377–402.

13. D. J. Pack. 1979. *"What Will Your Time Series Analysis Computer Package Do?"* Oak Ridge, Tenn.: Computer Science Division at Oak Ridge National Laboratory.

14. D. P. Reilly. 1981. "Recent Experiences with an Automatic Box-Jenkins Modelling Algorithm." In *Time Series Analysis,* edited by O. D. Anderson and M. R. Perryman. North-Holland Publishing.

References on the Practical Application of Time Series Forecasting and Forecasting in General

Many of these references also have subsections or chapters devoted to the Box-Jenkins method or ARIMA modeling, both univariate and multivariate.

15. W. K. Benton. 1972. *Forecasting for Management.* Reading, Mass.: Addison-Wesley.

16. J. C. Chambers, S. K. Mullick, and D. D. Smith. 1974. *An Executive's Guide to Forecasting.* New York: Wiley.

17. J. C. Chambers, S. K. Mullick, and D. D. Smith. July–August 1971. "How to Choose the Right Forecasting Technique." *Harvard Business Review.*

18. J. P. Cleary and H. Levenbach. 1982. *The Professional Forecaster—The Forecasting Process Through Data Analysis.* Belmont, Calif.: Lifetime Learning.

19. C. W. Gross and R. T. Peterson. 1976. *Business Forecasting.* Boston: Houghton Mifflin.

20. J. C. Hoff. 1979. *TIMEPACK II Concepts—Part I (Time Series Analysis and Forecasting).* Control Data Business Information Services, Form No. 65-2698-1.

21. J. C. Hoff. 1982. *TIMEPACK II Reference Manual.* Control Data Business Information Services, Form No. 65-2700-4.

22. H. Levenbach and J. P. Cleary. 1981. *The Beginning Forecaster—The Forecasting Process Through Data Analysis.* Belmont, Calif.: Lifetime Learning.

23. S. Makridakis and S. C. Wheelwright. 1978. *Interactive Forecasting—Univariate and Multivariate Methods.* San Francisco: Holden-Day.

24. D. C. Montgomery and L. A. Johnson. 1976. *Forecasting and Time Series Analysis.* New York: McGraw-Hill.

25. W. G. Sullivan and W. W. Claycombe. 1977. *Fundamentals of Forecasting*. Reston, Va.: Reston Publishing.

26. S. C. Wheelwright and S. Makridakis. 1973. *Forecasting Methods for Management*. New York: Wiley.

27. S. C. Wheelwright and S. Makridakis. 1978. *Forecasting Methods and Applications*. New York: Wiley.

References on the Theoretical Aspects of Time Series Analysis and Forecasting

28. T. W. Anderson. 1971. *The Statistical Analysis of Time Series*. New York: Wiley.

29. W. Gilchrist. 1976. *Statistical Forecasting*. London: Wiley.

30. C. W. J. Granger. 1980. *Forecasting in Business and Economics*. New York: Academic Press.

31. C. W. J. Granger and P. Newbold. 1977. *Forecasting Economic Time Series*. New York: Academic Press.

32. M. G. Kendall. 1973. *Time-Series*. London: Griffin.

33. M. Nerlove, D. M. Grether, and J. L. Carvalho. 1979. *Analysis of Economic Time Series—A Synthesis*. New York: Academic Press.

Miscellaneous References

34. N. R. Draper and H. Smith. 1966. *Applied Regression Analysis*. New York: Wiley.

35. G. G. C. Parker and E. L. Segura. March–April 1971. "How to Get a Better Forecast." *Harvard Business Review.*

36. R. S. Pindyck and D. L. Rubinfeld. 1970. *Econometric Models and Economic Forecasts*. New York: McGraw-Hill.

37. J. Johnston. 1963. *Econometric Methods*. New York: McGraw-Hill.

38. J. Shiskin, A. H. Young, and J. C. Musgrave. 1976. *The X-11 Variant of the Census Method II Seasonal Adjustment Program*. U.S. Department of Commerce, Bureau of Economic Analysis. Washington, D.C.: National Technical Information Service.

Appendix G

Answers for Examples

CHAPTER 8

Example 8.1: 1 MA, order 1
Example 8.2: 1 AR, order 1
Example 8.3: 2 MAs, orders 1 and 2
Example 8.4: 1 AR, order 1
Example 8.5: 1 MA, order 1
Example 8.6: 2 ARs, orders 1 and 2
Example 8.7: 1 MA, order 1; 1 AR, order 1
Example 8.8: 1 AR, order 3
Example 8.9: 1 MA, order 6
Example 8.10: 2 ARs, orders 1 and 2

CHAPTER 9

Example 9.1: 1 MA, order 1

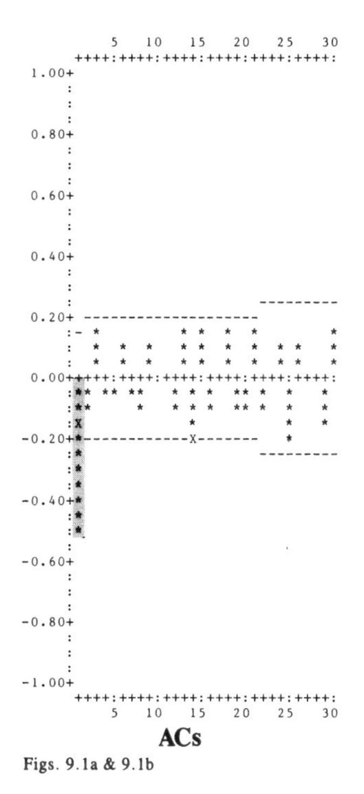

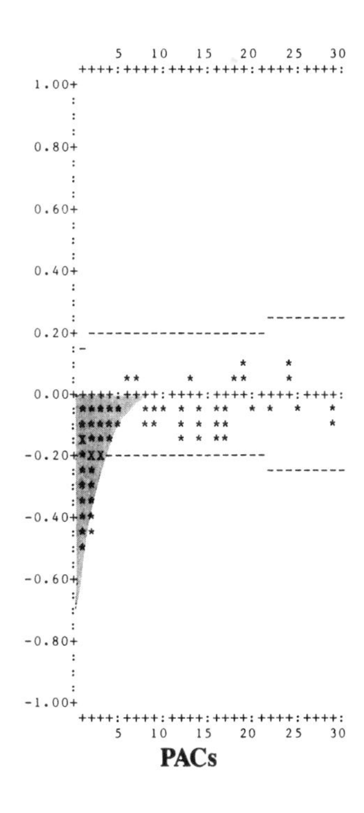

Figs. 9.1a & 9.1b

Example 9.2: 2 ARs, orders 1 and 2

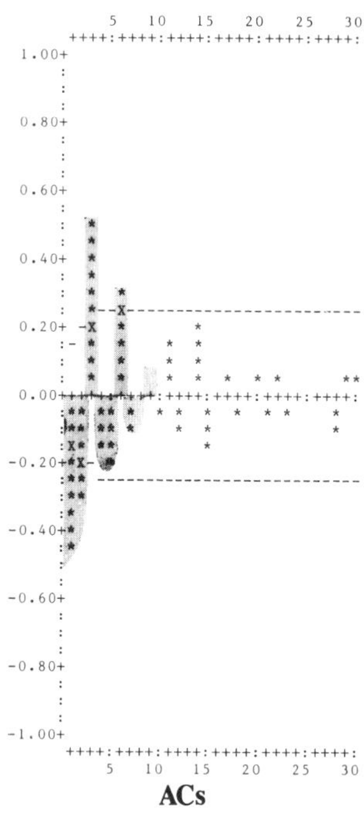

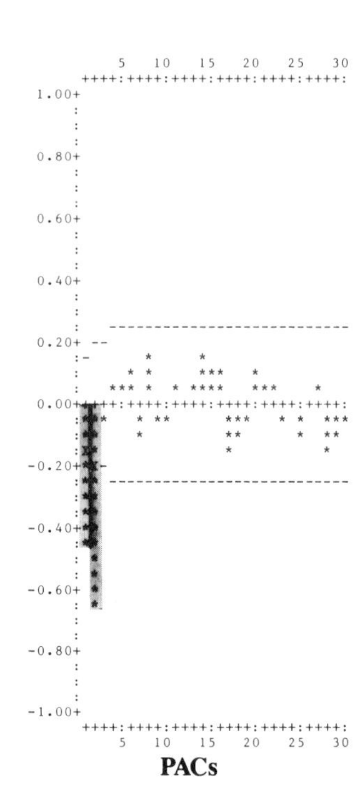

Figs. 9.2a & 9.2b

Example 9.3: 1 MA, order 6

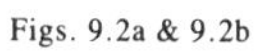

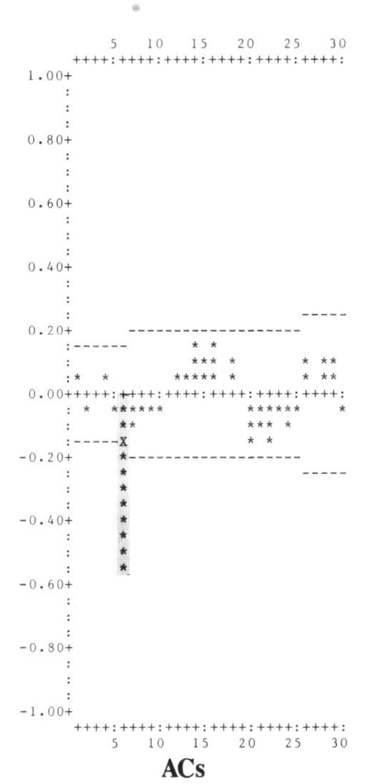

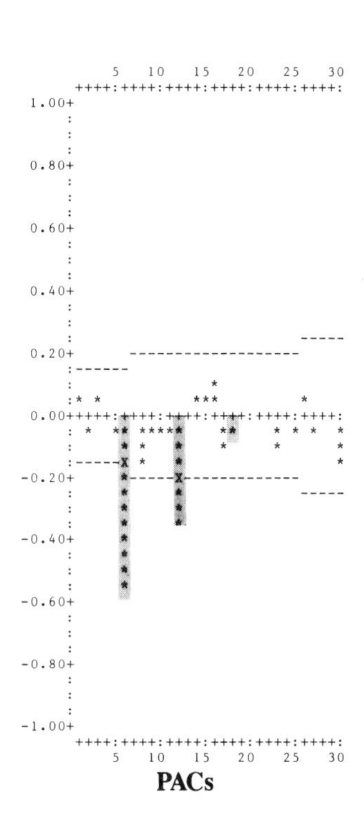

Figs. 9.3a & 9.3b

Example 9.4: 1 AR, order 1

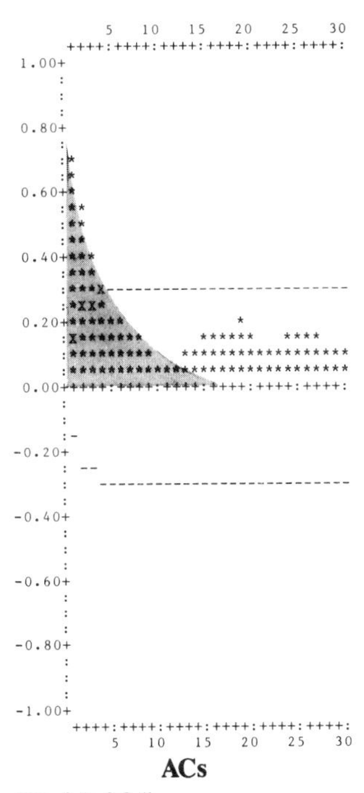

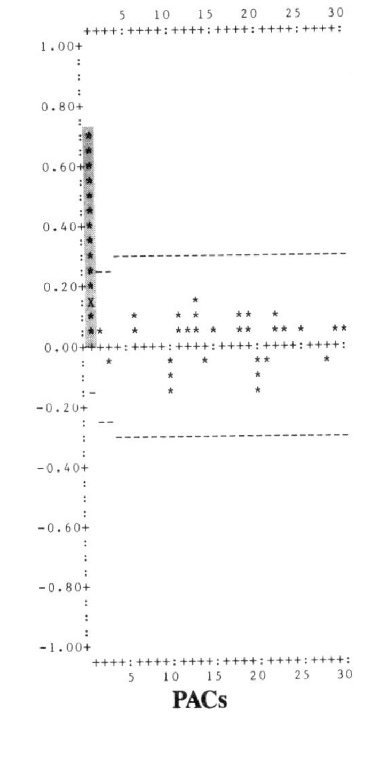

Figs. 9.4a & 9.4b

Example 9.5: 2 MAs, orders 1 and 2

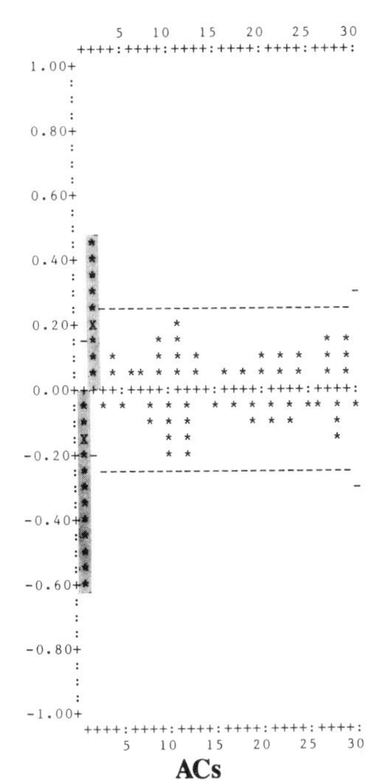

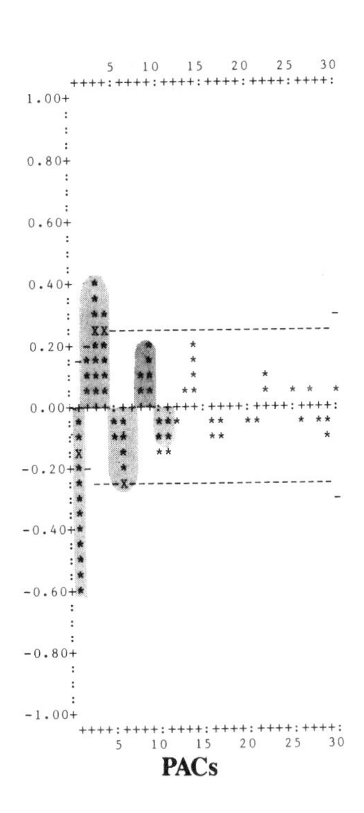

Figs. 9.5a & 9.5b

Example 9.6: 1 MA, order 1; 1 AR, order 1

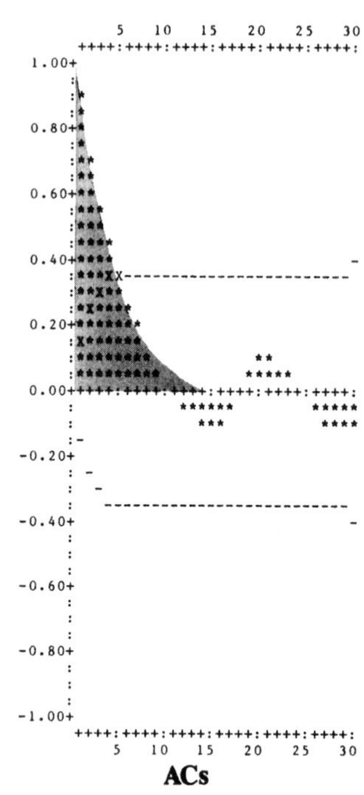

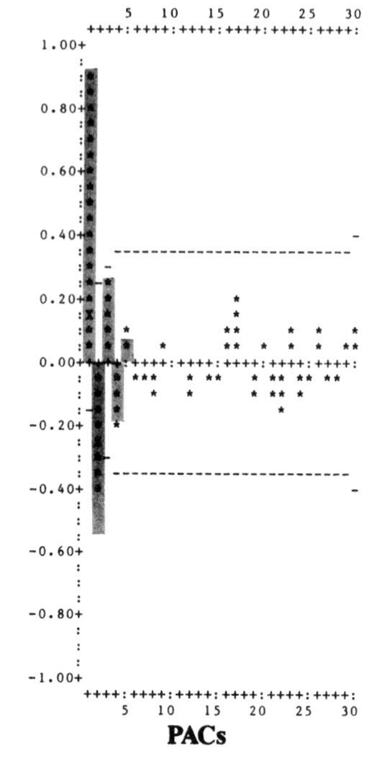

Figs. 9.6a & 9.6b

Example 9.7: 1 AR, order 3

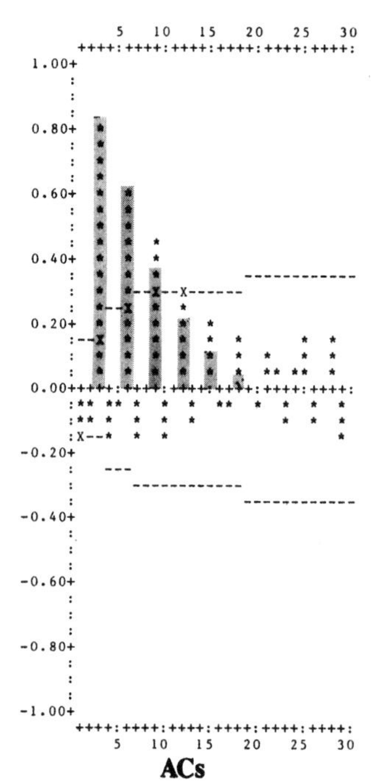

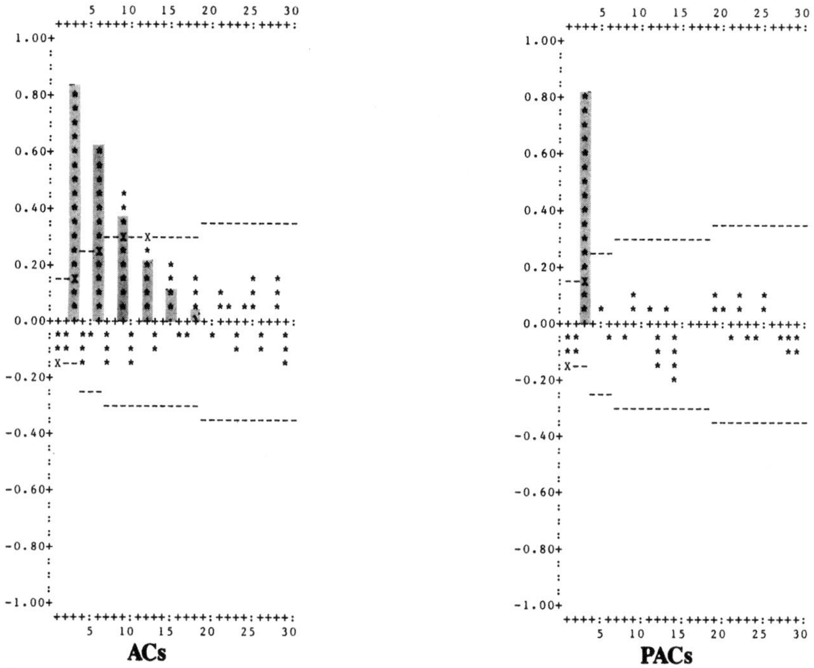

PACs

Figs. 9.7a & 9.7b

Example 9.8: 1 MA, order 1

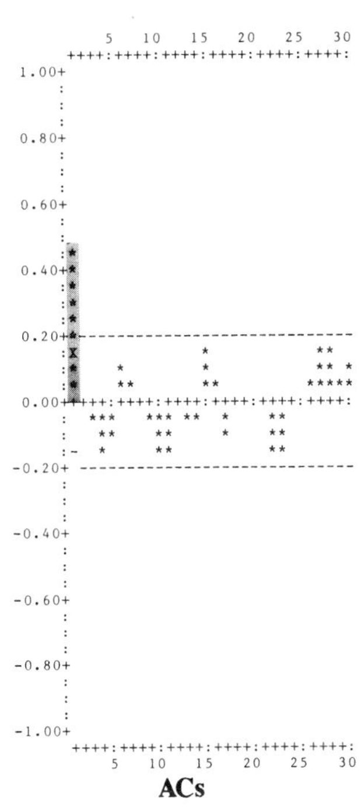

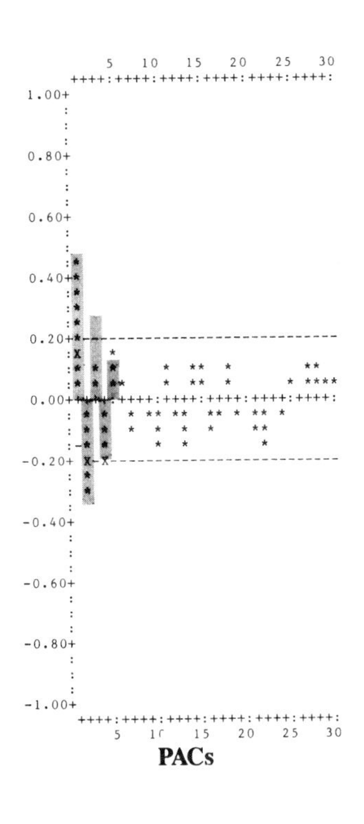

Figs. 9.8a & 9.8b

Example 9.9: 2 ARs, orders 1 and 2

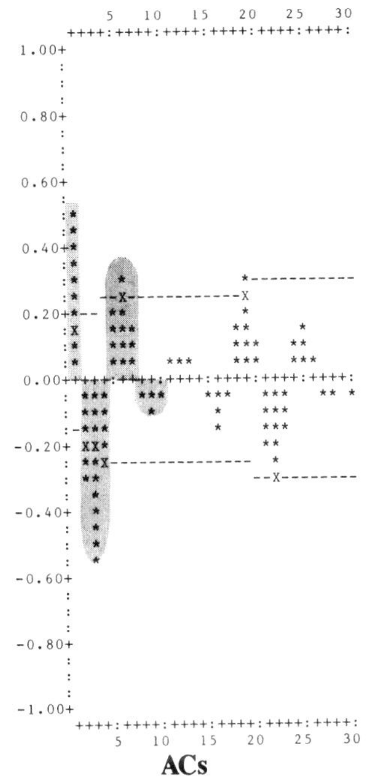

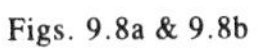

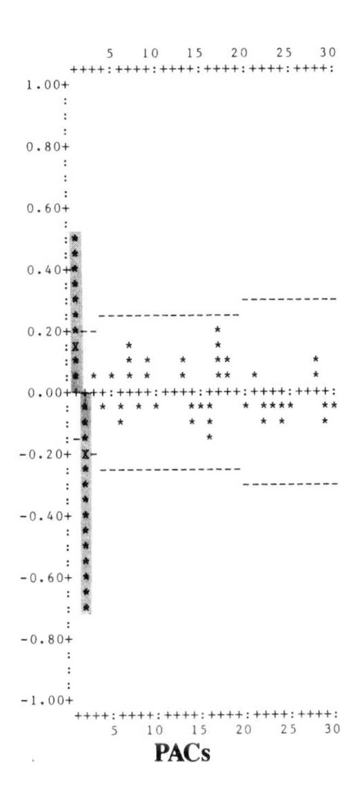

Figs. 9.9a & 9.9b

Example 9.10: 1 AR, order 1

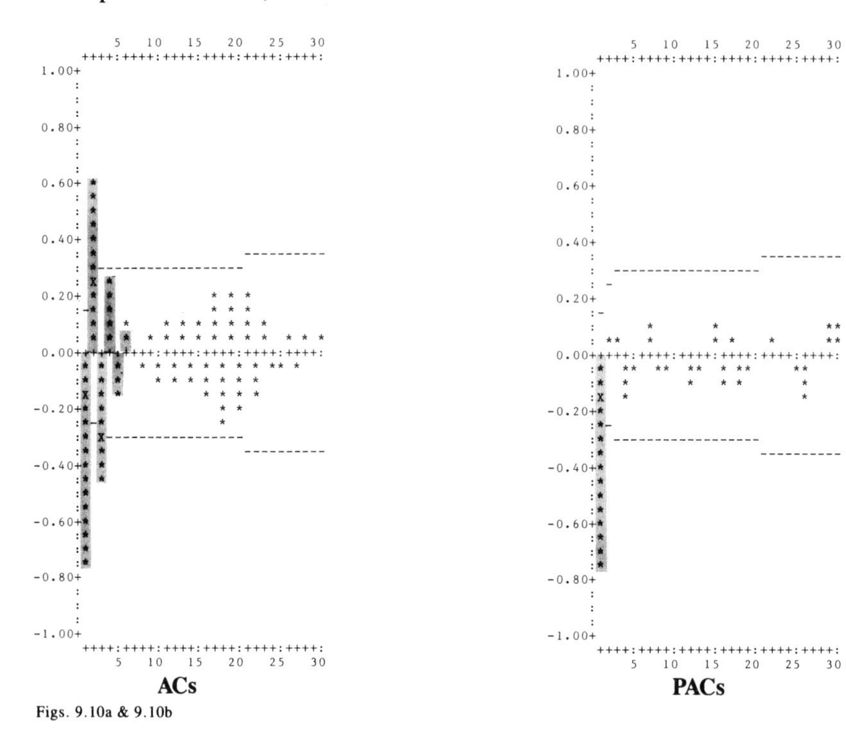

Figs. 9.10a & 9.10b

CHAPTER 15 Example 15.7: 1 RMA and 1 SMA (order 12)

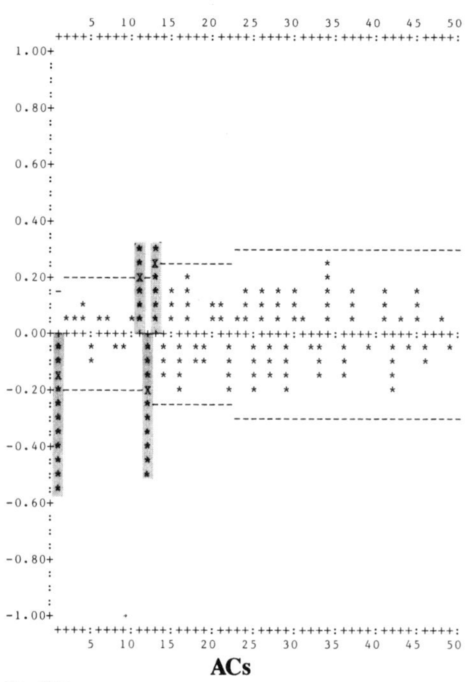

Fig. 15.7

Example 15.8: 2 RMAs and 1 SMA (order 12)

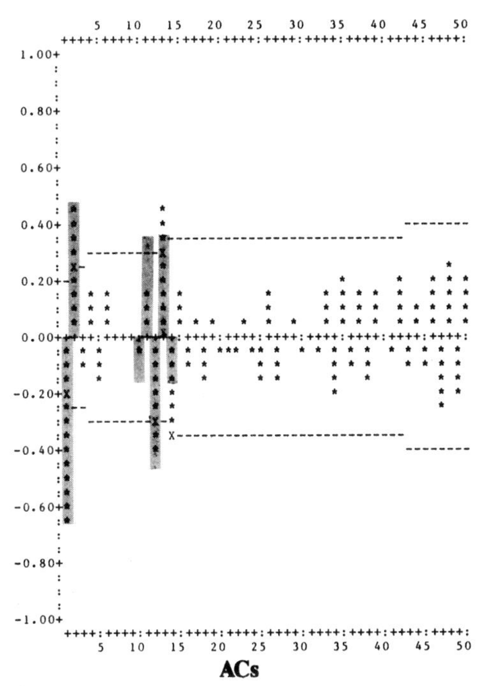

Fig. 15.8

Example 15.9: 1 RMA and 1 SMA (order 4)

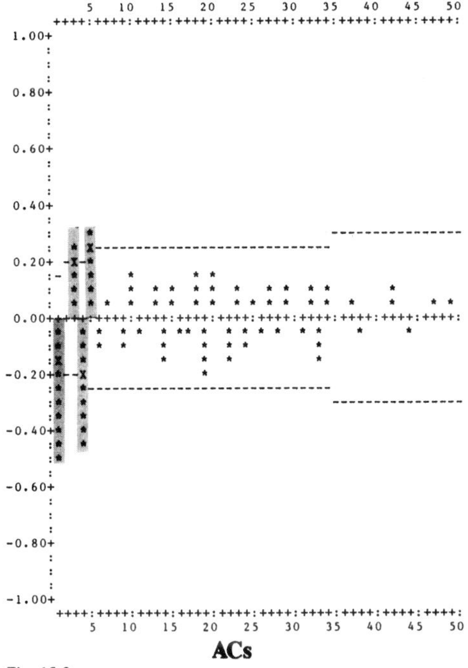

Fig. 15.9

Example 15.10: 1 RMA and 1 SAR (order 12); or 1 RMA and 1 SMA (order 12) and 1 SAR (order 12)

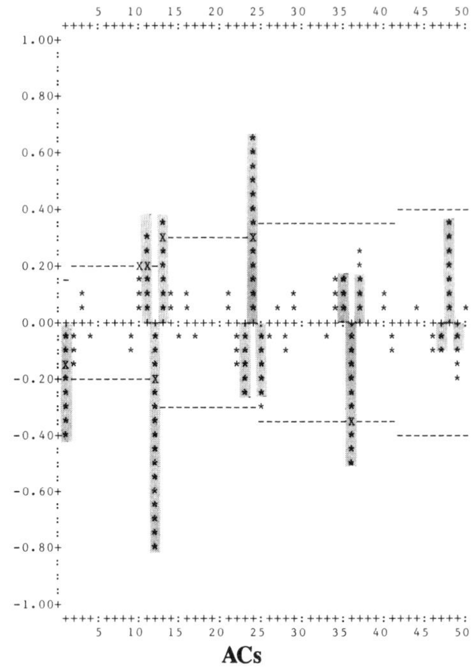

Fig. 15.10

Example 15.11: 1 RAR and 1 SMA (order 12)

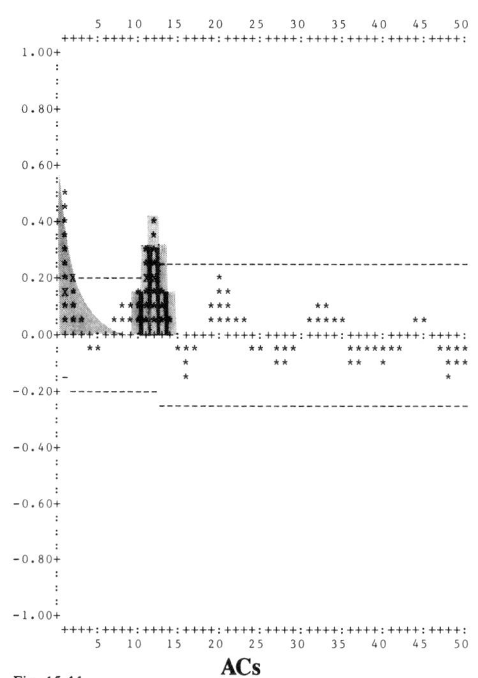

Fig. 15.11

Index